Inequalities in Geometry and Applications

Inequalities in Geometry and Applications

Editor

Gabriel-Eduard Vîlcu

MDPI • Basel • Beijing • Wuhan • Barcelona • Belgrade • Manchester • Tokyo • Cluj • Tianjin

Editor
Gabriel-Eduard Vîlcu
Department of Cybernetics,
Economic Informatics,
Finance and Accountancy,
Petroleum-Gas University of Ploieşti
Romania

Editorial Office
MDPI
St. Alban-Anlage 66
4052 Basel, Switzerland

This is a reprint of articles from the Special Issue published online in the open access journal *Mathematics* (ISSN 2227-7390) (available at: https://www.mdpi.com/journal/mathematics/special_issues/Inequalities_Geometry_Applications).

For citation purposes, cite each article independently as indicated on the article page online and as indicated below:

LastName, A.A.; LastName, B.B.; LastName, C.C. Article Title. *Journal Name* **Year**, *Volume Number*, Page Range.

ISBN 978-3-0365-0298-4 (Hbk)
ISBN 978-3-0365-0299-1 (PDF)

© 2021 by the authors. Articles in this book are Open Access and distributed under the Creative Commons Attribution (CC BY) license, which allows users to download, copy and build upon published articles, as long as the author and publisher are properly credited, which ensures maximum dissemination and a wider impact of our publications.

The book as a whole is distributed by MDPI under the terms and conditions of the Creative Commons license CC BY-NC-ND.

Contents

About the Editor vii

Preface to "Inequalities in Geometry and Applications" ix

Pablo Alegre, Joaquín Barrera and Alfonso Carriazo
A Closed Form for Slant Submanifolds of Generalized Sasakian Space Forms
Reprinted from: *Mathematics* **2019**, *7*, 1238, doi:10.3390/math7121238 1

Rifaqat Ali, Fatemah Mofarreh, Nadia Alluhaibi, Akram Ali and Iqbal Ahmad
On Differential Equations Characterizing Legendrian Submanifolds of Sasakian Space Forms
Reprinted from: *Mathematics* **2020**, *8*, 150, doi:10.3390/math8020150 17

Nadia Alluhaibi, Fatemah Mofarreh, Akram Ali and Wan Ainun Mior Othman
Geometric Inequalities of Warped Product Submanifolds and Their Applications
Reprinted from: *Mathematics* **2020**, *8*, 759, doi:10.3390/math8050759 27

Mohd. Aquib, Michel Nguiffo Boyom, Mohammad Hasan Shahid and Gabriel-Eduard Vîlcu
The First Fundamental Equation and Generalized Wintgen-Type Inequalities for Submanifolds in Generalized Space Forms
Reprinted from: *Mathematics* **2019**, *7*, 1151, doi:10.3390/math7121151 39

Joana Cirici and Scott O. Wilson
Almost Hermitian Identities
Reprinted from: *Mathematics* **2020**, *8*, 1357, doi:10.3390/math8081357 59

Simona Decu, Stefan Haesen and Leopold Verstraelen
Inequalities for the Casorati Curvature of Statistical Manifolds in Holomorphic Statistical Manifolds of Constant Holomorphic Curvature
Reprinted from: *Mathematics* **2020**, *8*, 251, doi:10.3390/math8020251 67

Yongping Deng, Muhammad Uzair Awan and Shanhe Wu
Quantum Integral Inequalities of Simpson-Type for Strongly Preinvex Functions
Reprinted from: *Mathematics* **2019**, *7*, 751, doi:10.3390/math7080751 81

Sharief Deshmukh and brahim Al-Dayel
A Note on Minimal Hypersurfaces of an Odd Dimensional Sphere
Reprinted from: *Mathematics* **2020**, *8*, 294, doi:10.3390/math8020294 95

Niufa Fang, and Zengle Zhang
The Minimal Perimeter of a Log-Concave Function
Reprinted from: *Mathematics* **2020**, *8*, 759, doi:10.3390/math8081365 105

Meraj Ali Khan, Ibrahim Aldayel
Ricci Curvature Inequalities for Skew CR-Warped Product Submanifolds in Complex Space Forms
Reprinted from: *Mathematics* **2020**, *8*, 1317, doi:10.3390/math8081317 117

Jian Liu
New Refinements of the Erdös–Mordell Inequality and Barrow's Inequality
Reprinted from: *Mathematics* **2019**, *7*, 726, doi:10.3390/math7080726 137

Gabriel Macsim, Adela Mihai and Ion Mihai
$\delta(2,2)$-Invariant for Lagrangian Submanifolds in Quaternionic Space Forms
Reprinted from: *Mathematics* **2020**, *8*, 480, doi:10.3390/math8040480 . . . . **149**

Moruz Marilena and Leopold Verstraelen
On the Extrinsic Principal Directions and Curvatures of Lagrangian Submanifolds
Reprinted from: *Mathematics* **2020**, *8*, 1533, doi:10.3390/math8091533 . . . . **165**

Vladimir Rovenski, Sergey Stepanov and Irina Tsyganok
On the Betti and Tachibana Numbers of Compact Einstein Manifolds
Reprinted from: *Mathematics* **2019**, *7*, 1210, doi:10.3390/math7121210 . . . . **171**

Aliya Naaz Siddiqui, Bang-Yen Chen and Oguzhan Bahadir
Statistical Solitons and Inequalities for Statistical Warped Product Submanifolds
Reprinted from: *Mathematics* **2019**, *7*, 797, doi:10.3390/math7090797 . . . . **177**

About the Editor

Gabriel-Eduard Vîlcu (Professor) obtained his Ph.D. in Mathematics at the University of Bucharest, Romania, in 2007. He is currently a full professor at the Petroleum-Gas University of Ploiești and also a senior researcher at the Research Center in Geometry, Topology and Algebra, at the Faculty of Mathematics and Computer Science, University of Bucharest. His main research interest is differential geometry and its applications. He has authored more than 70 articles in several renowned international journals and conference proceedings, and has also written and edited books in this field.

Preface to "Inequalities in Geometry and Applications"

Geometric inequalities have fascinated the mathematical world and not only since ancient times, as this field of research is in fact as old as mathematics itself. Over time, such inequalities have proven to be an excellent tool in investigating and solving basic problems in pure and applied sciences, including some that were apparently unrelated to geometric inequalities.

The aim of this book was to present recent developments in the field of geometric inequalities and their applications. The volume covers a vast range of topics, such as isoperimetric problem, Erdös–Mordell inequality, Barrow's inequality, Simpson inequality, Chen inequalities, q-integral inequalities, complex geometry, contact geometry, statistical manifolds, Riemannian submanifolds, optimization theory, topology of manifolds, log-concave functions, Obata differential equation, o-invariants, Einstein spaces, warped products, and solitons. By exposing new concepts, techniques and ideas, this book will certainly stimulate further research in this field.

Reviewed by leading experts, the chapters in this book were written by scientists from 13 different countries, most of them being outstanding researchers in the field. I am thankful to all the contributors and also to the journal *Mathematics* for giving me the opportunity to publish this book.

Gabriel-Eduard Vîlcu
Editor

Article

A Closed Form for Slant Submanifolds of Generalized Sasakian Space Forms

Pablo Alegre [1],*, Joaquín Barrera [2] and Alfonso Carriazo [2],†

[1] Departamento de Economía, Métodos Cuantitativos e Historia Económica. Área de Estadística e Investigación Operativa, Universidad Pablo de Olavide. Ctra. de Utrera, km. 1. 41013 Sevilla, Spain
[2] Department of Geometry and Topology, Faculty of Mathematics, University of Sevilla, Apdo. Correos 1160, 41080 Sevilla, Spain; barreralopezjoaquin@gmail.com (J.B.); carriazo@us.es (A.C.)
* Correspondence: psalerue@upo.es
† First and third authors are partially supported by the PAIDI group FQM-327 (Junta de Andalucía, Spain) and the MEC-FEDER grant MTM2011-22621. The third author is member of IMUS (Instituto de Matemáticas de laUniversidda de Sevilla).

Received: 4 November 2019; Accepted: 9 December 2019; Published: 13 December 2019

Abstract: The Maslov form is a closed form for a Lagrangian submanifold of $\mathbb{C}^m$, and it is a conformal form if and only if M satisfies the equality case of a natural inequality between the norm of the mean curvature and the scalar curvature, and it happens if and only if the second fundamental form satisfies a certain relation. In a previous paper we presented a natural inequality between the norm of the mean curvature and the scalar curvature of slant submanifolds of generalized Sasakian space forms, characterizing the equality case by certain expression of the second fundamental form. In this paper, first, we present an adapted form for slant submanifolds of a generalized Sasakian space form, similar to the Maslov form, that is always closed. And, in the equality case, we studied under which circumstances the given closed form is also conformal.

Keywords: slant submanifolds; generalized Sasakian space forms; closed form; conformal form; Maslov form

---

1. Introduction

It was proven by V. Borrelli, B.-Y. Chen and J. M. Morvan [1], and independently by A. Ros and F. Urbano [2], that if M is a Lagrangian submanifold, with $\dim(M) = m$, of $\mathbb{C}^m$, with mean curvature vector H and scalar curvature τ, then $\|H\|^2 \geq \frac{2(m+2)}{m^2(m-1)}\tau$, with equality if and only if M is either totally geodesic or a (piece of a) Whitney sphere. Moreover, they proved that M satisfies the equality case at every point if and only if its second fundamental form σ is given by

$$\sigma(X,Y) = \frac{m}{m+2}\{g(X,Y)H + g(JX,H)JY + g(JY,H)JX\}, \quad (1)$$

for any tangent vector fields X and Y. Thus, they found a simple relationship between one of the main intrinsic invariants, τ, and the main extrinsic invariant H.

It was also proven in [2], that the Maslov form, which is a closed form for a Lagrangian submanifold of $\mathbb{C}^m$, is a conformal form if and only if M satisfies (1).

Later, D. E. Blair and A. Carriazo [3] established an analogue inequality for anti-invariant submanifolds in $\mathbb{R}^{2m+1}$ with its standard Sasakian structure and characterized the equality case with a specific expression of the second fundamental form, similar to Equation (1). In a previous paper [4], we studied the corresponding inequality for slant submanifolds of generalized Sasakian space forms; we also characterized the equality case with an specific expression of the second fundamental form; and finally, we presented some examples satisfying the equality case.

Both B.-Y. Chen, [5] and A. Carriazo, [6], have studied the existence of closed forms for slant submanifolds in different environments. The existence of closed forms is particularly interesting, as they provide conditions about submanifolds admitting an immersion in a certain environment.

The purpose of this paper was to obtain some results similar to those of [2] for slant submanifolds of a generalized Sasakian space form. After a section with the main preliminaries, we show that for a slant submanifold of a generalized Sasakian manifold, the Maslov form is not always closed. Therefore, in the following section, we present a form that is always closed for a slant submanifold, so it really plays the role of the Maslov form in the cited papers. Later, if the submanifold satisfies the equality case in the corresponding inequality, that is, if the second fundamental form takes a particular expression [4], we study if the vector field associated with the given form is a conformal vector field.

2. Preliminaries

Given a Riemannian manifold $(\widetilde{M}, g)$, a tangent vector field X on $\widetilde{M}$ is called *closed* if its dual 1-form is closed. That is equivalent to

$$g(Y, \widetilde{\nabla}_Z X) = g(Z, \widetilde{\nabla}_Y X), \qquad (2)$$

for all Y and Z on $\widetilde{M}$, where $\widetilde{\nabla}$ is the Levi–Civita connection.

Moreover, X is called *conformal* if $L_X g = \rho g$, for ρ a function on $\widetilde{M}$, where L is the Lie derivative. A closed vector field X is conformal in and only if

$$\widetilde{\nabla}_Y X = fY, \qquad (3)$$

for any tangent vector field Y on $\widetilde{M}$ and for certain function f on $\widetilde{M}$.

In such a case, considering an orthonormal basis $\{e_1, \ldots, e_m\}$ on $\widetilde{M}$, it holds that $\widetilde{\nabla}_{e_i} X = fe_i$, for $i = 1, \ldots, m$.

Now, we will recall some notions about almost-contact Riemannian geometry. For more details about this subject, we recommend the book [7].

An odd-dimensional Riemannian manifold $(\widetilde{M}, g)$ is said to be an *almost contact metric manifold* if there exists on $\widetilde{M}$, a $(1,1)$ tensor field ϕ, a unit vector field ξ (called *the structure or Reeb vector field*) and a 1-form η, such that

$$\eta(\xi) = 1, \qquad \phi^2(X) = -X + \eta(X)\xi$$

and

$$g(\phi X, \phi Y) = g(X, Y) - \eta(X)\eta(Y),$$

for any vector fields X and Y on $\widetilde{M}$. In particular, in an almost contact metric manifold we also have

$$\phi \xi = 0, \qquad \eta \circ \phi = 0 \quad \text{and} \quad \eta(X) = g(X, \xi).$$

Such a manifold is said to be a *contact metric manifold* if $d\eta = \Phi$, where $\Phi(X,Y) = g(X, \phi Y)$ is called *the fundamental 2-form* of $\widetilde{M}$. The almost contact metric structure of M is said to be *normal* if $[\phi, \phi](X, Y) = -2d\eta(X,Y)\xi$, for any X and Y. A normal contact metric manifold is called a *Sasakian manifold*. It can be proven that an almost contact metric manifold is Sasakian if an only if

$$(\widetilde{\nabla}_X \phi)Y = g(X,Y)\xi - \eta(Y)X,$$

for any X and Y on M.

In [8], J.A. Oubiña introduced the notion of a *trans-Sasakian manifold*. An almost contact metric manifold $\widetilde{M}$ is a trans-Sasakian manifold if there exists two functions α and β on $\widetilde{M}$ such that

$$(\widetilde{\nabla}_X \phi)Y = \alpha(g(X,Y)\xi - \eta(Y)X) + \beta(g(\phi X, Y)\xi - \eta(Y)\phi X), \qquad (4)$$

for any X and Y on $\widetilde{M}$. If $\beta = 0$, $\widetilde{M}$ is said to be an α-*Sasakian manifold*. Sasakian manifolds appear as examples of α-Sasakian manifolds, with $\alpha = 1$. If $\alpha = 0$, $\widetilde{M}$ is said to be a β-*Kenmotsu manifold*. Kenmotsu manifolds are particular examples with $\beta = 1$. If both α and β vanish, then $\widetilde{M}$ is a *cosymplectic manifold*. In particular, from (4) it is easy to see that the following equation holds for a trans-Sasakian manifold:

$$\widetilde{\nabla}_X \xi = -\alpha \phi X + \beta (X - \eta(X)\xi). \tag{5}$$

It was proven by J.C. Marrero that, for dimensions greater or equal than 5, the only existing trans-Sasakian manifolds are α-Sasakian and β-Kenmotsu ones [9].

In [10], P. Alegre, D.E. Blair and A. Carriazo introduced the notion of a *generalized Sasakian space form* as an almost contact metric manifold $(\widetilde{M}, \phi, \xi, \eta, g)$ whose curvature tensor is given by

$$\begin{aligned}\widetilde{R}(X,Y)Z &= f_1 \{g(Y,Z)X - g(X,Z)Y\} \\ &+ f_2 \{g(X,\phi Z)\phi Y - g(Y,\phi Z)\phi X + 2g(X,\phi Y)\phi Z\} \\ &+ f_3 \{\eta(X)\eta(Z)Y - \eta(Y)\eta(Z)X + g(X,Z)\eta(Y)\xi - g(Y,Z)\eta(X)\xi\},\end{aligned} \tag{6}$$

where f_1, f_2 and f_3 are differential functions on $\widetilde{M}$. These manifolds are denoted by $\widetilde{M}(f_1, f_2, f_3)$; generalize the notion of *Sasakian space form*, $\widetilde{M}(c)$, whose curvature tensor satisfies the expression (6), with

$$f_1 = \frac{c+3}{4}, \quad f_2 = f_3 = \frac{c-1}{4},$$

where c is the constant ϕ-sectional curvature.

Now we recall some general definitions and facts about submanifolds. Let M be a submanifold isometrically immersed in a Riemannian manifold $(\widetilde{M}, g)$. We denote by ∇ the induced Levi–Civita connection on M. Thus, *the Gauss and Weingarten formulas* are respectively given by

$$\begin{aligned}\widetilde{\nabla}_X Y &= \nabla_X Y + \sigma(X,Y), \\ \widetilde{\nabla}_X V &= -A_V X + D_X V,\end{aligned}$$

for vector fields X and Y tangent to M and a vector field V normal to M, where σ denotes the second fundamental form, A_V the shape operator in the direction of V and D the normal connection. The second fundamental form and the shape operator are related by

$$g(A_V X, Y) = g(\sigma(X,Y), V). \tag{7}$$

M is called a *totally geodesic submanifold* if σ vanishes identically.

We denote by R and $\widetilde{R}$, the curvature tensors of M and $\widetilde{M}$, respectively. They are related by *Gauss and Codazzi's equations*

$$\begin{aligned}\widetilde{R}(X,Y;Z,W) &= R(X,Y;Z,W) \\ &+ g(\sigma(X,Z), \sigma(Y,W)) - g(\sigma(X,W), \sigma(Y,Z)),\end{aligned} \tag{8}$$

$$(\widetilde{R}(X,Y)Z)^\perp = (\widetilde{\nabla}_X \sigma)(Y,Z) - (\widetilde{\nabla}_Y \sigma)(X,Z), \tag{9}$$

where $(\widetilde{R}(X,Y)Z)^\perp$ denotes the normal component of $\widetilde{R}(X,Y)Z$ and

$$(\widetilde{\nabla}_X \sigma)(Y,Z) = D_X(\sigma(Y,Z)) - \sigma(\nabla_X Y, Z) - \sigma(Y, \nabla_X Z),$$

is the *derivative of Van der Waerden-Bortolotti*.

On the other hand, the *mean curvature vector* H is defined by

$$H = (1/\dim M) \text{ trace } \sigma, \tag{10}$$

and M is said to be *minimal* if H vanishes identically.

The *scalar curvature* τ of M at $p \in M$ is defined by

$$\tau = \sum_{1 \leq i < j \leq \dim M} K(e_i, e_j), \tag{11}$$

where $K(e_i, e_j)$ denotes the sectional curvature of M associated with the plane section spanned by e_i and e_j, for any tangent vector fields e_i and e_j in a local orthonormal frame of M.

For a submanifold of an almost contact manifold, we denote

$$\phi X = TX + NX \quad \text{and} \quad \phi V = tV + nV$$

the tangent and normal part of ϕX and ϕV for any X tangent vector field and V normal vector field. If the ambient space is trans-Sasakian, taking the tangent and normal part at (4) we obtain:

$$(\nabla_X T)Y - t\sigma(X,Y) - A_{NY}X = \alpha(g(X,Y)\xi - \eta(Y)X) \\ + \beta(g(TX,Y)\xi - \eta(Y)TX), \tag{12}$$

$$(\nabla_X N)Y + \sigma(X, TY) - n\sigma(X,Y) = -\beta\eta(Y)NX, \tag{13}$$

$$(\nabla_X t)V - A_{nV}X + TA_VX = \beta g(NX, V)\xi, \tag{14}$$

$$(\nabla_X n)V + \sigma(X, tV) + NA_VX = 0. \tag{15}$$

And from (5):

$$\nabla_X \xi = -\alpha TX + \beta(X - \eta(X)\xi), \tag{16}$$

$$\sigma(X, \xi) = -\alpha NX. \tag{17}$$

Now, we recall the definition of *slant submanifolds*. These submanifolds were defined by B.-Y. Chen in [5] on almost Hermitian geometry. Later, A. Lotta defined slant submanifolds on the almost contact metric setting in [11]: given a submanifold M tangent to ξ, for each nonzero vector X tangent to M at p, such that X is not proportional to ξ_p, we denote by $\theta(X)$ as the angle between ϕX and $T_p M$. Then, M is said to be slant if the angle $\theta(X)$ is a constant, which is independent of the choice of $p \in M$ and $X \in T_p M - <\xi_p>$. The angle θ of a slant immersion is called the slant angle of the immersion. Invariant and anti-invariant immersions are slant immersions with slant angles $\theta = 0$ and $\theta = \pi/2$, respectively. A slant immersion, which is neither invariant nor anti-invariant, is called a proper slant immersion. Slant submanifolds of Sasakian manifolds were studied by J.L. Cabrerizo, A. Carriazo, L.M. Fernández and M. Fernández in [12,13].

From now on, we denote by $m + 1 = 2n + 1$ the dimension of M and $2m + 1 = 4n + 1$ the dimension of $\widetilde{M}$. We assume $m \geq 2$. Then, for a slant submanifold holds:

$$T^2 X = \cos^2 \theta(-X + \eta(X)\xi), \tag{18}$$

$$tNX = \sin^2 \theta(-X + \eta(X)\xi), \tag{19}$$

$$NTX + nNX = 0, \tag{20}$$

and because of the dimensions,

$$n^2 V = -\cos^2 \theta V, \quad NtV = -\sin^2 \theta V \quad \text{and} \quad TtV + tnV = 0,$$

for any X, Y tangent vector fields and V normal vector field.

Given a proper slant submanifold M^{2n+1}, with slant angle θ, immersed in an almost contact manifold $\widetilde{M}^{4n+1}$, we considered an adapted slant reference, [6]; it was built as follows. Given e_1 a unit tangent vector field, orthogonal to ξ, we took:

$$e_2 = (\sec\theta)Te_1, \quad e_{1*} = (\csc\theta)Ne_1, \quad e_{2*} = (\csc\theta)Ne_2.$$

For $k > 1$, then proceeding by induction, for each $l = 1, \ldots, n-1$, we chose a unit tangent vector field e_{2l+1} of M, such as e_{2l+1}, which is orthogonal to $e_1, e_2, \ldots, e_{2l-1}, e_{2l}, \xi$ and took:

$$e_{2l+2} = (\sec\theta)Te_{2l+1}, \quad e_{(2l+1)*} = (\csc\theta)Ne_{2l+1}, \quad e_{(2l+2)*} = (\csc\theta)Ne_{2l+2}.$$

In this way

$$\{e_1, \ldots, e_m, \xi, e_{1*}, \ldots, e_{m*}\} \tag{21}$$

is an orthonormal reference such that $e_1, \ldots, e_m$ belong to the contact distribution, $\mathcal{D}$ and $e_{1*}, \ldots, e_{m*}$ are normal to M. Moreover, it can be directly computed that:

$$\begin{array}{lll} Te_{2j-1} = (\cos\theta)e_{2j}, & Te_{2j} = -(\cos\theta)e_{2j-1}, & j = 1, \ldots, k; \\ Ne_i = (\sin\theta)e_{i*}, & te_{i*} = -(\sin\theta)e_i, & i = 1, \ldots, m; \\ ne_{(2j-1)*} = -(\cos\theta)e_{(2j)*}, & ne_{(2j)*} = (\cos\theta)e_{(2j-1)*}, & j = 1, \ldots, k. \end{array}$$

Finally, a slant submanifold of an (α, β) trans-Sasakian generalized Sasakian space form $\widetilde{M}^{2m+1}(f_1, f_2, f_3)$, is called *-slant submanifold, [4], if its second fundamental form σ is given by the following expression:

$$\sigma(X, Y) = \frac{m+1}{m+2}\Big\{(g(X,Y) - \eta(X)\eta(Y))H$$
$$+ \left(\frac{1}{\sin^2\theta}g(\phi X, H) - \alpha\frac{m+2}{m+1}\eta(X)\right)NY \tag{22}$$
$$+ \left(\frac{1}{\sin^2\theta}g(\phi Y, H) - \alpha\frac{m+2}{m+1}\eta(Y)\right)NX\Big\}.$$

They are specially interesting because it was proven in [4] that this expression of the second fundamental form characterizes the equality case of the following inequality involving the squared mean curvature $\|H\|^2$ and the scalar curvature τ:

$$(m+1)^2\|H\|^2 - 2\frac{m+2}{m-1}\tau \geq -\frac{m(m+2)}{m-1}((m+1)f_1 + 3f_2\cos^2\theta - 2f_3 - 2\alpha\sin^2\theta). \tag{23}$$

3. The Maslov Form

For any submanifold of any almost contact manifold, we consider the Maslov form ω_H as the dual form of ϕH; that is

$$\omega_H(X) = g(X, \phi H),$$

for any X tangent vector field in the submanifold. We can also define a canonical 1-form on M by

$$\Theta = \sum_{1=1}^{m} \omega_i^{i*},$$

where ω_i^{i*} are the connection forms given by Cartan's structure equations.

We can relate these two forms for certain slant submanifolds. In [12], proper slant submanifolds such as for any tangent vector fields X and Y were studied with:

$$(\nabla_X T)Y = \cos^2\theta(g(X,Y)\xi - \eta(Y)X). \tag{24}$$

They were called *slant Sasakian submanifolds* in [6]; however, we can point that they are α-Sasakian manifolds with the induced structure $\overline{\phi} = \sec\theta T$. That aims us to defined *slant trans-Sasakian submanifolds* as those verifying:

$$(\nabla_X \overline{\phi})Y = \overline{\alpha}(g(X,Y)\xi - \eta(Y)X) + \overline{\beta}(g(\overline{\phi}X,Y)\xi - \eta(Y)\overline{\phi}X). \tag{25}$$

For a slant trans-Sasakian submanifold of a trans-Sasakian manifold the relation between the structure functions is given by

$$\sec\theta \overline{\alpha} = \alpha \quad \text{and} \quad \overline{\beta} = \beta. \tag{26}$$

From (25) and (12) it is deduced that

$$A_{NY}X = A_{NX}Y + \alpha\sin^2\theta(\eta(Y)X - \eta(X)Y), \tag{27}$$

for any X, Y tangent vector fields.

Then, for such a submanifold, the relation between Θ and the Maslov form is given in the following theorem.

Theorem 1. *Let M^{m+1} be a slant trans-Sasakian submanifold of a generalized Sasakian space form $\widetilde{M}^{2m+1}(f_1, f_2, f_3)$ endowed with an (α, β) trans-Sasakian structure. Then:*

$$\omega_H = -\frac{\sin\theta}{m+1}(\Theta + m\alpha\sin\theta\eta). \tag{28}$$

Proof. Considering an adapted slant basis, it holds

$$\omega_H(e_i) = g(e_i, \phi H) = -g(Ne_i, H) = -\sin\theta g(e_{i^*}, H), \tag{29}$$

for $i = 1, \ldots, m$. Moreover,

$$\Theta = \sum_{l=1}^{2n}\sum_{i=1}^{2n} \sigma_{li}^{l^*}\omega^i + \sum_{l=1}^{2n} \sigma_{l\xi}^{l^*}\eta. \tag{30}$$

But,

$$\sigma_{l\xi}^{l^*} = g(\sigma(e_l,\xi), e_{l^*}) = -\csc\theta g(Ne_l, Ne_l) = -\sin\theta, \tag{31}$$

and

$$\begin{aligned}\sigma_{li}^{l^*} &= g(\sigma(e_l, e_i), e_{l^*}) = \csc\theta g(\sigma(e_l, e_i), Ne_l) \\ &= \csc\theta g(A_{Ne_l}e_i, e_l) = \csc\theta g(A_{Ne_i}e_l, e_l) \\ &= g(\sigma(e_l, e_l), e_{i^*}) = \sigma_{ll}^{i^*},\end{aligned} \tag{32}$$

where we have used (27); that is, M is a slant trans-Sasakian submanifold.

And therefore, from (30)–(32),

$$\Theta + m\alpha\sin\theta\eta = \sum_{i=1} 2n(tr\sigma^{i^*})\omega^i.$$

As $\sigma(\xi,\xi) = 0$:

$$H = \frac{1}{m+1}\sum_{j=1}^{m}\sigma(e_j,e_j). \tag{33}$$

Now, from (29) and (33), it holds that

$$\omega_H(e_i) = -\frac{\sin\theta}{m+1}\sum_{j=1}^{2k}\sigma_{jj}^{i*} = -\frac{\sin\theta}{m+1}(\Theta + m\alpha\sin\theta\eta)(e_i),$$

for $i = 1,\ldots,m$. Finally, as $\omega_H(\xi) = g(tH,\xi) = 0$, the proof is finished. □

Following the same steps that [5] did for slant submanifolds of an almost Hermitian manifold or [6] for an almost contact manifold, and after a long computation, the differentials of θ and η can be proven. The proof is straightforward so we have omitted it.

Lemma 1. *Let M^{m+1}, a proper slant submanifold of a generalized Sasakian space form $\widetilde{M}^{2m+1}$ endowed with an (α,β) trans-Sasakian structure, with M tangent to ξ and $m \geq 2$. Then, the 1-forms Θ and η satisfy:*

$$d\Theta = -2\sin\theta\cos\theta(\alpha^2 + f_2(m+1))\left(\sum_{j=1}^{k}\omega^{2j-1}\wedge\omega^{2j} - \sum_{j=1}^{k}\omega^{(2j-1)*}\wedge\omega^{(2j)*}\right)$$
$$+(-2\sin^2\theta(\alpha^2 + f_2(m+1))+\alpha^2 + f_2 - f_1 - \beta^2) \tag{34}$$
$$\left(\sum_{j=1}^{k}\omega^{2j-1}\wedge\omega^{(2j-1)*} + \sum_{j=1}^{k}\omega^{2j}\wedge\omega^{(2j)*}\right),$$

and

$$d\eta = -2\alpha\cos\theta\sum_{j=1}^{k}\omega^{2j-1}\wedge\omega^{2j} - 2\alpha\sin\theta\sum_{j=1}^{k}\omega^{2j-1}\wedge\omega^{(2j-1)*} -$$
$$-2\alpha\sin\theta\sum_{j=1}^{k}\omega^{2j}\wedge\omega^{(2j)*} + 2\alpha\cos\theta\sum_{j=1}^{k}\omega^{(2j-1)*}\wedge\omega^{(2j)*}, \tag{35}$$

where θ is the slant angle of M.

As we are considering a trans-Sasakian manifold with a dimension greater or equal than 5, from [9], it must be an α-Sasakian or a β-Kenmotsu manifold. So we distinguish both two cases in the following theorems.

Theorem 2. *Let M^{m+1} be a proper slant trans-Sasakian submanifold of a connected generalized Sasakian space form $\widetilde{M}^{2m+1}(f_1,f_2,f_3)$ endowed with an α-Sasakian structure. Then, the Maslov form is closed if and only if $f_1 = 0$. In such a case, it holds $f_2 = f_3 = -\alpha^2$.*

Proof. As $\widetilde{M}^{2m+1}$ is α-Sasakian, from Proposition 4.1 of [14], α is constant. From (28),

$$d\omega_H = -\frac{\sin\theta}{m+1}(d\Theta + m\alpha\sin\theta d\eta).$$

Then, from (34) and (35), it is deduced that $d\omega_H = 0$ if and only if it holds $\alpha^2 + f_2 = 0$ and $f_1 = 0$. Moreover, Theorem 4.2 of [14] establishes that both conditions are equivalent, as $f_1 - \alpha^2 = f_2 = f_3$. □

Remark 1. *If the ambient space is a Sasakian space form $\widetilde{M}^{2m+1}(c)$, the Maslov form is closed if and only if $c = -3$, as it was proved in [6].*

Theorem 3. *Let M^{m+1} be a proper slant trans-Sasakian submanifold of a generalized Sasakian space form $\widetilde{M}^{2m+1}(f_1, f_2, f_3)$ endowed with a β-Kenmotsu structure. Then, the Maslov form is closed if and only if*

$$f_1 = -\beta^2 \quad \text{and} \quad f_2 = 0.$$

In such a case, it holds $f_3 = \xi(\beta)$.

Proof. Again from (28), (34) and (35), $d\omega_H = 0$ if and only if $f_2 = 0$ and $f_1 + \beta^2 = 0$. The last condition is obtained from Proposition 4.3 in [14], where it was proven that $f_1 - f_3 + \xi(\beta) + \beta^2 = 0$. □

Remark 2. *We note that on the opposite that for Lagrangian submanifold of $\mathbb{C}^n$, [2], or totally real submanifolds of $\mathbb{R}^{2m+1}$, [3], the Maslov is not always closed. That aims us to look for an adapted form that is closed in more cases.*

4. An Adapted Closed Form

As the Maslov form is not always closed for slant submanifolds it is necessary to find a new form related with this Maslov form but including the special slant character of the submanifold.

Both the Maslov form and Θ can be considered forms at $\widetilde{M}$ or M. As both η and Θ vanish at $TM^\perp$, it is the same defining them on $\widetilde{M}$ or M; however, it is not the same considering $d\eta$ or $d\eta]_M$ and $d\Theta$ or $d\Theta]_M$. Although both B.-Y. Chen and A. Carriazo, [5] and [6], studied conditions for $d\omega_H$ and $d\Theta$ vanishing at the manifold; their real interest was finding a closed form at the submanifold, not at the manifold.

Therefore, we consider the restrictions of Θ and η at the submanifold. From (34) and (35) it is deduced:

$$d\eta]_M = -2\alpha \cos\theta \sum_{j=1}^{m} \omega^{2j-1} \wedge \omega^{2j} \tag{36}$$

and

$$d\Theta]_M = -2\sin\theta \cos\theta (\alpha^2 + f_2(m+1)) \sum_{j=1}^{m} \omega^{2j-1} \wedge \omega^{2j}. \tag{37}$$

So we find that, for obtaining a closed form, the relation between Θ and η is not the given by the Maslov form at (28).

Again, we particularize to α-Sasakian or a β-Kenmotsu manifolds. Firstly, we consider an α-Sasakian manifold. It was proven in [14], that if $\alpha \neq 0$ and $\widetilde{M}(f_1, f_2, f_3)$ is connected, then α is constant, and the functions are constant and related by $f_1 - \alpha^2 = f_2 = f_3$. We can write:

$$f_1 = \frac{c + 3\alpha^2}{4}, \quad f_2 = f_3 = \frac{c - \alpha^2}{4}.$$

From now on, we suppose $\widetilde{M}$ is connected.

Lemma 2. *Let M^{m+1} be a slant submanifold of an α-Sasakian generalized Sasakian space form $\widetilde{M}^{2m+1}(f_1, f_2, f_3)$, with $\alpha \neq 0$. Then, the form $\Theta - \sin\theta \dfrac{\alpha^2 + f_2(m+1)}{\alpha} \eta$ is closed at M.*

Proof. It is directly deduced from (36) and (37) that $\alpha d\Theta - \sin\theta(\alpha^2 + f_2(m+1))d\eta = 0$, and as α is constant, the result is proven. □

Moreover, the field associated to the closed form is

$$-\frac{m+1}{\sin\theta} tH - \sin\theta \left(m + \frac{\alpha^2 + f_2(m+1)}{\alpha}\right) \zeta, \tag{38}$$

so we already have the following theorem.

Theorem 4. *Let M^{m+1} be a slant submanifold of an α-Sasakian generalized Sasakian space form $\widetilde{M}^{2m+1}(f_1, f_2, f_3)$, with $\alpha \neq 0$. Then, the field $tH + \dfrac{\sin^2 \theta}{m+1} \dfrac{m\alpha + \alpha^2 + f_2(m+1)}{\alpha} \xi$ is closed.*

Corollary 1. *Let M^{m+1} a slant submanifold of a Sasakian space form $\widetilde{M}^{2m+1}(c)$; the field $tH + \sin^2 \theta \dfrac{c+3}{4} \xi$ is closed.*

Note that this result improves the one obtained by A. Carriazo in [6] giving a closed form for a slant submanifold of any Sasakian space form.

Corollary 2. *Let M^{2m+1} be a compact and simply connected manifold. Then, M can not be immersed in a generalized Sasakian space form, $\widetilde{M}^{4m+1}(0, -\alpha^2, -\alpha^2)$, endowed with an α-Sasakian structure, $\alpha \neq 0$, like a slant submanifold.*

Proof. If M^{m+1} is a slant submanifold of $\widetilde{M}^{4m+1}(0, -\alpha^2, -\alpha^2)$, with an α-Sasakian structure. By Theorem 4 the vector field

$$tH + \dfrac{\sin^2 \theta}{m+1} \dfrac{m\alpha + \alpha^2 + f_2(m+1)}{\alpha} \xi \neq 0,$$

is closed, and the corresponding form is also closed. Therefore it represents a cohomology class in $H^1(M; \mathbb{R})$. But, as M is compact, it can not be an exact form. So $H^1(M; \mathbb{R})$ is a nontrivial cohomology class and M could not be simply connected what is a contradiction. □

On the other hand, for a β-Kenmotsu manifold $d\eta = 0$ and from Theorem 1, $\omega_H = -\dfrac{\sin \theta}{m+1} \Theta$. The following lemma studies when it is a closed form.

Lemma 3. *Let M^{m+1} be a proper slant submanifold of a β-Kenmotsu generalized Sasakian space form $\widetilde{M}^{2m+1}$, with M tangent to ξ and $m \geq 2$. Then, the Maslov form at M is closed if and only if $f_2 = 0$.*

Proof. For a β-Kenmotsu manifold $\omega_H = -\dfrac{\sin \theta}{m+1} \Theta$. And writing (37) for $\alpha = 0$,

$$d\Theta]_M = -2 \sin \theta \cos \theta f_2 (m+1) \sum_{j=1}^{m} \omega^{2j-1} \wedge \omega^{2j}. \tag{39}$$

Therefore, the Maslov form is closed in M if and only $f_2 = 0$. □

Note, that in such a case $f_1 - f_3 + \xi(\beta) + \beta^2 = 0$ ([14], Proposition 4.3). Moreover, we observe that, on the opposite that for α-Sasakian manifolds, we cannot find a closed form for a slant submanifold of any generalized Sasakian space form with a β-Kenmotsu structure.

However, for $f_2 = 0$, we have obtained a closed vector field as follows.

Theorem 5. *Let M^{m+1} be a slant submanifold of an β-Kenmotsu generalized Sasakian space form $\widetilde{M}^{2m+1}(f_1, 0, f_3)$. Then, the field $tH + \sin^2 \theta \dfrac{m}{m+1} \xi$ is closed.*

Again, we can present a topological obstruction for slant immersions:

Corollary 3. *Let M^{2m+1} be a compact and simply connected manifold. Then, M cannot be immersed in a generalized Sasakian space form, $\widetilde{M}^{4m+1}(f_1, 0, f_3)$, endowed with an β-Kenmotsu structure, as a slant submanifold.*

From now on, we will write $tH + a\xi$ and $tH + b\xi$, with

$$a = \frac{\sin^2\theta}{m+1} \frac{m\alpha + \alpha^2 + f_2(m+1)}{\alpha} \quad \text{and} \quad b = \sin^2\theta \frac{m}{m+1},$$

for the correspondent closed vector fields.

5. About Conformal Forms for α-Sasakian Space Forms

As we said in the Introduction, for those Lagrangian submanifolds of $\mathbb{C}^m$ verifying the equality case, the Maslov form, that is closed, is also conformal. Now we study if the closed form presented in the previous section is conformal for those slant submanifolds verifying the equality case at (23).

We are considering a connected manifold, so α, f_1, f_2 and f_3 are constant functions.

We want to compute $\nabla_X(tH + a\xi)$, for any X tangent vector field. It is a long computation. Firstly, we compute ∇N for later use. Using the expression of the second fundamental form of a *-slant submanifold, (22), and (20) in (13):

$$(\nabla_X N)Y = \frac{m+1}{m+2} \Big\{ (g(X,Y) - \eta(X)\eta(Y))nH - g(X,TY)H \\
+ 2\left(\frac{1}{\sin^2\theta}g(\phi X, H) - \frac{m+2}{m+1}\eta(X)\right)nNY \\
+ \left(\frac{1}{\sin^2\theta}g(\phi Y, H) - \frac{m+2}{m+1}\eta(Y)\right)nNX \\
- \frac{1}{\sin^2\theta}g(\phi TY, H)NX \Big\}. \tag{40}$$

Lemma 4. *Let M be *-slant submanifold of an generalized Sasakian space form $\widetilde{M}(f_1, f_2, f_3)$ endowed with an α-Sasakian structure. For every X tangent vector field belonging to the contact distribution it holds:*

$$\nabla_X(tH + a\xi) = -g(D_X H, NX)X + \left(\frac{1}{\sin^2\theta}\frac{m+1}{m+2}g^2(H,NX) - a\right)TX \\
- \frac{3}{\sin^2\theta}\frac{m+1}{m+2}g(H,NX)tnH \\
+ \frac{1}{\sin^2\theta}\frac{m+1}{m+2}g(H,nNX)tH + g(H,nNX)\xi. \tag{41}$$

Proof. Firstly, from Codazzi's equation we will compute $D_X H$, and after, $\nabla_X(tH + a\xi)$. Writing Codazzi's equation, (9), for a generalized Sasakian space form, for any unit orthogonal X, Y tangent vector fields in the contact distribution, using (3) (6), $R(X,Y)Y^\perp$ gives:

$$3\frac{m+2}{m+1}f_2 g(X,TY)NY = D_X H + 3\frac{m+2}{m+1}g(Y,TX)NY \\
+ \frac{2}{\sin^2\theta}\{g((\nabla_X N)Y, H)NY + g(NY, D_X H)NY + g(NY, H)(\nabla_X N)Y\} \\
+ \frac{1}{\sin^2\theta}\{-g((\nabla_Y N)X, H)NY - g((\nabla_Y N)Y, H)NX - g(NX, H)(\nabla_Y N)Y \\
- g(NX, D_Y H)NY - g(NY, D_Y H)NX - g(NY, H)(\nabla_Y N)X\}. \tag{42}$$

Then, using (40), we obtain:

$$
\begin{aligned}
3\frac{m+2}{m+1}&(f_2+1)g(X,TY)NY = D_XH \\
&+ \frac{1}{\sin^2\theta}\{2g(NY,D_XH)NY - g(NX,D_YH)NY - g(NY,D_YH)NX\} \\
&+ \frac{1}{\sin^2\theta}\frac{m+1}{m+2}\Big\{-g(NX,H)nH - \frac{2}{\sin^2\theta}g(NY,H)g(nNY,H)NX - 3g(NY,H)g(X,TY)H \\
&\Big(-3g(X,TY)\|H\|^2 + \frac{4}{\sin^2\theta}g(NX,H)g(nNY,H) - \frac{2}{\sin^2\theta}g(NY,H)g(nNX,H)\Big)NY\Big\}.
\end{aligned}
\tag{43}
$$

At this point, we use that, taking into account Corollary 1, $tH + a\zeta$ is a closed vector field.

$$g(\nabla_X(tH+a\zeta),Y) = g(\nabla_Y(tH+a\zeta),X).$$

Then, using (16)

$$g(\nabla_X tH, Y) = g(\nabla_Y tH, X) - 2ag(TY, X),$$

and therefore, (14) gives

$$
\begin{aligned}
g(NX, D_YH) &= -g(X, tD_YH) = -g(X, \nabla_Y tH - A_{nH}Y + TA_HY) \\
&= g(NY, D_XH) + 2ag(TX, Y) - g(\sigma(TY, X), H) + g(\sigma(Y, TX), H).
\end{aligned}
\tag{44}
$$

Now, using (22) carries to

$$
\begin{aligned}
g(NX, D_YH) =& g(NY, D_XH) + 2ag(TX, Y) + \frac{m+1}{m+2}2g(Y, TX)\|H\|^2 \\
&+ \frac{m+1}{m+2}\frac{2}{\sin^2\theta}(g(NY, H)g(NTX, H) - g(NX, H)g(NTY, H)).
\end{aligned}
\tag{45}
$$

So (43) gives

$$
\begin{aligned}
3\frac{m+2}{m+1}&(f_2+1)g(X,TY)NY = D_XH + \frac{1}{\sin^2\theta}\{g(NY,D_XH)NY - g(NY,D_YH)NX\} \\
&+ \frac{1}{\sin^2\theta}\frac{m+1}{m+2}\Big\{\Big(-2a\frac{m+2}{m+1}g(TX,Y) + g(Y,TX)\|H\|^2 + \frac{2}{\sin^2\theta}g(NX,H)g(nNY,H)\Big)NY \\
&-g(NX,H)nH - \frac{2}{\sin^2\theta}g(NY,H)g(nNY,H)NX + 3g(NY,H)g(Y,TX)H\Big\}.
\end{aligned}
\tag{46}
$$

Now, for dimensions over or equal than 5, we can consider X orthogonal to Y and TY. Multiplying by NX,

$$
\begin{aligned}
0 =& g(D_XH, NX) - g(NY, D_YH) \\
&+ \frac{1}{\sin^2\theta}\frac{m+1}{m+2}\{-g(NX,H)g(nNX,H) - 2g(NY,H)g(nNY,H)\}.
\end{aligned}
\tag{47}
$$

Interchanging X and Y at (47), and adding it to the previous equation:

$$g(NX,H)g(nNX,H) = -g(NY,H)g(nNY,H). \tag{48}$$

For TY, that is also orthogonal to X, TX,

$$g(NX,H)g(nNX,H) = -g(NTY,H)g(nNTY,H) = \cos^2\theta g(NY,H)g(nNY,H). \tag{49}$$

From (48) and (51), we get $g(NX, H)g(nNX, H) = 0$ for every X unit vector field in the contact distribution. Moreover, developing $0 = g(N(X+Y), H)g(nN(X+Y), H)$, we obtain

$$g(NX, H)g(nNY, H) = -g(NY, H)g(nNX, H). \tag{50}$$

Also, at (47), we get

$$g(D_X H, NX) = g(NY, D_Y H), \tag{51}$$

so $g(D_X H, NX)$ is independent of X unit vector field in the contact distribution.

Now, multiplying (46) by NY,

$$0 = 2g(D_X H, NY) + \frac{3}{\sin^2 \theta} \frac{m+1}{m+2} g(NX, H)g(nNY, H). \tag{52}$$

But (44) for any X, a unitary vector field orthogonal to Y, TY, in the contact distribution it states:

$$g(NX, D_Y H) - g(\sigma(TX, Y), H) = g(NY, D_X H) - g(\sigma(TY, X), H). \tag{53}$$

Using (52) and (22) at (53)

$$\frac{-7}{2\sin^2 \theta} \frac{m+1}{m+2} g(NX, H)g(nNY, H) = \frac{-7}{2\sin^2 \theta} \frac{m+1}{m+2} g(NY, H)g(nNX, H), \tag{54}$$

where X, Y can be interchanged. Comparing (50) with (54) it is proven that

$$g(NX, H)g(nNY, H) = 0, \tag{55}$$

and consequently, by (52),

$$g(D_X H, NY) = 0, \tag{56}$$

for each X orthogonal to Y and TY at the contact distribution.

It only rests on us to compute $g(D_X H, NTX)$ in order to know $D_X H$. Multiplying (46) by NTX we obtain:

$$g(D_X H, NTX) = -\frac{\cos^2 \theta}{\sin^2 \theta} \frac{m+1}{m+2} g(H, NX)^2. \tag{57}$$

Therefore, taking an orthogonal basis $\{e_1^*, ..., e_n^*\}$ at $T^\perp M$,

$$D_X H = \sum g(D_X H, e_j^*) e_j^* = $$
$$= \frac{1}{\sin^2 \theta} g(D_X H, NX) NX - \frac{1}{\sin^4 \theta} \frac{m+1}{m+2} g(H, NX)^2 NTX, \tag{58}$$

for any X unit tangent field orthogonal to ξ.

Finally, for any X at the contact distribution, and any Z tangent vector field,

$$g(\nabla_X(tH + a\xi), Z) = g(\nabla_X tH, Z) - ag(TX, Z)$$
$$= g(t\nabla_X H + A_{nH} X - TA_H X, Z) - ag(TX, Z)$$
$$= -g(D_X H, NZ) + g(nH, h(X, Z)) + g(H, h(X, TZ)) - ag(TX, Z)$$
$$= -g\left(g(D_X H, NX)NX - \frac{1}{\sin^4 \theta} \frac{m+1}{m+2} g(H, NX)^2 NTX, NZ\right)$$
$$+ \frac{m+1}{m+2} \left(\frac{1}{\sin^2 \theta} g(NX, H)g(NZ, nH) + \left(\frac{1}{\sin^2 \theta} g(NZ, H) - \frac{m+2}{m+1} \eta(Y)\right) g(NX, nH)\right)$$

$$+\frac{m+1}{m+2}\left(\frac{1}{\sin^2\theta}g(NX,H)g(NTZ,H)+\frac{1}{\sin^2\theta}g(NTZ,H)g(NX,H)\right)$$
$$-ag(TX,Z), \tag{59}$$

where we used (22). This last equation, using (19), direct gives the desired expression of $\nabla_X tH + \frac{c-1}{4}\cos^2\theta\xi$. □

The quid point of the above proof is to deduce, from Codazzi's equation and the expression of the second fundamental form, that $g(D_X H, NX)$ is independent of X and also that $g(D_X H, NY) = 0$ for Y orthogonal to X, TX. This is the same sketch than A. Ros and F. Urbano did in [2].

Now, we repeat the same steps in order to obtain $\nabla_\xi(tH + a\xi)$.

Lemma 5. *Let M be *-slant submanifold of a Sasakian space form $\widetilde{M}(c)$; it holds:*

$$\nabla_\xi(tH + a\xi) = -TtH. \tag{60}$$

Proof. Using that, from Corollary 1, $tH + a\xi$ is a closed vector field,

$$g(\nabla_\xi(tH + a\xi), Y) = g(\nabla_Y(tH + a\xi), \xi),$$

so using (16),

$$g(\nabla_\xi tH, X) = g(\nabla_X tH, \xi) = -g(tH, \nabla_X \xi) = g(tH, TX) = -g(TtH, X),$$

for any X tangent vector field, which finishes the proof. □

Theorem 6. *Let M be *-slant submanifold of a generalized Sasakian space form $\widetilde{M}(f_1, f_2, f_3)$, endowed with an α-Sasakian structure. Then, for every X tangent vector field it holds:*

$$\nabla_X(tH + a\xi) = (-g(D_X H, NX) + \eta(X)g(H, nNX))(X - \eta(X)\xi)$$
$$+ \left(\frac{1}{\sin^2\theta}\frac{m+1}{m+2}g(H, NX)^2 - a\right)TX$$
$$+ \left(\frac{-3}{\sin^2\theta}\frac{m+1}{m+2}g(H, NX) + \eta(X)\right)tnH$$
$$+ \frac{1}{\sin^2\theta}\frac{m+1}{m+2}g(H, nNX)tH + g(H, nNX)\xi.$$

Proof. It is a direct consequence of Lemmas 4 and 5. □

So, in general, for a *-slant submanifold of a generalized Sasakian space form, the closed form is not conformal. However, for the corresponding vector field, the covariant derivative with respect to X is in the direction of X, TX, tnH and ξ.

6. About Conformal Forms for β-Kenmotsu Space Forms

At Section 4 we obtained that, for a β-Kenmotsu generalized Sasakian space form $\widetilde{M}(f_1, 0, f_3)$, the vector field $tH + \sin^2\theta \frac{m}{m+1}\xi = tH + b\xi$ is always closed. So, the associated form plays the role of the Maslov form for Lagrangian submanifolds of Kaehler manifolds. In this section we study if it is conformal for a *-slant submanifold.

The study is similar to the one made at Section 5, so we omit the proofs.

Lemma 6. *Let M be *-slant submanifold of a β-Kenmotsu generalized Sasakian space form $\widetilde{M}(f_1, 0, f_3)$. Then, for every X tangent vector field belonging to the contact distribution it holds:*

$$\nabla_X(tH + b\xi) = \left(-g(D_XH, NX) + \beta \sin^2\theta \frac{m}{m+1}\right) X$$
$$+ \frac{m+1}{m+2}\left(\frac{1}{\sin^2\theta}g(H, NX)^2 - \|H\|^2\right) TX$$
$$- \frac{3}{\sin^2\theta}\frac{m+1}{m+2}g(H, NX)tnH$$
$$+ \frac{1}{\sin^2\theta}\frac{m+1}{m+2}g(H, nNX)tH + \beta g(NX, H)\xi.$$
(61)

Again, the quid point of the proof is to deduce, from Codazzi's equation and the expression of the second fundamental form, that $g(D_XH, NX)$ is independent of X and that $g(D_XH, NY) = 0$ for X orthogonal to Y.

Now, we repeat the same steps in order to obtain $\nabla_\xi(tH + b\xi)$.

Lemma 7. *Let M be *-slant submanifold of a β-Kenmotsu space form $\widetilde{M}(f_1, 0, f_3)$; it holds:*

$$\nabla_\xi tH = -\beta tH.$$
(62)

Finally, we get:

Theorem 7. *Let M be *-slant submanifold of a β Kenmotsu space form $\widetilde{M}(f_1, 0, f_3)$. Then, for every X tangent vector field it holds:*

$$\nabla_X(tH + b\xi) =$$
$$= \left(-g(D_XH, NX) + \beta\eta(X)g(H, nNX) + \beta\sin^2\theta\frac{m}{m+1}\right)(X - \eta(X)\xi)$$
$$+ \frac{m+1}{m+2}\left(\frac{1}{\sin^2\theta}g(H, NX)^2 - \|H\|^2\right)TX - \frac{3}{\sin^2\theta}\frac{m+1}{m+2}g(H, NX)tnH$$
$$+ \frac{1}{\sin^2\theta}\frac{m+1}{m+2}g(H, nNX)tH - \beta\eta(X)tH + \beta g(NX, H)\xi.$$

Proof. It is a direct consequence from Lemmas 6 and 7. □

Again, for a *-slant submanifold of a β-Kenmotsu generalized Sasakian space form, the closed form is not conformal. However, for the corresponding vector field, the covariant derivative with respect to X is in the direction of X, TX, tH, tnH and ξ.

Author Contributions: All the authors contributed equally to this work.

Conflicts of Interest: The authors declare no conflict of interest.

References

1. Borrelli, V.; Chen, B.Y.; Morvan, J.M. Une caractérisation géométrique de la sphère de Whitney. *C. R. Acad. Sci. Paris* **1995**, *321*, 1485–1490.
2. Ros, A.; Urbano, F. Lagrangian submanifolds of $\mathbf{C}^n$ with conformal Maslov form and the Whitney sphere. *J. Math. Soc. Jpn.* **1998**, *50*, 203–226. [CrossRef]
3. Blair, D.E.; Carriazo, A. The contact Whitney sphere. *Note Mat.* **2000**, *20*, 125–133.
4. Alegre, P. Barrera, J.; Carriazo, A. A new class of slant submanifolds in generalized Sasakian space forms. *Appear Med. J. Math.* **2020**, under review.
5. Chen, B.Y. *Geometry of Slant Submanifolds*; Katholieke Universiteit Leuven: Leuven, Belgium, 1990.

6. Carriazo, A. Obstructions to slant immersions in contact manifolds. *Ann. Mat. Pura Appl.* **2001**, *179*, 459–470. [CrossRef]
7. Blair, D.E. *Contact Manifolds in Riemannian Geometry, Lecture Notes in Mathematics 509*; Springer: New York, NY, USA, 1976.
8. Oubi na, J.A. New classes of almost contact metric structures. *Publ. Math. Debrecen* **1985**, *32*, 187–193.
9. Marrero, J.C. The Local Structure of Trans-Sasakian Manifolds. *Ann. Mat. ura Appl. CLXII* **1992**, *162*, 77–86. [CrossRef]
10. Alegre, P.; Blair, D.; Carriazo, A. Generalized Sasakian-space-forms. *Israel J. Math.* **2004**, *141*, 157–183. [CrossRef]
11. Lotta, A. Slant submanifolds in contact geometry. *Bull. Math. Soc. Roum.* **1996**, *39*, 183–198.
12. Cabrerizo, J.L.; Carriazo, A.; Fernández, L.M.; Fernández, M. Slant submanifolds in Sasakian manifolds. *Glasgow Math. J.* **2000**, *42*, 125–138. [CrossRef]
13. Cabrerizo, J.L.; Carriazo, A.; Fernández, L.M.; Fernández, M. Structure on a slant submanifold of a contact manifold. *Indian J. Pure Appl. Math.* **2000**, *31*, 857–864.
14. Alegre, P.; Carriazo, A. Structures on generalized Sasakian-space-forms. *Differ. Geom. Appl.* **2008**, *26*, 656–666. [CrossRef]

© 2019 by the authors. Licensee MDPI, Basel, Switzerland. This article is an open access article distributed under the terms and conditions of the Creative Commons Attribution (CC BY) license (http://creativecommons.org/licenses/by/4.0/).

Article

On Differential Equations Characterizing Legendrian Submanifolds of Sasakian Space Forms

Rifaqat Ali [1], Fatemah Mofarreh [2], Nadia Alluhaibi [3], Akram Ali [4,*] and Iqbal Ahmad [5]

1. Department of Mathematics, College of Sciences and Arts, Muhayil, King Khalid University, Abha 9004, Saudi Arabia; rifaqat.ali1@gmail.com
2. Mathematical Science Department, Faculty of Science, Princess Nourah bint Abdulrahman University, Riyadh 11546, Saudi Arabia; fyalmofarrah@pnu.edu.sa
3. Department of Mathematics, Science and Arts College, Rabigh Campus, King Abdulaziz University, Jeddah 21911, Saudi Arabia; nallehaibi@kau.edu.sa
4. Department of Mathematics, College of Science, King Khalid University, Abha 9004, Saudi Arabia
5. College of Engineering, Qassim University, Buraidah 51452, Al-Qassim, Saudi Arabia; iqbal@qec.edu.sa
* Correspondence: akali@kku.edu.sa

Received: 10 December 2019; Accepted: 15 January 2020; Published: 21 January 2020

Abstract: In this paper, we give an estimate of the first eigenvalue of the Laplace operator on minimally immersed Legendrian submanifold N^n in Sasakian space forms $\widetilde{N}^{2n+1}(\epsilon)$. We prove that a minimal Legendrian submanifolds in a Sasakian space form is isometric to a standard sphere $\mathbb{S}^n$ if the Ricci curvature satisfies an extrinsic condition which includes a gradient of a function, the constant holomorphic sectional curvature of the ambient space and a dimension of N^n. We also obtain a Simons-type inequality for the same ambient space forms $\widetilde{N}^{2n+1}(\epsilon)$.

Keywords: legendrian submanifolds; sasakian space forms; obata differential equation; isometric immersion

MSC: 58C40; 53C42; 35P15

1. Introduction and Motivations

In 1959, Yano and Nagano [1] proved that if a complete Einstein space of dimension strictly greater than 2 admits a 1-parameter group of non-homothetic conformal transformations, then it is isometric to a sphere. Later, Obata [2] gave a simplified proof of the result of Yano and Nagano by analyzing a differential equation, nowadays known as Obata equation. Recall that a complete manifold (N^n, g) admits a non-constant function ψ satisfying the Obata differential equation

$$Hess(\psi) + \psi g = 0, \qquad (1)$$

if and only if (N^n, g) is isometric to the standard sphere $\mathbb{S}^n$. Such characterizations of complete spaces are of great interest and they were investigated by many geometers (see [3–12]). For example, Tashiro [13] has shown that the Euclidean spaces $\mathbb{R}^n$ are characterized by a differential equation $\nabla^2 \psi = cg$, where c is a positive constant. Utilizing Obata Equation (1), Barros et al. [14] have shown that a compact gradient almost Ricci soliton $(N^n, g, \nabla \psi, \lambda)$ with the Codazzi Ricci tensor and constant sectional curvature is isometric to the Euclidean sphere, and then ψ is a height function in this case. For more terminologies related to the Obata equation, see [8]. In [15], Lichnerowicz proved that, if the first non-zero eigenvalue μ_1 of the Laplacian on a compact manifold (M^n, g) with $Ric \geq n-1$, is not less than n, while $\mu_1 = n$, then (M^n, g) is isometric to the sphere $\mathbb{S}^n$. This means that the Obata's rigidity theorem could be used to analyze the equality case of Lichnerowicz's eigenvalue estimates in [15]. In the sequel, inspired by ideas developed in [16–18], we derive some rigidity theorems in the present paper.

On the other hand, by considering N^n as a compact submanifold immersed in Euclidean space $\mathbb{R}^{n+p}$ or the standard Euclidean sphere $\mathbb{S}^{n+p}$, Jiancheng Zhang [17] derived the Simons-type [18] inequalities of the first eigenvalue μ_1 and the squared norm of the second fundamental form S without need of minimallty. In addition, a lower bound of S can be provided if it is constant. Similar results can be found in [14,16]. As a generalization in the case of an odd-dimensional sphere, a minimally immersed Legendrian submanifold into a Sasakian space form of constant holomorphic sectional curvature ϵ should be considered in order to obtain Simon's-like inequality theorem.

2. Preliminaries and Notations

An odd-dimensional C^∞-manifold $(\widetilde{N}, g)$ is said to be an *almost contact metric manifold* if it is equipped with almost contact structure (ϕ, η, ζ) satisfying following properties:

$$\phi^2 = -I + \zeta \otimes \eta, \quad \eta(\zeta) = 1, \quad \phi(\zeta) = 0, \quad \eta \circ \phi = 0, \tag{2}$$

$$g(\phi V_1, \phi V_2) = g(V_1, V_2) - \eta(V_1)\eta(V_2), \quad \& \quad \eta(V_1) = g(V_1, \zeta), \tag{3}$$

$\forall V_1, V_2 \in \Gamma(T\widetilde{N})$, where ϕ, ζ and η are a tensor field of type $(1,1)$, a structure vector field and a dual 1-form, respectively. Moreover, an almost contact metric manifold $\widetilde{N}^{2m+1}$ is referred to as a *Sasakian manifold* if it fulfills the following relation

$$(\widetilde{\nabla}_{V_1} \phi) V_2 = g(V_1, V_2)\zeta - \eta(V_2)V_1. \tag{4}$$

It follows that

$$\widetilde{\nabla}_{V_1} \zeta = -\phi V_1, \tag{5}$$

for any $V_1, V_2 \in \Gamma(T\widetilde{N})$, where $\widetilde{\nabla}$ stands for the Riemannian connection in regard to g. A Sasakian manifold $\widetilde{N}^{2m+1}$ equipped with constant ϕ-sectional curvature ϵ is referred to as *Sasakian space form* and denoted by $\widetilde{N}^{2m+1}(\epsilon)$. Then, the following formula for the curvature tensor $\widetilde{R}$ of $\widetilde{N}^{2m+1}(\epsilon)$ can be expressed as:

$$\begin{aligned}\widetilde{R}(V_1, V_2, V_3, V_4) =& \frac{\epsilon+3}{4}\Big\{g(V_2, V_3)g(V_1, V_4) - g(V_1, V_3)g(V_2, V_4)\Big\} \\&+ \frac{\epsilon-1}{4}\Big\{\eta(V_1)\eta(V_3)g(V_2, V_4) + \eta(V_4)\eta(V_2)g(V_1, V_3) \\&\quad - \eta(V_2)\eta(V_3)g(V_1, V_4) - \eta(V_1)g(V_2, V_3)\eta(V_4) \\&\quad + g(\phi V_2, V_3)g(\phi V_1, V_4) - g(\phi V_1, V_3)g(\phi V_2, V_4) \\&\quad + 2g(V_1, \phi V_2)g(\phi V_3, V_4)\Big\},\end{aligned} \tag{6}$$

$\forall V_1, V_2, V_3, V_4 \in \Gamma(T\widetilde{N})$. Moreover, $\mathbb{R}^{2m+1}$ and $\mathbb{S}^{2m+1}$ with standard Sasakian structures can be given as typical examples of Sasakian space forms. An n-dimensional Riemannian submanifold N^n of $\widetilde{N}^{2m+1}(\epsilon)$ is referred to as totally real if the standard almost contact structure ϕ of $\widetilde{N}^{2m+1}(\epsilon)$ maps any tangent space of N^n into its corresponding normal space (see [4,19–21]). Now, let N^n be an isometric immersed submanifold of dimension n in $\widetilde{N}^{2m+1}(\epsilon)$. Then N^n is referred to as a Legendrian submanifold if ζ is a normal vector field on N^n, i.e., N^n is a C- totally real submanifold, and $m = n$ [22]. Legendrian submanifolds play a substantial role in contact geometry. From Riemannian geometric perspective, studying Legendrian submanifolds of Sasakian manifolds was initiated in

1970's. Many geometers have drawn significant attention to minimal Legendrian submanifolds in particular. In order to proceed let us recall the definition of the curvature tensor $\tilde{R}$ for Legendrian submanifold in $\tilde{N}^{2n+1}(\epsilon)$ which is given by

$$\tilde{R}(V_1, V_2, V_3, V_4) = \left(\frac{\epsilon+3}{4}\right)\left\{g(V_2,V_3)g(V_1,V_4) - g(V_1,V_3)g(V_2,V_4)\right\}. \tag{7}$$

Let $\{e_1, \cdots, e_n\}$ be an adapted orthogonal frame to N^n. Then, the second fundamental from h associated to N^n is defined as

$$h(e_i, e_j) = \sum_{\gamma=1}^{n} \sigma_{ij}^{\gamma} e_{\gamma},$$

where $\sigma_{ij}^{\gamma} = \langle A_{\gamma} e_i, e_j \rangle$ and A_{γ} is the shape operator in the direction of e_{γ}. Hence, the Gauss formula for Legendrian submanifold N^n in $\tilde{N}^{2n+1}(\epsilon)$ in the local coordinates has the form

$$R_{jkl}^{i} = \left(\delta_{ii}\delta_{jj} - \delta_{ij}\delta_{ji}\right)\left(\frac{\epsilon+3}{4}\right) + \sum_{\gamma=1}^{n}(\sigma_{ik}^{\gamma}\sigma_{jl}^{\gamma} - \sigma_{il}^{\gamma}\sigma_{jk}^{\gamma}).$$

Therefore, we have

$$R_{jij}^{i} = \left(\delta_{ii}\delta_{jj} - \delta_{ij}\delta_{ji}\right)\left(\frac{\epsilon+3}{4}\right) + \sum_{\gamma=1}^{n}(\sigma_{ii}^{\gamma}\sigma_{jj}^{\gamma} - \sigma_{ij}^{\gamma}\sigma_{ji}^{\gamma}). \tag{8}$$

We should note that Ψ is a C-totally real minimal immersion. Then, (8) yields

$$Ric(e_i, e_j) = (n-1)\left(\frac{\epsilon+3}{4}\right)\delta_{ij} - \sum_{\gamma=1}^{n}\sigma_{ir}^{\gamma}\sigma_{jr}^{\gamma}. \tag{9}$$

Now, we recall that Bochner formula [4] as follows: if $\psi : N^n \to \mathbb{R}$ is a function defined on a Riemannian manifold N^n, then we have

$$\frac{1}{2}\Delta|\nabla\psi|^2 = |Hess(\psi)|^2 + Ric_{N^n}(\nabla\psi, \nabla\psi) + g(\nabla\psi, \nabla(\Delta\psi)), \tag{10}$$

where, Ric denotes the Ricci tensor of N^n and $|A|$ stands for the norm of an operator A which is given by $|A|^2 = tr(AA^*); A^*$ is the transpose of A.

3. The Main Results

Now, we give a proof of the following essential proposition that we need later to prove our main Theorems 1 and 2.

Proposition 1. *Let $\Psi : N^n \to \tilde{N}^{2n+1}(\epsilon)$ be a minimal immersion of a compact Legendrian submanifold into the Sasakian space form $\tilde{N}^{2n+1}(\epsilon)$ and ψ be a first eigenfunction associated to the Laplacian of N^n. Then if $\{e_1, \cdots, e_n\}$ is an orthonormal tangent basis on N^n, we have*

$$\left\{(n-1)\left(\frac{\epsilon+3}{4}\right) - \mu_1\right\}\int_N |\nabla\psi|^2 dV + \int_N |Hess(\psi)|^2 dV = \int_N \sum_{i=1}^{n}|h(\nabla\psi, e_i)|^2 dV, \tag{11}$$

and particularly, we get

$$\int_N \sum_{i=1}^n |h(\nabla\psi, e_i)|^2 dV = \int_N \left|\text{Hess}(\psi) + \frac{\mu_1}{n}\psi I\right|^2 dV$$
$$+ \left\{(n-1)\left(\frac{\epsilon+3}{4}\right) - \frac{\mu_1}{n}\right\} \int_N |\nabla\psi|^2 dV, \quad (12)$$

where I denotes the identity operator on TN, μ_1 is an eigenvalue of the eigenfunction ψ such that $\Delta\psi + \mu_1\psi = 0$, and $\text{Hess}(\psi)$ is the squared norm of the Hessian of ψ.

Proof. Let I be the identity operator on TN. Then we have

$$\left|\text{Hess}(\psi) - t\psi I\right|^2 = |\text{Hess}(\psi)|^2 - 2t\psi g(I, \text{Hess}(\psi)) + |I|^2 t^2 \psi^2. \quad (13)$$

It should be noted that $|I|^2 = \text{trace}(II^*) = n$, and

$$g(\text{Hess}(\psi), I) = \text{trace}(\text{Hess}(\psi)I^*) = \text{trace}(\text{Hess}(\psi)) = \Delta\psi.$$

Therefore, if $\Delta\psi + \mu_1\psi = 0$, we derive it for any $t \in \mathbb{R}$. Integrating Equation (13), and using the above equation and Stokes theorem, we get

$$\int_N \left|\text{Hess}(\psi) - t\psi I\right|^2 dV = \int_N |\text{Hess}(\psi)|^2 dV + \left(2t + \frac{n}{\mu_1}t^2\right) \int_N |\nabla\psi|^2 dV. \quad (14)$$

Setting $t = -\frac{\mu_1}{n}$ in (14), we get

$$\int_N |\text{Hess}(\psi)|^2 dV = \int_N \left|\text{Hess}(\psi) + \frac{\mu_1}{n}\psi I\right|^2 dV + \frac{\mu_1}{n} \int_N |\nabla\psi|^2 dV. \quad (15)$$

On other hand, Equation (9) yields

$$\text{Ric}(\psi_i e_i, \psi_j e_j) = (n-1)\left(\frac{\epsilon+3}{4}\right)\delta_{ij}\psi_i\psi_j - \sum_{\gamma=1}^{2n+1}\sum_{r=1}^{n} \sigma_{ir}^\gamma \sigma_{jr}^\gamma \psi_i\psi_j.$$

Tracing the above equation, we obtain

$$\text{Ric}(\nabla\psi, \nabla\psi) = \left(\frac{\epsilon+3}{4}\right)(n-1)|\nabla\psi|^2 - \sum_{i=1}^n |h(\nabla\psi, e_i)|^2. \quad (16)$$

As we consider that $\Delta\psi = -\mu_1\psi$, combining the integration of Bochner formula with utilizing Stokes theorem, one arrives

$$\int_N |\text{Hess}(\psi)|^2 dV + \int_N \text{Ric}_{N^n}(\nabla\psi, \nabla\psi) dV = \mu_1 \int_N |\nabla\psi|^2 dV. \quad (17)$$

From (16) and (17), we conclude

$$\left\{\left(\frac{\epsilon+3}{4}\right)n - \mu_1\right\} \int_N |\nabla\psi|^2 dV = \int_N \sum_{i=1}^n |h(\nabla\psi, e_i)|^2 dV$$
$$+ \left(\frac{\epsilon+3}{4}\right) \int_N |\nabla\psi|^2 - \int_N |\text{Hess}(\psi)|^2 dV.$$

This is the first result (11) of proposition. On the other hand, using (15) in the last equality, we obtain

$$\left\{\left(\frac{\epsilon+3}{4}\right)n - \mu_1\right\}\int_N |\nabla\psi|^2 dV = \int_N \sum_{i=1}^n |h(\nabla\psi, e_i)|^2 dV + \left(\frac{\epsilon+3}{4}\right)\int_N |\nabla\psi|^2 dV$$
$$- \int_N \left|Hess(\psi) + \frac{\mu_1}{n}\psi I\right|^2 dV - \frac{\mu_1}{n}\int_N |\nabla\psi|^2 dV.$$

The above formula can written as

$$\left\{\left(\frac{\epsilon+3}{4}\right)n - \left(\frac{\epsilon+3}{4}\right) - \mu_1 + \frac{\mu_1}{n}\right\}\int_N |\nabla\psi|^2 dV = \int_N \sum_{i=1}^n |h(\nabla\psi, e_i)|^2 dV$$
$$- \int_N \left|Hess(\psi) + \frac{\mu_1}{n}\psi I\right|^2 dV$$

After some computation, we get

$$\int_N \sum_{i=1}^n |h(\nabla\psi, e_i)|^2 dV = \int_N \left|Hess(\psi) + \frac{\mu_1}{n}\psi I\right|^2 dV$$
$$+ \left\{\frac{(n-1)}{n}\left(\left(\frac{\epsilon+3}{4}\right)n - \mu_1\right)\right\}\int_N |\nabla\psi|^2 dV,$$

which completes the proof of the proposition. □

The first result of our study can be given as follows.

Theorem 1. *Suppose that $\Psi : N^n \to \tilde{N}^{2n+1}(\epsilon)$ is a minimal immersion of a compact Legedrian submanifold into Sasakian space form $\tilde{N}^{2n+1}(\epsilon)$ and ψ is a first eigenfunction of the Laplacian of N^n associated to the first eigenvalue μ_1. Then, we have*

(i) *The second fundamental form satisfies the following*

$$\int_N |Hess(\psi)|^2 dV \leq \int_N \sum_{i=1}^n |h(\nabla\psi, e_i)|^2 dV + \left(\frac{\epsilon+3}{4}\right)\int_N |\nabla\psi|^2 dV, \tag{18}$$

provided that the inequality $n\left(\frac{\epsilon+3}{4}\right) \geq \mu_1$ holds, where $Hess(\psi)$ denotes the squared norm of the Hessian of ψ and $\{e_1, \cdots, e_n\}$ is an orthonormal frame tangent to N^n. Moreover, the equality holds if and only if

$$\mu_1 = \left(\frac{\epsilon+3}{4}\right)n. \tag{19}$$

(ii) *Furthermore, if the inequality*

$$\int_N |Hess(\psi)|^2 dV \geq \int_N \sum_{i=1}^n |h(\nabla\psi, e_i)|^2 dV \tag{20}$$

holds, then we have lower bound for eigenvalue μ_1, that is,

$$\mu_1 \geq \left(\frac{\epsilon+3}{4}\right)(n-1).$$

(iii) *In particular, if the following inequality*

$$\frac{\mu_1}{n}\int_N |\nabla\psi|^2 dV \geq \left(\frac{\epsilon+3}{4}\right)\int_N \sum_{i=1}^n |h(\nabla\psi, e_i)|^2 dV \tag{21}$$

holds, then the eigenvalue μ_1 satisfies the following inequality

$$\mu_1 \geq \left(\frac{\epsilon+3}{4}\right)(n-1).$$

Proof. We proceed as follows. Let

$$n\left(\frac{\epsilon+3}{4}\right) \geq \mu_1.$$

We point out that (11) of Proposition 1 is non-negative. Therefore, we can write

$$\int_N \sum_{i=1}^n |h(\nabla\psi, e_i)|^2 dV + \left(\frac{\epsilon+3}{4}\right)\int_N |\nabla\psi|^2 dV \geq \int_N |Hess(\psi)|^2 dV.$$

Furthermore, the equality sign of the above inequality holds if and only if

$$\mu_1 = n\left(\frac{\epsilon+3}{4}\right).$$

Moreover, the first equation of Proposition 1 can take the form

$$\int_N |Hess(\psi)|^2 dV = \int_N \sum_{i=1}^n |h(\nabla\psi, l_i)|^2 dV$$
$$+ \left\{\mu_1 - \left(\frac{\epsilon+3}{4}\right)(n-1)\right\}\int_N |\nabla\psi|^2 dV. \quad (22)$$

Now, if we consider the following inequality

$$\int_N |Hess(\psi)|^2 dV \geq \int_N \sum_{i=1}^n |h(\nabla\psi, l_i)|^2 dV,$$

then Equation (22) yields that

$$\left\{\mu_1 - \left(\frac{\epsilon+3}{4}\right)(n-1)\right\} \geq 0.$$

Finally, we note that

$$\int_N |\nabla\psi|^2 dV \geq \frac{n}{\mu_1}\left(\frac{\epsilon+3}{4}\right)\int_N \sum_{i=1}^n |h(\nabla\psi, l_i)|^2 dV.$$

This implies that

$$\int_N |Hess(\psi)|^2 dV \geq \int_N \sum_{i=1}^n |h(\nabla\psi, l_i)|^2 dV,$$

which completes the proof of the theorem. □

Now, we recall the following lemma which would help us to prove the next Theorem.

Lemma 1 ([16]). *Let $T : U \to U$ be a trace-less non-null symmetric linear operator defined over a finite dimensional vector space U. Let $\{e_1, \cdots, e_n\}$ be an orthonormal frame diagonalizing T, i.e., $Te_i = \mu_i e_i$. If $\dim \operatorname{Ker} T = q$, then we get*

$$\mu_j^2 \leq \frac{(n-q-1)|T|^2}{(n-q)}, \quad \forall j.$$

Now, we give the second result of the study as follows.

Theorem 2. *Let $\Psi : N^n \to \widetilde{N}^{2n+1}(\epsilon)$ be a minimal immersion of a compact Legendrian submanifold into a Sasakian space form $\widetilde{N}^{2n+1}(\epsilon))$, μ_1 be the first eigenvalue of the Laplacian of N^n and $\dim \operatorname{Ker}(h) = q$. Then, we have*

$$\int_N S |\operatorname{Hess}(\psi)|^2 dV \geq \left\{ \frac{(n-q)(n\beta-1)(n\beta-\mu_1)}{(n-q-1)n\beta} \right\} \int_N |\nabla \psi|^2 dV,$$

where $\beta = \frac{\epsilon+3}{4}$ and S is the squared norm of the second fundamental form h. Moreover, if S is constant, we get

$$S \geq \frac{(n-q)(n\beta-1)}{n\beta(n-q-1))(n\beta-\mu_1)},$$

where $\Delta \psi + \mu_1 \psi = 0$.

Proof. Let $\{e_1, \cdots, e_n\}$ be an orthogonal referential diagonalizing T, i.e., $Te_i = k_i e_i$ and let θ_i be the angle between $\nabla \psi$ and e_i. Then, we have

$$|h(\nabla \psi, e_i)|^2 = g(T \nabla \psi, e_i)^2 = g(\nabla \psi, T e_i)^2 = k_i^2 \cos^2 \theta_i |\nabla \psi|^2.$$

By virtue of (11) in Proposition 1, we obtain

$$\int_N \left(\sum_{i=1}^n k_i^2 \cos^2 \theta_i \right) |\nabla \psi|^2 dV = \int_N |\operatorname{Hess}(\psi)|^2 dV$$

$$+ \left\{ \left(\frac{\epsilon+3}{4} \right)(n-1) - \mu_1 \right\} \int_N |\nabla \psi|^2 dV.$$

Utilizing Lemma 1, the above equation gives

$$\left(\frac{n-q-1}{n-q} \right) \int_N S |\nabla \psi|^2 dV \geq \int_N |\operatorname{Hess}(\psi)|^2 dV$$

$$+ \left\{ \left(\frac{\epsilon+3}{4} \right)(n-1) - \mu_1 \right\} \int_N |\nabla \psi|^2 dV. \quad (23)$$

Let us assume the following inequality

$$\int_N |\operatorname{Hess}(\psi)|^2 dV \geq \left(\frac{4\mu_1}{\epsilon+3} \right) \int_N |\nabla \psi|^2 dV,$$

holds. Using this assumption with fixing $\beta = \frac{\epsilon+3}{4}$, then (23) becomes

$$\left(\frac{n-q-1}{n-q} \right) \int_N S |\nabla \psi|^2 dV \geq \left(\frac{n^2 \beta^2 - n\beta \mu_1 - n\beta^2 + \mu_1}{n\beta} \right) \int_N |\nabla \psi|^2 dV.$$

After some computations, we get

$$\int_N S|Hess(\psi)|^2 dV \geq \left\{ \frac{(n-q)(n\beta-1)(n\beta-\mu_1)}{(n-q-1)n\beta} \right\} \int_N |\nabla \psi|^2 dV.$$

This completes the proof. □

The following theorem gives the characterization Theorem as follows.

Theorem 3. *Let $\Psi : N^n \to \widetilde{N}^{2n+1}(\epsilon)$ be a minimal immersion of a compact Legendrian submanifold into Sasakian space form $\widetilde{N}^{2n+1}(\epsilon)$ and ψ be a first eigenfunction associated to the Laplacian of N^n. Then, we have*

(i) *If $\nabla \psi \in Ker(h)$, then $\Psi(N^n)$ is isometric to the standard sphere $\mathbb{S}^n$ with $\mu_1 > 0$ and $n = 1$.*
(ii) *If following Ricci inequality holds*

$$Ric_{N^n}(\nabla \psi, \nabla \psi) \geq (n-1)\left(\frac{\epsilon+3}{4}\right)|\nabla \psi|^2,$$

then $\Psi(N^n)$ is isometric to a sphere $\mathbb{S}^n$ with $\epsilon > -3$ and $n \geq 2$.

Proof. At first, we provide the state of Obata Theorem [2] as follows: a Riemannian manifold M^n is isometric to a unit sphere $\mathbb{S}^n$ if and only if it is equipped with a differentiable function ψ such that $Hess(\psi) = -\psi$, where $Hess(\psi)$ is the Hessian form. Now, we assume that $\nabla \psi \in \ker(h)$, i.e.,

$$h(\nabla \psi, e_i) = 0, \quad \forall e_i.$$

Then by using Equation (12), we attain

$$\int_N \left| Hess(\psi) + \frac{\mu_1}{n}\psi \right|^2 dV = \frac{(n-1)(\mu_1 - n\beta)}{n} \int_N |\nabla \psi|^2 dV.$$

Using the fact that the right-hand side of the above equation is non-positive leads to

$$0 < \mu_1 = n\left(\frac{\epsilon+3}{4}\right).$$

Therefore, $Hess(\psi) = -\mu_1 \psi$, as $\mu_1 > 0$ and $n = 1$. Now, utilizing Obata Theorem [2], we conclude that $\Phi(N^n)$ is isometric to $\mathbb{S}^n$ with $\mu_1 = n$. Thus, we have gotten the first part of Theorem 3. To prove the second statement of the theorem, let us consider that

$$Ric_{N^n}(\nabla \psi, \nabla \psi) \geq \left(\frac{\epsilon+3}{4}\right)(n-1)|\nabla \psi|^2.$$

According to Equation (16), we find that

$$\int_N (n-1)\left(\frac{\epsilon+3}{4}\right)|\nabla \psi|^2 dV \geq \sum_{i=1}^n |h(\nabla \psi, e_i)|^2 dV + \left(\frac{\epsilon+3}{4}\right)(n-1)\int_N |\nabla \psi|^2 dV.$$

This leads to

$$\sum_{i=1}^n |h(\nabla \psi, e_i)|^2 dV \leq 0. \tag{24}$$

Hence, we conclude that $h(\nabla \psi, e_i) = 0$, i.e., $\nabla \psi \in \ker(h)$. The proof is now complete. □

Tashiro [13] has proved more general results than of Obata and Kanai. The following theorem is of interest in characterizing the Euclidean space in terms of a certain differential equation. Therefore, we are able to prove the following result.

Theorem 4. Let $\Psi : N^n \to \tilde{N}^{2n+1}(\epsilon)$ be a minimal immersion of a compact Legendrian submanifold into Sasakian space form $\tilde{N}^{2n+1}(\epsilon)$. Then N^n is isometric to Eculidean space $\mathbb{R}^n$ if and only if the following equation is satisfied

$$\int_N \sum_{i=1}^n |h(\nabla \psi, e_i)|^2 dV + \int_N \frac{\mu_1^2}{n} dV = \left\{ \mu_1 - \left(\frac{\epsilon + 3}{4} \right)(n-1) \right\} \int_N |\nabla \psi|^2 dV, \qquad (25)$$

where ψ is a first eigenfunction associated to the Laplacian of N^n with first non-zero eigenvalue μ_1.

Proof. Let us consider the equation

$$|Hess(\psi) + tI|^2 = |Hess(\psi)|^2 + t^2 |I|^2 + 2t g(Hess(\psi), I),$$

which implies that

$$|Hess(\psi) + tI|^2 = |Hess(\psi)|^2 + t^2 n - 2t \Delta \psi.$$

Putting $t = -\frac{\mu_1}{n}$ and integrating the above equation along volume element dV, we obtain

$$\int_N \left| Hess(\psi) - \frac{\mu_1}{n} I \right|^2 dV = \int_N \left(|Hess(\psi)|^2 + \frac{\mu_1^2}{n} \right) dV.$$

Using (16) and (17), we get

$$\int_N \left| Hess(\psi) - \frac{\mu_1}{n} I \right|^2 dV = \int_N \sum_{i=1}^n |h(\nabla \psi, e_i)|^2 dV - \left\{ \mu_1 - \left(\frac{\epsilon + 3}{4} \right)(n-1) \right\} \int_N |\nabla \psi|^2 dV$$

$$+ \int_N \frac{\mu_1^2}{n} dV. \qquad (26)$$

If (25) is satisfied, then (26) implies that

$$\left| Hess(\psi) - \frac{\mu_1}{n} I \right|^2 = 0.$$

Hence, we get

$$Hess(\psi)(X, X) = \frac{\mu_1}{n} g(X, X), \qquad (27)$$

for any $X \in \Gamma(N)$. Therefore, by applying Tashiro Theorem [13], we conclude that N^n is isometric to the Euclidean space $\mathbb{R}^n$. The converse part can be proved easily from (26) if N^n is isometric to Euclidean space $\mathbb{R}^n$. □

We provide an interesting application of Theorem 3 in the following corollary by choosing $\epsilon = 1$ (see [19]).

Corollary 1. Let $\Psi : N^n \to \mathbb{S}^{2n+1}$ be a minimal immersion of a compact Legendrian submanifold into the sphere $\mathbb{S}^{2n+1}$ and ψ be a first eigenfunction associated to the Laplacian of N^n. Then, we get the following

(i) If $\nabla \psi \in Ker(h)$, then $\Psi(N^n)$ is isometric to standard sphere $\mathbb{S}^n$.
(ii) If $Ric_{N^n}(\nabla \psi, \nabla \psi) \geq (n-1) |\nabla \psi|^2$, then $\Psi(N^n)$ is isometric to the sphere $\mathbb{S}^n$.

Author Contributions: Writing and original draft, R.A.; funding acquisition, editing and draft, F.M.; review and editing, N.A.; methodology, project administration, A.A.; formal analysis, resources, I.A. All authors have read and agreed to the published version of the manuscript.

Funding: The first and fourth authors would like to express their gratitude to Deanship of Scientific Research at King Khalid University, Saudi Arabia for providing funding research groups under the research grant R. G. P. 2/57/40.

Acknowledgments: This research was funded by the Deanship of Scientific Research at Princess Nourah bint Abdulrahman University through the Fast-track Research Funding Program.

Conflicts of Interest: The authors declare no conflict of interest.

References

1. Yano, K.; Nagano, T. Einstein spaces admitting a one-parameter group of conformal transformations. *Ann. Math.* **1959**, *69*, 451–461. [CrossRef]
2. Obata, M. Certain conditions for a Riemannian manifold to be isometric with a sphere. *J. Math. Soc. Jpn.* **1962**, *14*, 333–340. [CrossRef]
3. Barros, A. Applications of Bochner formula to minimal submanifold of the sphere. *J. Geom. Phys.* **2002**, *44*, 196–201. [CrossRef]
4. Berger, M.; Gauduchon, P.; Mazet, E. Le spectre d'une variété riemannienne. In *Le Spectre d'Une Variété Riemannienne*; Springer: Berlin/Heidelberg, Germany, 1971; pp. 141–241.
5. Chavel, I. *Eigenvalues in Riemannian Geometry*; Academic Press: New York, NY, USA, 1984.
6. Choi, H.I.; Wang, A.N. A first eigenvalue estimate for minimal hypersurfaces. *J. Differ. Geom.* **1983**, *18*, 559–562. [CrossRef]
7. Deshmukh, S. Conformal vector fields and eigenvectors of Laplace operator. *Math. Phys. Anal. Geom.* **2012**, *15*, 163–172. [CrossRef]
8. Deshmukh, S.; Al-Solamy, F. A note on conformal vector fields on a Riemannian manifold. *Colloq. Math.* **2014**, *136*, 65–73. [CrossRef]
9. Deshmukh, S. Characterizing Spheres and Euclidean Spaces by Conformal Vector Fields. *Ann. Math. Pura Appl.* **2017**, *196*, 2135–2145. [CrossRef]
10. Erkekoğlu, F.; García-Río, E.; Kupeli, D.N.; Ünal, B. Characterizing specific Riemannian manifolds by differential equations. *Acta Appl. Math.* **2003**, *76*, 195–219. [CrossRef]
11. García-Río, E.; Kupeli, D.N.; Unal, B. On a differential equation characterizing Euclidean sphere. *J. Differ. Equ.* **2003**, *134*, 287–299. [CrossRef]
12. Jamali, M.; Shahid, M.H. Application of Bochner formula to generalized Sasakian space forms. *Afrika Matematika* **2018**, *29*, 1135–1139. [CrossRef]
13. Tashiro, Y. Complete Riemannian manifolds and some vector fields. *Trans. Am. Math. Soc.* **1965**, *117*, 251–275. [CrossRef]
14. Barbosa, J.N.; Barros, A. A lower bound for the norm of the second fundamental form of minimal hypersurfaces of $\mathbb{S}^{n+1}$. *Arch. Math.* **2003**, *81*, 478–484. [CrossRef]
15. Lichnerowicz, A. *Geometrie des Groupes de Transformations*; Dunod: Malakoff, France, 1958.
16. Barros, A.; Gomes, J.N.; Ernani, J.R. A note on rigidity of the almost Ricci soliton. *Arch. Math.* **2013**, *100*, 481–490. [CrossRef]
17. Liu, J.; Zhang, Q. Simons-type inequalities for the compact submanifolds in the space of constant curvature. *Kodai Maths J.* **2007**, *30*, 344–351. [CrossRef]
18. Simons, J. Minimal Varieties in Riemannian Manifolds. *Ann. Math.* **1968**, *88*, 62–105. [CrossRef]
19. Ali, A.; Laurian-Ioan, P. Geometric classification of warped products isometrically immersed in Sasakian space forms. *Math. Nachr.* **2018**, *292*, 234–251.
20. Mihai, I.; Presură, I. An improved first Chen inequality for Legendrian submanifolds in Sasakian space forms. *Period. Math. Hung.* **2017**, *74*, 220–226. [CrossRef]
21. Sasahara, T. A class of biminimal Legendrian submanifolds in Sasakian space forms. *Math. Nachr.* **2014**, *287*, 79–90. [CrossRef]
22. Aquib, M.; Boyom, M.N.; Shahid, M.H.; Vîlcu, G.-E. The First Fundamental Equation and Generalized Wintgen-Type Inequalities for Submanifolds in Generalized Space Forms. *Mathematics* **2019**, *7*, 1151. [CrossRef]

© 2020 by the authors. Licensee MDPI, Basel, Switzerland. This article is an open access article distributed under the terms and conditions of the Creative Commons Attribution (CC BY) license (http://creativecommons.org/licenses/by/4.0/).

Article

Geometric Inequalities of Warped Product Submanifolds and Their Applications

Nadia Alluhaibi [1], Fatemah Mofarreh [2], Akram Ali [3,*] and Wan Ainun Mior Othman [4]

[1] Department of Mathematics, Science and Arts College, Rabigh Campus, King Abdulaziz University, Jeddah 21589, Saudi Arabia; nallehaibi@kau.edu.sa
[2] Mathematical Science Department, Faculty of Science, Princess Nourah bint Abdulrahman University, Riyadh 11546, Saudi Arabia; fyalmofarrah@pnu.edu.sa
[3] Department of Mathematics, College of Science, King Khalid University, Abha 62529, Saudi Arabia
[4] Institute of Mathematical Sciences, Faculty of Science, University of Malaya, Kuala Lumpur 50603, Malaysia; wanainun@um.edu.my
* Correspondence: akali@kku.edu.sa

Received: 17 April 2020; Accepted: 7 May 2020; Published: 11 May 2020

Abstract: In the present paper, we prove that if Laplacian for the warping function of complete warped product submanifold $\mathbb{M}^m = \mathbb{B}^p \times_h \mathbb{F}^q$ in a unit sphere $\mathbb{S}^{m+k}$ satisfies some extrinsic inequalities depending on the dimensions of the base $\mathbb{B}^p$ and fiber $\mathbb{F}^q$ such that the base $\mathbb{B}^p$ is minimal, then $\mathbb{M}^m$ must be diffeomorphic to a unit sphere $\mathbb{S}^m$. Moreover, we give some geometrical classification in terms of Euler–Lagrange equation and Hamiltonian of the warped function. We also discuss some related results.

Keywords: warped product; sphere theorem; Laplacian; inequalities; diffeomorphic

1. Introduction and Main Results

We will use the following acronyms throughout the paper: 'WP' for Warped product, 'WF' for warping function, 'RM' for Riemannian manifold, and 'SFF' for second fundamental form. The idea of the warped product was initiated by Bishop and O'Neil [1] when they gave an example of complete Riemannian manifold with negative curvature. If $(\mathbb{B}, g_\mathbb{B})$ and $(\mathbb{F}, g_\mathbb{F})$ are two Riemannian manifolds (RMs), and h is a positive differentiable function defined on the base manifold $\mathbb{B}$, then we define the metric $g = \pi^* g_\mathbb{B} + h^2 \sigma^* g_\mathbb{F}$ on the product manifold $\mathbb{B} \times \mathbb{F}$, where π and σ are the projection maps on B and F, respectively. Under such stipulations, the product manifold is referred to as warped product (WP) of $\mathbb{B}$ and $\mathbb{F}$, and written as $\mathbb{M} = \mathbb{B} \times_h \mathbb{F}$. Here, h is referred to as warping function (WF).

We observe that $\mathbb{M}$ is a Riemannian product, or trivial warped product, when h is constant. Notice that there has been a great interest in the study of warped products over the recent years. For example, S. Nolker [2] derived the decompositions of the standard spaces of an isometric immersion of warped products and D.K. Kim and Y.H. Kim in [3] proved that if the scalar is non-constant then there is no non-trivial compact Einstein warped product. Recently, an interesting fundamental result proved by Djaczer in [4] showed that an isometric immersion of warped products into space forms must be product of isometric immersions under extrinsic conditions. Moreover, by using DDVV conjecture, Roth [5] obtained an inequality for submanifold of WP $I \times_h \mathbb{M}^m(c)$ where I is an interval and $\mathbb{M}^m(c)$ is a real space form and also provided some rigidity results based on submanifolds of $\mathbb{R} \times_{e^{\lambda t}} \mathbb{H}^m(c)$, where λ is a real constant. Salavessa in [6] obtained that the Heinz mean curvature $m\|\mathbb{H}\|^2 \leq \frac{A_\Psi(\partial \mathcal{D})}{V_\Psi(\mathcal{D})}$ holds in WP spaces of type $M \times_{e^\Psi} N$ in case that a graph of submanifold $(x, h(x))$ of Riemannian WP $M \times_{e^\Psi} N$ is immersed with parallel mean curvature, where $A_\Psi(\partial \mathcal{D})$ and $V_\Psi(\mathcal{D})$ are Ψ−weighted area and volume, respectively.

On the other hand, the investigation of the relations between curvature invariants and topology is an important problem in Riemannian geometry as well as in global differential geometry. For instance, a beautiful and classical theorem established by Myers [7] states that "if $\mathbb{M}$ is a complete Riemannian manifold with Ricci curvature $Ric(\mathbb{M}) > 1$, then the diameter $d(\mathbb{M})$ of $\mathbb{M}$ is not greater than π, and, therefore, $\mathbb{M}$ is compact and its fundamental group $\pi_1((\mathbb{M}))$ is finite". Due to the distinctive work of Rauch [8], Berger [9] proved the rigidity theorem for a simply connected and complete manifold $\mathbb{M}$ of even dimension and the sectional curvature satisfying $\frac{1}{4} \leq K_\mathbb{M} \leq 1$. Furthermore, Grove and Shiohama in [10] has generalized the sphere theorem. There are lots of interesting and well-known results regarding the topology of complete manifolds of positive Ricci curvature. The curvature and topology of manifolds play a substantial role in global differential geometry. Later on, a splitting theorem, resulting from the work of Cheeger and Gromoll in [11], states that "if $\mathbb{M}$ is a complete non-compact manifold of non-negative Ricci curvature and if $\mathbb{M}$ contains a straight line, then M is isometric to the Riemannian product $\mathbb{M} \times \mathbb{R}$". In the sequel, Schoen and Yau [12] proved that a complete non-compact $\mathbb{M}$ of dimension 3 and positive Ricci curvature is diffeomorphic to $\mathbb{R}^3$. Using the first eigenvalue of the Laplacian operator, the result stating that "if $\mathbb{M}$ is complete such that if $Ric(\mathbb{M}) > 1$ and if $d(\mathbb{M}) = \pi$, then $\mathbb{M}$ is isometric to the standard unit sphere" has been proven by Cheng in [13].

The non-existence of a compact stable minimal submanifold or stable currents is sharply associated with the topology and geometric function theory on Riemannian structure of the whole manifold. Recently, it has been shown in [14] that if the sectional curvature of a compact oriented minimal submanifold $\mathbb{M}$ of dimension m in the unit sphere $\mathbb{S}^{m+k}$ with codimension p satisfies some pinching condition $K_\mathbb{M} \geq \frac{p.sign(p-1)}{2(p+1)}$, then $\mathbb{M}$ is either a totally geodesic sphere, one of the Clifford minimal hypersurface $\mathbb{S}^k(\frac{k}{m}) \times \mathbb{S}^{m-k}(\frac{m-k}{m})$ in $\mathbb{S}^{m+1}$ for $k = 1, \ldots, m-1$, or the Veronese surface in $\mathbb{S}^4$. Later on, some new results for the non-existence of the stable currents, vanishing homology groups, topological and differential theorems are well known (see [15–23] and references therein). Therefore, it was an objective for mathematicians to understand geometric function theory and topological invariant of Riemannian submanifolds as well as in Riemannian space forms. Surely, this is a fruitful problem in Riemannian geometry. Using the result of Lawson and Simon [24] and following Leung [20] homotopic sphere theorem for compact oriented submanifolds in a sphere, also motivated by the idea of complete Riemannian manifold and without assumption that $\mathbb{M}^m$ is simply connected, Xu and Zao (Theorem 1.2 in [21]) concluded the following result:

Theorem 1. *[21] Let $\mathbb{M}^m$ be an oriented complete submanifold of dimension m in the unit sphere $\mathbb{S}^{m+k}$ satisfying the following inequality*

$$\|\mathbf{B}(\mathbb{X}, \mathbb{X})\|^2 < \frac{1}{3}, \quad \forall \mathbb{X} \in \Gamma(TM), \tag{1}$$

where $\mathbb{X}$ is a unit vector at any point of $\mathbb{M}^m$ and $\mathbf{B}$ is SFF, the second fundamental form. Then $\mathbb{M}^m$ is diffeomorphic to the sphere $\mathbb{S}^m$.

This is one of the motivations to study—the differential and topological manifolds, and their direct relations with warped product submanifolds theory. In this way, a natural question arises: Is it possible to extend Theorem 1 to the warped product submanifolds to the cases with base manifold is minimal in a sphere? What is the best pinching constant for the differentiable rigidity sphere theorem of complete minimal warped product submanifold in a unite sphere under pinching conditions using the Laplace operator for the warping function?

The main goal of this note is to extend the rigidity Theorem 1 to a complete warped product submanifolds and find the solution for our proposed problem where motivation comes from the Nash embedding theorem [25] which states that "every Riemannian manifold has an isometric immersion into Euclidean space of sufficient high codimension". To prove our findings we shall use the technique of Chen [26] for an isometric minimal immersion from warped products to the ambient manifold, where he proved the following relation as:

$$\sum_{\alpha=1}^{p}\sum_{\beta=1}^{q} K(e_\alpha \wedge e_\beta) = \frac{q\Delta h}{h}. \qquad (2)$$

Therefore, using Theorem 1 and formula (2), we announce our main finding of this study as follows:

Theorem 2. *Let $\ell : \mathbb{M}^{p+q} = \mathbb{B}^p \times_h \mathbb{F}^q \longrightarrow \mathbb{S}^{p+q+k}$ be an isometric immersion from a WP submanifold $\mathbb{M}^{p+q}$ of dimension $(p+q)$ into a unit sphere $\mathbb{S}^{p+q+k}$ of dimension $(p+q+k)$ such that the base manifold $\mathbb{B}^p$ is minimal. Assume that $\mathbb{M}^{p+q}$ is an oriented complete WP submanifold satisfying the following inequality*

$$\frac{\Delta h}{h} > \left(\frac{2(3pq-1)}{3q} \right), \qquad (3)$$

where Δh is the Laplace operator for the warping function h defined on base manifold $\mathbb{B}^p$. Then $\mathbb{M}^{p+q}$ is diffeomorphic to a sphere $\mathbb{S}^{p+q}$.

In particular, if we follows the statement of Theorem D in [21], then we give another topological sphere theorem which is a consequence of Theorem 2, i.e.,

Theorem 3. *Let $\ell : \mathbb{M}^{p+q} = \mathbb{B}^p \times_h \mathbb{F}^q \longrightarrow \mathbb{S}^{p+q+k}$ be an isometric immersion from an $(p+q)$-dimensional oriented complete WP submanifold $\mathbb{M}^{p+q}$ into a $(p+q+k)$-dimensional unit sphere $\mathbb{S}^{p+q+k}$ such that the base manifold $\mathbb{B}^p$ is minimal. If the following inequality holds*

$$\frac{\Delta h}{h} > \left(\frac{2(3pq-1)}{3q} \right),$$

where Δf is the Laplace of f defined on base manifold $\mathbb{B}^p$, then $\mathbb{M}^{p+q}$ is homeomorphic to the sphere $\mathbb{S}^{p+q}$.

Hence, we noticed that Theorems 2 and 3 are differentiable sphere theorems for complete warped product submanifolds without assumption that $\mathbb{M}^n$ is simply connected.

2. Preliminaries and Notations

Let $\mathbb{S}^{m+k}$ denote the sphere with constant sectional curvature $c = 1 > 0$ and dimension $(m+k)$. We use the fact that $\mathbb{S}^{m+k}$ admits a canonical isometric embedding in $\mathbb{R}^{m+k+1}$ as

$$\mathbb{S}^{m+k} = \{\mathbb{X} \in \mathbb{R}^{m+k+1} : ||\mathbb{X}||^2 = 1\}.$$

Thus, the Riemannian curvature tensor $\widetilde{R}$ of a sphere $\mathbb{S}^{m+k}$ fulfils

$$\widetilde{R}(Z_1, Z_2, Z_3, Z_4) = g(Z_1, Z_4)g(Z_2, Z_3) - g(Z_2, Z_4)g(Z_1, Z_4), \qquad (4)$$

$\forall Z_1, Z_2, Z_3, Z_4 \in \Gamma(T\widetilde{M})$, where $T\widetilde{M}$ is a tangent bundle of $\mathbb{S}^{m+k}$. Hence, $\mathbb{S}^{m+k}$ is a manifold with constant sectional curvature 1 and codimension k.

Let $\nabla^\perp$ and ∇ be the induced connections on normal bundle $T^\perp \mathbb{M}$ and the tangent bundle $T\mathbb{M}$ of $\mathbb{M}$, respectively, where $\mathbb{M}$ is a m-dimensional RM in a Riemannian $\widetilde{M}^n$ of dimension n with induced metric g. The Weingarten and Gauss formulae are defined as

$$\widetilde{\nabla}_{Z_1} \xi = -A_\xi Z_1 + \nabla^\perp_{Z_1} \xi,$$

and
$$\tilde{\nabla}_{Z_1} Z_2 = \nabla_{Z_1} Z_2 + \mathbf{B}(Z_1, Z_2),$$

$\forall\, Z_1, Z_2 \in \Gamma(TM)$ and $\xi \in \Gamma(T^\perp M)$, where A_ξ and $\mathbf{B}$ are, respectively, shape operator (corresponding to ξ) and the second fundamental form as $\mathbb{M}^m$ immersed into $\tilde{M}$, and they verify the relation

$$g(\mathbf{B}(Z_1, Z_2), \xi) = g(A_\xi Z_1, Z_2).$$

If the curvature tensors of $\tilde{M}^n$ and $\mathbb{M}^m$ are denoted by $\tilde{R}$ and R, then the Gauss equation is given by

$$R(Z_1, Z_2, Z_3, Z_4) = \tilde{R}(Z_1, Z_2, Z_3, Z_4) + g(\mathbf{B}(Z_1, Z_4), \mathbf{B}(Z_2, Z_3)) - g(\mathbf{B}(Z_1, Z_3), \mathbf{B}(Z_2, Z_4)), \quad (5)$$

$\forall\, Z_1, Z_2, Z_3, Z_4 \in \Gamma(T\tilde{M})$.

Let $\{e_1, \cdots e_m\}$ be an orthonormal basis of $T_x \mathbb{M}$ and $e_s = (e_{m+1}, \cdots e_{m+k})$ belongs to an orthonormal basis of $T^\perp \mathbb{M}$, then the squared norm of $\mathbf{B}$ is

$$\mathbf{B}^s_{\alpha\beta} = g(\mathbf{B}(e_\alpha, e_\beta), e_s), \qquad (6)$$

and

$$||\mathbf{B}(e_\alpha, e_\beta)||^2 = \sum_{s=m+1}^{m+k} \sum_{\alpha=1}^{p} \sum_{\beta=1}^{q} (\mathbf{B}^s_{\alpha\beta})^2. \qquad (7)$$

The squared norm of the mean curvature vector $\mathbb{H}$ of a Riemannian submanifold $\mathbb{M}^m$ is defined by

$$||\mathbb{H}||^2 = \frac{1}{m^2} \sum_{s=m+1}^{m+k} \left(\sum_{\alpha=1}^{m} \mathbf{B}^s_{\alpha\alpha} \right)^2. \qquad (8)$$

A submanifold $\mathbb{M}^m$ of a RM, $\tilde{M}^{m+k}$, is referred to as *totally geodesic* and *totally umbilical* if

$$\mathbf{B}(Z_1, Z_2) = 0 \quad \text{and} \quad \mathbf{B}(Z_1, Z_2) = g(Z_1, Z_2)\mathbb{H},$$

$\forall\, Z_1, Z_2 \in \Gamma(TM)$, respectively, where $\mathbb{H}$ is the mean curvature vector of $\mathbb{M}^m$. Moreover, if $\mathbb{H} = 0$, then $\mathbb{M}^m$ is *minimal* in $\tilde{M}^{m+k}$.

Now, we give a definition of the scalar curvature of Riemannian submanifold $\mathbb{M}^m$, which is denoted by $\tau(T_x \mathbb{M}^m)$, at some x in $\mathbb{M}^m$, as

$$\tau(T_x \mathbb{M}^m) = \sum_{1 \leq \alpha < \beta \leq m} K_{\alpha\beta}, \qquad (9)$$

where $K_{\alpha\beta} = K(e_\alpha \wedge e_\beta)$. The first equality (9) is equal to the following equation:

$$2\tau(T_x \mathbb{M}^m) = \sum_{1 \leq \alpha < \beta \leq m} K_{\alpha\beta}, \quad 1 \leq \alpha, \beta \leq m.$$

The above equation will be considerably used in subsequent proofs throughout the paper. In similar way, the scalar curvature $\tau(L_x)$ of an L−plane is defined as

$$\tau(L_x) = \sum_{1 \leq \alpha < \beta \leq m} K_{\alpha\beta}.$$

If the plane section spanned by e_α and e_β at x, then the sectional curvatures of the submanifold $\mathbb{M}^m$ and Riemannian manifold $\tilde{M}^{m+k}$ are denoted by $K_{\alpha\beta}$ and $\tilde{K}_{\alpha\beta}$, respectively. Thus, $\tilde{K}_{\alpha\beta}$ and $K_{\alpha\beta}$ are considered to be the extrinsic and intrinsic sectional curvature of the span $\{e_\alpha, e_\beta\}$ at x. Using Gauss Equation (5), and using (9), we conclude that

$$\sum_{1\leq\alpha<\beta\leq m+k} K_{\alpha\beta} = \sum_{1\leq\alpha<\beta\leq m+k} \widetilde{K}_{\alpha\beta} + \sum_{r=m+1}^{n+k} \left(\mathbf{B}^r_{\alpha\alpha}\mathbf{B}^r_{\beta\beta} - (\mathbf{B}^r_{\alpha\beta})^2 \right). \tag{10}$$

Now, we provide the proofs of the main findings of the study.

3. Proof of Main Findings

3.1. Proof of Theorem 2

Assume that $\mathbb{M}^m = \mathbb{B}^p \times_h \mathbb{F}^q \to \mathbb{S}^{m+k}$ is a warped product in which the base $\mathbb{B}^p$ is minimal. Let $\{e_1 \ldots e_p, e_{p+1} \ldots e_m\}$ be a local orthonormal frame fields of $\mathbb{M}^m$ such that $\{e_1 \ldots e_p\}$ are tangents to $\mathbb{B}^p$ and $\{e_{p+1} \ldots e_m\}$ are tangents to $\mathbb{F}^q$. First, we define the two unit vectors $\mathbb{X}$ and $\mathbb{Y}$ to estimate the upper bound of the terms $||\mathbf{B}(e_\alpha, e_\beta)||^2$. We can define these two unit vectors as follows:

$$\mathbb{X} = \frac{1}{\sqrt{2}}(e_\alpha + e_\beta), \text{ and } \mathbb{Y} = \frac{1}{\sqrt{2}}(e_\alpha - e_\beta), \quad 1\leq \alpha \leq p \ \& \ 1\leq \beta \leq q.$$

Eliminating e_α and e_β from the above equation, one obtains:

$$e_\alpha = \frac{1}{\sqrt{2}}(\mathbb{X} + \mathbb{Y}), \text{ and } e_\beta = \frac{1}{\sqrt{2}}(\mathbb{X} - \mathbb{Y}), \quad 1\leq \alpha \leq p \ \& \ 1\leq \beta \leq q.$$

Then we derive

$$||\mathbf{B}(e_\alpha, e_\beta)||^2 = \left\|\mathbf{B}\left(\frac{\mathbb{X}+\mathbb{Y}}{\sqrt{2}}, \frac{\mathbb{X}-\mathbb{Y}}{\sqrt{2}}\right)\right\|^2$$
$$= \frac{1}{4}\|\mathbf{B}(\mathbb{X},\mathbb{X}) - \mathbf{B}(\mathbb{Y},\mathbb{Y})\|^2$$
$$= \frac{1}{4}\Big\{\|\mathbf{B}(\mathbb{X},\mathbb{X})\|^2 + \|\mathbf{B}(\mathbb{Y},\mathbb{Y})\|^2 - 2g(\mathbf{B}(\mathbb{X},\mathbb{X}), \mathbf{B}(\mathbb{Y},\mathbb{Y}))\Big\}.$$

Using the Cauchy–Schwartz inequality for orthonormal vector fields, we conclude that

$$||\mathbf{B}(e_\alpha, e_\beta)||^2 \leq \frac{1}{4}\Big\{\|\mathbf{B}(\mathbb{X},\mathbb{X})\|^2 + \|\mathbf{B}(\mathbb{Y},\mathbb{Y})\|^2 + 2\|\mathbf{B}(\mathbb{X},\mathbb{X})\|\|\mathbf{B}(\mathbb{Y},\mathbb{Y})\|\Big\}.$$

In virtue of (1), the above equation implies that

$$||\mathbf{B}(e_\alpha, e_\beta)||^2 < \frac{1}{4}\left(\frac{1}{3} + \frac{2}{3} + \frac{1}{3}\right) = \frac{1}{3}. \tag{11}$$

Next, from curvature tensor Equation (4) of the sphere $\mathbb{S}^{m+k}$ and the Gauss Equation (5), we find that

$$m^2||\mathbb{H}||^2 + m(m-1) = ||\mathbf{B}||^2 + \sum_{1\leq A<B\leq m} K(e_A \wedge e_B).$$

The above equation can be written for warped product manifold $\mathbb{M}^n$ and from the viewpoint of (8) and (6) as:

$$\sum_{s=m+1}^{m+k}\left(\sum_{A=1}^{m}\mathbf{B}_{AA}^{s}\right)^{2}+m(m-1)=\sum_{s=m+1}^{m+k}\sum_{i,j=1}^{p}(\mathbf{B}_{ij}^{s})^{2}+\sum_{s=m+1}^{m+k}\sum_{a,b=1}^{q}(\mathbf{B}_{ab}^{s})^{2}$$
$$+2\sum_{s=m+1}^{m+k}\sum_{\alpha=1}^{p}\sum_{\beta=1}^{q}(\mathbf{B}_{\alpha\beta}^{s})^{2}+\sum_{\alpha=1}^{p}\sum_{\beta=1}^{q}K(e_{\alpha}\wedge e_{\beta})$$
$$+\sum_{1\leq i<j\leq p}K(e_{i}\wedge e_{j})+\sum_{1\leq a<b\leq q}K(e_{a}\wedge e_{b}).$$

Using (10) and (2) in the above equation, we derive

$$\sum_{s=m+1}^{m+k}\left(\sum_{A=1}^{m}\mathbf{B}_{AA}^{s}\right)^{2}+m(m-1)=\sum_{s=m+1}^{m+k}\sum_{i,j=1}^{p}(\mathbf{B}_{ij}^{s})^{2}+\sum_{s=m+1}^{m+k}\sum_{a,b=1}^{q}(\mathbf{B}_{ab}^{s})^{2}$$
$$+2\sum_{s=m+1}^{m+k}\sum_{\alpha=1}^{p}\sum_{\beta=1}^{q}(\mathbf{B}_{\alpha\beta}^{s})^{2}+\frac{q\Delta f}{f}$$
$$+\sum_{1\leq i<j\leq p}\widetilde{K}(e_{i}\wedge e_{j})+\sum_{1\leq a<b\leq q}\widetilde{K}(e_{a}\wedge e_{b})$$
$$+\sum_{s=m+1}^{m+k}\sum_{1\leq i<j\leq p}\left(\mathbf{B}_{ii}^{s}\mathbf{B}_{jj}^{s}-(\mathbf{B}_{ij}^{s})^{2}\right)$$
$$+\sum_{s=m+1}^{m+k}\sum_{1\leq a<b\leq q}\left(\mathbf{B}_{aa}^{s}\mathbf{B}_{bb}^{s}-(\mathbf{B}_{ab}^{s})^{2}\right).$$

Thus, from (4) and some rearrangements in the last equation, one obtains:

$$\sum_{s=m+1}^{m+k}\left(\sum_{A=1}^{m}\mathbf{B}_{AA}^{s}\right)^{2}=\sum_{s=m+1}^{m+k}\sum_{i,j=1}^{p}(\mathbf{B}_{ij}^{s})^{2}+\sum_{s=m+1}^{m+k}\sum_{a,b=1}^{q}(\mathbf{B}_{ab}^{s})^{2}-2pq$$
$$+2\sum_{s=m+1}^{m+k}\sum_{\alpha=1}^{p}\sum_{\beta=1}^{q}(\mathbf{B}_{\alpha\beta}^{s})^{2}+\frac{q\Delta h}{h}-\sum_{s=m+1}^{m+k}\sum_{1\leq i<j\leq p}(\mathbf{B}_{ij}^{s})^{2}$$
$$+\sum_{s=m+1}^{m+k}\sum_{1\leq i<j\leq p}\mathbf{B}_{ii}^{s}\mathbf{B}_{jj}^{s}+\sum_{s=m+1}^{m+k}\left((\mathbf{B}_{11}^{s})^{2}+\cdots+(\mathbf{B}_{pp}^{s})^{2}\right)$$
$$-\sum_{s=m+1}^{m+k}\left((\mathbf{B}_{11}^{s})^{2}+\cdots+(\mathbf{B}_{pp}^{s})^{2}\right)+\sum_{s=m+1}^{m+k}\sum_{1\leq a<b\leq q}\mathbf{B}_{aa}^{s}h_{bb}^{s}$$
$$-\sum_{s=m+1}^{m+k}\sum_{1\leq a<b\leq q}(\mathbf{B}_{ab}^{s})^{2}+\sum_{s=m+1}^{m+k}\left((\mathbf{B}_{p+1p+1}^{s})^{2}+\cdots+(\mathbf{B}_{mm}^{s})^{2}\right)$$
$$-\sum_{s=m+1}^{m+k}\left((\mathbf{B}_{p+1p+1}^{s})^{2}+\cdots+(\mathbf{B}_{mm}^{s})^{2}\right).$$

This can take the form

$$\sum_{s=m+1}^{m+k}\left(\sum_{A=1}^{m} \mathbf{B}_{AA}^{s}\right)^{2} = \sum_{s=m+1}^{m+k}\sum_{i,j=1}^{p}(\mathbf{B}_{ij}^{s})^{2} + \sum_{s=m+1}^{m+k}\sum_{a,b=1}^{q}(\mathbf{B}_{ab}^{s})^{2} - 2pq$$

$$+ 2\sum_{s=m+1}^{m+k}\sum_{\alpha=1}^{p}\sum_{\beta=1}^{q}(\mathbf{B}_{\alpha\beta}^{s})^{2} + \frac{q\Delta h}{h}$$

$$+ \sum_{s=m+1}^{m+k}\left\{\sum_{1\leq i<j\leq p} \mathbf{B}_{ii}^{s}\mathbf{B}_{jj}^{s} + (\mathbf{B}_{11}^{s})^{2} + \cdots + (\mathbf{B}_{pp}^{s})^{2}\right\}$$

$$- \sum_{s=m+1}^{m+k}\left\{\sum_{1\leq i<j\leq p} (\mathbf{B}_{ij}^{s})^{2} + (\mathbf{B}_{11}^{s})^{2} + \cdots + (\mathbf{B}_{pp}^{s})^{2}\right\}$$

$$+ \sum_{s=m+1}^{m+k}\left\{\sum_{1\leq a<b\leq q} \mathbf{B}_{aa}^{s}\mathbf{B}_{bb}^{s} + (\mathbf{B}_{p+1p+1}^{s})^{2} + \cdots + (\mathbf{B}_{mm}^{s})^{2}\right\}$$

$$- \sum_{s=m+1}^{m+k}\left\{\sum_{1\leq a<b\leq q} (\mathbf{B}_{ab}^{s})^{2} + (\mathbf{B}_{p+1p+1}^{s})^{2} + \cdots + (\mathbf{B}_{mm}^{s})^{2}\right\}$$

Using the binomial theorem and the fact that the base manifold $\mathbb{B}^p$ is minimal, then it not hard to check that

$$\sum_{s=m+1}^{m+k}\left(\sum_{A=p+1}^{m} \mathbf{B}_{AA}^{s}\right)^{2} = \sum_{s=m+1}^{m+k}\sum_{i,j=1}^{p}(\mathbf{B}_{ij}^{s})^{2} + \sum_{s=m+1}^{m+k}\sum_{a,b=1}^{q}(\mathbf{B}_{ab}^{s})^{2} - 2pq$$

$$+ 2\sum_{s=m+1}^{m+k}\sum_{\alpha=1}^{p}\sum_{\beta=1}^{q}(\mathbf{B}_{\alpha\beta}^{s})^{2} + \frac{q\Delta h}{h}$$

$$+ \sum_{s=m+1}^{m+k}\left((\mathbf{B}_{11}^{s})^{2} + \cdots + (\mathbf{B}_{pp}^{s})^{2}\right) - \sum_{s=m+1}^{m+k}\sum_{i,j=1}^{p}(\mathbf{B}_{ij}^{s})^{2}$$

$$+ \sum_{s=m+1}^{m+k}\left((\mathbf{B}_{p+1p+1}^{s})^{2} + \cdots + (\mathbf{B}_{mm}^{s})^{2}\right) - \sum_{s=m+1}^{m+k}\sum_{a,b=1}^{q}(\mathbf{B}_{ab}^{s})^{2}. \quad (12)$$

From the hypothesis of the theorem, we know that $\mathbb{B}^p$ is minimal and using this, we get that the fifth term of the right hand side in Equation (12) is equal to zero and seventh the term is equal to the first term of left hand side. Thus, we have:

$$2pq = 2\sum_{s=m+1}^{m+k}\sum_{\alpha=1}^{p}\sum_{\beta=1}^{q}(\mathbf{B}_{\alpha\beta}^{s})^{2} + \frac{q\Delta h}{h}.$$

From (7), it implies that

$$\|\mathbf{B}(e_\alpha, e_\beta)\|^2 = \frac{q}{2}\left(-\frac{\Delta h}{h}\right) + pq. \quad (13)$$

From assumption(3), we find that

$$-\frac{\Delta h}{h} < \left(\frac{2 - 6pq}{3q}\right) \quad (14)$$

Combining (13) with (14), one obtains:

$$||\mathbf{B}(e_\alpha, e_\beta)||^2 < \frac{q}{2}\left(\frac{2-6pq}{3q}\right) + pq = \frac{1}{3} - \frac{3pq}{3} + pq$$
$$< \frac{1}{3}. \tag{15}$$

Therefore, the proof follows from Theorem 1 and pinching condition (1) together with (15).

Remark 1. *The proofs of Theorems 2 and 3 follow easily using the same technique.*

3.2. Some Applications

Assume that $\{e_1, \ldots, e_m\}$ is a local orthonormal basis of vector field $\mathbb{M}^m$. Then the gradient of function φ and its squared norm is defined as:

$$\nabla \varphi = \sum_{i=1}^{m} e_i(\varphi) e_i,$$

and

$$||\nabla \varphi||^2 = \sum_{i=1}^{m} (e_i(\varphi))^2. \tag{16}$$

Let φ be a differentiable function defined on $\mathbb{M}^m$ such that $\varphi \in \mathcal{F}(\mathbb{M}^m)$, then the *Lagrangian* of the function φ is given in (p. 44, [27]).

$$L_\varphi = \frac{1}{2}||\nabla \varphi||^2. \tag{17}$$

The Euler–Lagrange formula of the *Lagrangian* (17) satisfies

$$\Delta \varphi = 0. \tag{18}$$

At point $x \in \mathbb{M}^n$ in a local orthonormal basis, the Hamiltonian would take the form (see [27] for more details):

$$H(p, x) = \frac{1}{2} \sum_{i=1}^{m} p(e_i)^2.$$

Put $p = d\varphi$, where d is a differential operator, and using (16), we get:

$$H(d\varphi, x) = \frac{1}{2} \sum_{i=1}^{m} d\varphi(e_i)^2 = \frac{1}{2} \sum_{i=1}^{m} e_i(\varphi)^2 = \frac{1}{2}||\nabla \varphi||^2. \tag{19}$$

Assuming that $\mathbb{M}^m = \mathbb{B}^p \times_f \mathbb{F}^q$ is a warped product, then $\forall\ Z_1 \in \Gamma(T\mathbb{B})$ and $Z_2 \in \Gamma(T\mathbb{F})$, we have

$$\nabla_{Z_2} Z_1 = \nabla_{Z_1} Z_2 = (Z_1 \ln h) Z_2.$$

Using the unit vector fields X and Z which are tangents to $\Gamma(T\mathbb{B})$ and $\Gamma(T\mathbb{F})$, resp.; then one obtains:

$$K(Z_1 \wedge Z_2) = g(R(Z_1, Z_2)Z_1, Z_2) = (\nabla_{Z_1} Z_1) \ln h g(Z_2, Z_2) - g(\nabla_{Z_1}((Z_1 \ln h) Z_2), Z_2)$$
$$= (\nabla_{Z_1} Z_1) \ln h g(Z_2, Z_2) - g(\nabla_{Z_1}(Z_1 \ln h) Z_2 + (Z_1 \ln h) \nabla_{Z_1} Z_2, Z_2)$$
$$= (\nabla_{Z_1} Z_1) \ln h g(Z_2, Z_2) - (Z_1 \ln h)^2 - Z_1(Z_1 \ln h).$$

If $\{e_1, \cdots e_m\}$ is an orthonormal basis for $\mathbb{M}^m$, then we can take a sum over the vector fields as follows

$$\sum_{\alpha=1}^{p}\sum_{\beta=1}^{q} K(e_\alpha \wedge e_\beta) = \sum_{\alpha=1}^{p}\sum_{\beta=1}^{q} \left((\nabla_{e_\alpha} e_\alpha) \ln h - e_\alpha(e_\beta \ln h) - (e_\alpha \ln f)^2 \right)$$
$$= q\left(\Delta(\ln h) - ||\nabla(\ln h)||^2 \right). \tag{20}$$

Thus, from (20) and (2), it follows that

$$\frac{\Delta h}{h} = \Delta(\ln h) - ||\nabla(\ln h)||^2. \tag{21}$$

Here, motivated by the historical development on the study of Lagrangian and Hamiltonian, we will give the following theorems as

Theorem 4. *Let $\ell : \mathbb{M}^m = \mathbb{B}^p \times_h \mathbb{F}^q \longrightarrow \mathbb{S}^{m+k}$ be an isometric immersion from an oriented complete WP submanifold $\mathbb{M}^m$ of dimension m into a sphere $\mathbb{S}^{m+k}$ of dimension $(m+k)$ such that the base manifold $\mathbb{B}^p$ is minimal and the function h satisfies the Euler–Lagrange equation with following inequality*

$$L_h < \left(\frac{1-3pq}{3q} \right) 2h^2, \tag{22}$$

where L_h is the Lagrangian of h. Then $\mathbb{M}^m$ is diffeomorphic to $\mathbb{S}^m$.

Proof. Using the fact that the warping function $\ln h$ satisfies the Euler–Lagrange equation, from the hypothesis of the theorem, and using (18), we have

$$\Delta \ln h = 0. \tag{23}$$

From (21) and (15), we derive

$$\Delta \ln h - \frac{||\nabla h||^2}{h^2} > 2p - \frac{2}{3q}. \tag{24}$$

It follows from (23) and (24) that

$$||\nabla h||^2 < 2ph^2 - \frac{2h^2}{3q}.$$

Using (17), we get desired result (22) which ends the proof. □

Theorem 5. *Suppose that $\ell : \mathbb{M}^m = \mathbb{B}^p \times_h \mathbb{F}^q \longrightarrow \mathbb{S}^{m+k}$ is an isometric immersion from an oriented complete WP submanifold $\mathbb{M}^m$ of dimension m into a sphere $\mathbb{S}^{m+k}$ of dimension $(m+k)$ such that the base manifold $\mathbb{B}^p$ is minimal and satisfies the relation*

$$H(dh, x) < \left\{ \frac{\Delta(lnh)}{2} + \left(\frac{1}{3q} - p \right) \right\} h^2. \tag{25}$$

Then $\mathbb{M}^m$ is diffeomorphic to $\mathbb{S}^m$.

Proof. Using Equation (19) in (24), we get required pinching condition (25). □

4. Conclusion Remark

We provide the characterization of a complete warped manifold to be diffeomorphically a unit sphere and some geometric classifications using Euler Lagrange formula along with Hamiltonian of the warping function. The topology of warped products and main extrinsic and intrinsic curvature invariants are emphatically related. Hence, our results may be seen as topological and differential sphere theorems from the viewpoint of warped product submanifolds theory. This paper shows the relation between the notion of warped product manifold and homotopy-homology theory. Therefore, we hope that this paper will be of great interest with respect to the topology of Riemannian geometry [28–35] which may find possible applications in physics.

Author Contributions: Conceptualization, A.A., F.M. and N.A.; methodology, A.A.; software, N.A.; validation, W.A.M.O., N.A. and F.M.; formal analysis, W.A.M.O.; investigation, A.A.; resources, N.A.; data curation, F.M.; writing–original draft preparation, N.M.; writing–review and editing, A.A. and F.M.; visualization, W.A.M.O.; supervision, A.A.; project administration, F.M.; funding acquisition, F.M. All authors have read and agreed to the published version of the manuscript.

Funding: This research was funded by the Deanship of Scientific Research at Princess Nourah bint Abdulrahman University through the Fast-track Research Funding Program and The APC was funded by Fast-track Research Funding Program.

Acknowledgments: This research was funded by the Deanship of Scientific Research at Princess Nourah bint Abdulrahman University through the Fast-track Research Funding Program.

Conflicts of Interest: The authors declare no conflict of interest.

References

1. Bishop, R.L.; O'Neil, B. Manifolds of negative curvature. *Trans. Am. Math. Soc.* **1969**, *145*, 1–9. [CrossRef]
2. Nolker, S. Isometric immersions of warped products. *Differ. Geom. Appl.* **1996**, *6*, 1–30. [CrossRef]
3. Kim, D.K.; Kim, Y.H. Compact Einstein warped product spaces with nonpositive scalar curvature. *Proc. Am. Math. Soc.* **2003**, *131*, 2573–2576. [CrossRef]
4. Dajczer, D.; Vlachos, T. Isometric immersions of warped products. *Proc. Am. Math. Soc.* **2013**, *141*, 1795–1803. [CrossRef]
5. Roth, J. A DDVV inequality for submanifold of warped products. *Bull. Aust. Math. Soc.* **2017**, *95*, 495–499. [CrossRef]
6. Salavessa, I.M.C. Heinz mean curvature estimates in warped product spaces $M \times_{e^\psi} N$. *Ann. Glob. Anal. Geom.* **2018**, *53*, 265. [CrossRef]
7. Myers, S.B. Riemannian manifolds with positive mean curvature. *Duke Math. J.* **1941**, *8*, 401–404. [CrossRef]
8. Rauch, H. A contribution to differential geometry in the large. *Ann. Math.* **1951**, *54*, 38–55. [CrossRef]
9. Berger, M. Les variétés riemanniennes ($\frac{1}{4}$)-pincées. *Ann. Scuola Norm. Sup. Pisa Cl. Sci.* **1960**, *14*, 161–170.
10. Grove, K.; Shiohama, K. A generalized sphere theorem. *Ann. Math.* **1977**, *106*, 201–211. [CrossRef]
11. Cheeger, J.; Gromoll, D. The splitting theorem for manifolds of nonnegative Ricci curvature. *J. Differ. Geom.* **1971**, *6*, 119–129. [CrossRef]
12. Schoen, B.; Yau, S.T. Complete three dimensional manifolds with positive Ricci curvature and scalar curvature. In *Seminar on Differential Geometry*; Princeton University Press: Princeton, NJ, USA, 1982; Volume 102, pp. 209–228.
13. Cheng, S.Y. Eigenvalue comparison theorem and its geometric applications. *Math. Z.* **1975**, *143*, 289–297. [CrossRef]
14. Gu, J.R.; Xu, X.W. On Yau rigidity theorem for minimal submanifolds in spheres. *Math. Res. Lett.* **2012**, *19*, 511–523. [CrossRef]
15. Andrews, B.; Baker, C. Mean curvature flow of pinching submanifolds in spheres. *J. Differ. Geom.* **2010**, *58*, 357–395. [CrossRef]
16. Ali, A.; Laurian, P.I.; Alkhaldi, A.H. Ricci curvature on warped product submanifolds in spheres with geometric applications. *J. Geom. Phys.* **2019**, *146*, 103510. [CrossRef]
17. Ali, A.; Alkhaldi, A.H.; Laurian, P.I. Stable currents and homology groups in a compact CR-warped product submanifold with negative constant sectional curvature. *J. Geom. Phys.* **2020**, *148*, 103566. [CrossRef]

18. Gauchman, H. Minimal submanifold of a sphere with bounded second fundamental form. *Trans. Am. Math. Soc.* **1986**, *298*, 779–791. [CrossRef]
19. Leung, P.F. On a relation between the topology and the intrinsic and extrinsic geometries of a compact submanifold. *Proc. Edinb. Math. Soc.* **1985**, *28*, 305–311. [CrossRef]
20. Leung, P.F. On the topology of a compact submanifold of a sphere with bounded second fundmental form. *Manuscr. Math.* **1993**, *79*, 183–185. [CrossRef]
21. Xu, X.W.; Zhao, E.T. Topological and differentiable sphere theorems for complete submanifolds. *Commun. Anal. Geom.* **2009**, *17*, 565–585. [CrossRef]
22. Xu, X.W.; Ye, F. Differentiable sphere theorems for submanifolds of positive k-th Ricci curvature. *Manuscr. Math.* **2012**, *138*, 529–543. [CrossRef]
23. Xu, H.W.; Gu, J.R. Geometric, topological and differentiable rigidity of submanifolds in space forms. *Geom. Funct. Anal.* **2013**, *23*, 1684. [CrossRef]
24. Lawson, H.W.; Simons, J. On stable currents and their application to global problems in real and complex geometry. *Ann. Math.* **1973**, *98*, 427–450. [CrossRef]
25. Nash, J. The imbedding problem for Riemannian manifolds. *Ann. Math.* **1956**, *63*, 20–63. [CrossRef]
26. Chen, B.Y. On isometric minimal immersions from warped products into real space forms. *Proc. Edinb. Math. Soc.* **2002**, *45*, 579–587. [CrossRef]
27. Calin, O.; Chang, D.C. *Geometric Mechanics on Riemannian Manifolds: Applications to Partial Differential Equations*; Springer Science & Business Media: Berlin/Heidelberg, Germany, 2006.
28. Alkhaldi, A.H.; Ali, A. Classification of warped product submanifolds in Kenmotsu space forms admitting gradient Ricci solitons. *Mathematics* **2019**, *7*, 112. [CrossRef]
29. Aquib, M.; Boyom, B.N.; Shahid, M.H.; Vîlcu, G.E. The first fundamental equation and generalized wintgen-type inequalities for submanifolds in generalized space forms. *Mathematics* **2019**, *7*, 1151. [CrossRef]
30. Ali, R.; Mofarreh, F.; Alluhaibi, N.; Ali, A.; Ahmad, I. On differential equations characterizing Legendrian submanifolds of Sasakian space forms. *Mathematics* **2020**, *8*, 150. [CrossRef]
31. Ali, A.; Alkhaldi, A.H.; Laurian, P.I. Ali. R (2020) Eigenvalue inequalities for the p-Laplacian operator on C-totally real submanifolds in Sasakian space forms. *Appl. Anal.* **2020**. [CrossRef]
32. Decu, S.; Haesen, S.; Verstraelen, L. Inequalities for the Casorati Curvature of Statistical Manifolds in Holomorphic Statistical Manifolds of Constant Holomorphic Curvature. *Mathematics* **2020**, *8*, 251. [CrossRef]
33. Macsim, G.; Mihai, A.; Mihai, I. *delta*(2, 2)-Invariant for Lagrangian Submanifolds in Quaternionic Space Forms. *Mathematics* **2020**, *8*, 480. [CrossRef]
34. Siddiqui, A.N.; Ali, A.; Alkhaldi, A.H. Chen optimal inequalities of CR-warped products of generalized Sasakian space forms. *J. Taibah Univ. Sci.* **2020**, *14*, 322–330. [CrossRef]
35. Siddiqui, A.N.; Chen, B.-Y.; Bahadir, O. Statistical solitons and inequalities for statistical warped product submanifolds. *Mathematics* **2019**, *7*, 797. [CrossRef]

© 2020 by the authors. Licensee MDPI, Basel, Switzerland. This article is an open access article distributed under the terms and conditions of the Creative Commons Attribution (CC BY) license (http://creativecommons.org/licenses/by/4.0/).

Article

The First Fundamental Equation and Generalized Wintgen-Type Inequalities for Submanifolds in Generalized Space Forms

Mohd. Aquib [1], Michel Nguiffo Boyom [2], Mohammad Hasan Shahid [1] and Gabriel-Eduard Vîlcu [3,*]

[1] Department of Mathematics, Faculty of Natural Sciences, Jamia Millia Islamia, New Delhi-110025, India; aquib80@gmail.com (M.A.); mshahid@jmi.ac.in (M.H.S.)
[2] Institut Montpelliérain Alexander Grothendieck, Université de Montpellier, CC051, Pl. E. Bataillon, F-34095 Montpellier, France; boyom@math.univ-montp2.fr
[3] Department of Cybernetics, Economic Informatics, Finance and Accountancy, Petroleum-Gas University of Ploieşti, Bd. Bucureşti 39, 100680 Ploieşti, Romania
* Correspondence: gvilcu@upg-ploiesti.ro; Tel.: +40-244-575-847

Received: 23 October 2019; Accepted: 25 November 2019; Published: 1 December 2019

Abstract: In this work, we first derive a generalized Wintgen type inequality for a Lagrangian submanifold in a generalized complex space form. Further, we extend this inequality to the case of bi-slant submanifolds in generalized complex and generalized Sasakian space forms and derive some applications in various slant cases. Finally, we obtain obstructions to the existence of non-flat generalized complex space forms and non-flat generalized Sasakian space forms in terms of dimension of the vector space of solutions to the first fundamental equation on such spaces.

Keywords: Wintgen inequality; generalized complex space form; generalized Sasakian space form; Lagrangian submanifold; Legendrian submanifold

MSC: 53B05; 53B20; 53C40

---

1. Introduction

The classical Wintgen inequality is a sharp geometric inequality established in [1], according to which the Gaussian curvature $\mathcal{K}$ of any surface $\mathcal{N}^2$ in the Euclidean space $\mathbb{E}^4$, the normal curvature $\mathcal{K}^\perp$, and also the squared mean curvature $\|\mathcal{H}\|^2$ of $\mathcal{N}^2$, satisfy

$$\|\mathcal{H}\|^2 \geq \mathcal{K} + |\mathcal{K}^\perp|$$

and the equality is attained only in the case when the ellipse of curvature of $\mathcal{N}^2$ in $\mathbb{E}^4$ is a circle. Later, this inequality was extended independently by Rouxel [2] and Gaudalupe and Rodriguez [3] for surfaces of arbitrary codimension m in real space forms $\overline{\mathcal{N}}^{m+2}(c)$ with constant sectional curvature c as

$$\|\mathcal{H}\|^2 + c \geq \mathcal{K} + |\mathcal{K}^\perp|.$$

The generalized Wintgen inequality, also known as the DDVV-inequality or the DDVV-conjecture, is a natural extension of the above inequalities that was conjectured in 1999 by De Smet, Dillen, Verstraelen and Vrancken [4] and settled in the general case independently by Ge and Tang [5] and Lu [6]. The

Mathematics **2019**, *7*, 1151; doi:10.3390/math7121151   www.mdpi.com/journal/mathematics

DDVV-conjecture generalizes the classical Wintgen inequality to the case of an isometric immersion $f: M^n \to N^{n+p}(c)$ from an n-dimensional Riemannian submanifold M^n into a real space form $N^{n+p}(c)$ of dimension $(n+p)$ and of constant sectional curvature c, stating that such an isometric immersion satisfies

$$\rho + \rho^\perp \leq \|\mathcal{H}\|^2 + c,$$

where ρ is the normalized scalar curvature, while $\rho^\perp$ denotes the normalized normal scalar curvature. Notice that there are many examples of submanifolds satisfying the equality case of the above inequality and these submanifolds are known as Wintgen ideal submanifolds [7].

Recently, the generalized Wintgen inequality was extended for several kinds of submanifolds in many ambient spaces, e.g., complex space forms [8], Sasakian space forms [9], quaternionic space forms [10], warped products [11], and Kenmotsu statistical manifolds [12]. In the first part of the present paper, we obtain generalized Wintgen-type inequalities for different types of submanifolds in generalized complex space forms and also in generalized Sasakian space forms, generalizing the main results in [8,9], and also discuss some applications. The last part of the paper is devoted to the investigation of the Hessian equation on both generalized complex space forms and generalized Sasakian space-forms. In particular, some obstructions to the existence of these spaces are established. Recall that the notion of generalized complex space form was introduced in differential geometry by Tricerri and Vanhecke [13], the authors proving that, if $n \geq 3$, a $2n$-dimensional generalized complex space form is either a real space form or a complex space form, a result partially extendable to four-dimensional manifolds. However, the existence of proper generalized complex space form in dimension 4 was obtained by Olszak [14], using some conformal deformations of four-dimensional flat Bochner–Kähler manifolds of non-constant scalar curvature. It is important to note that the generalized complex space forms are a particular kind of almost Hermitian manifolds with constant holomorphic sectional curvature and constant type in the sense of Gray [15].

On the other hand, Alegre, Blair and Carriazo [16] generalized the notions of Sasakian space form, Kenmotsu space form and cosymplectic space form, by introducing the concept of generalized Sasakian space form. Notice that several examples of non-trivial generalized Sasakian space-forms are given in [16] using different geometric constructions, such as Riemannian submersions, warped products, and D-conformal deformations. Afterwards, many interesting results have been proved in these ambient spaces (see, e.g., [17–27]). We only recall that, very recently, Bejan and Güler [28] obtained an unexpected link between the class of generalized Sasakian space-forms and the class of Kähler manifolds of quasi-constant holomorphic sectional curvature, providing conditions under which each of these structures induces the other one.

2. Preliminaries

An almost Hermitian manifold consists in a smooth manifold $\overline{\mathcal{N}}$ endowed with an almost complex structure J and a Riemannian metric g that is compatible with the structure J. Such a manifold is called Kähler if $\overline{\nabla} J = 0$, where $\overline{\nabla}$ is the Levi–Civita connection of the metric g.

On the other hand, an almost Hermitian manifold $\overline{\mathcal{N}}$ is called a generalized complex space form [13], denoted by $\overline{\mathcal{N}}(f_1, f_2)$, if the Riemannian curvature tensor $\overline{R}$ satisfies

$$\begin{aligned}\overline{R}(X,Y)Z &= f_1\{g(Y,Z)X - g(X,Z)Y\} + f_2\{g(X,JZ)JY \\ &\quad - g(Y,JZ)JX + 2g(X,JY)JZ\}\end{aligned} \quad (1)$$

for all vector fields X, Y and Z on $\overline{\mathcal{N}}$, where f_1 and f_2 are smooth functions on $\overline{\mathcal{N}}$. This name is motivated by the fact that, in the case of a complex space form, viz. a Kähler manifold with constant holomorphic sectional curvature $4c$, the curvature tensor field of the manifold satisfies Equation (1) with $f_1 = f_2 = c$.

Let $\mathcal{N}$ be a submanifold of real dimension n in a generalized complex space form $\overline{\mathcal{N}}(f_1,f_2)$ of complex dimension m. If ∇ and $\overline{\nabla}$ are the Levi–Civita connections on $\mathcal{N}$ and $\overline{\mathcal{N}}(f_1,f_2)$, respectively, then the fundamental formulas of Gauss and Weingarten are [29]

$$\overline{\nabla}_X Y = \nabla_X Y + h(X,Y),$$

$$\overline{\nabla}_X \xi = -S_\xi X + \nabla^\perp_X Y,$$

where X,Y are vector fields tangent to $\mathcal{N}$, ξ is a vector field normal to $\mathcal{N}$, and $\nabla^\perp$ represents the normal connection. Recall that, in the above basic formulas, h denotes the second fundamental form and S is the shape operator, they being connected by

$$g(h(X,Y),\xi) = g(S_\xi X, Y).$$

On the other hand, the Gauss' equation is expressed by [30]

$$\begin{aligned}\overline{R}(X,Y,Z,W) &= R(X,Y,Z,W) + g(h(X,Z),h(Y,W))\\ &\quad - g(h(X,W),h(Y,Z))\end{aligned} \quad (2)$$

for all vector fields X,Y,Z,W tangent to $\mathcal{N}$, where $\overline{R}$ denotes the curvature tensor of $\overline{\mathcal{N}}(f_1,f_2)$, while R represents the curvature tensors of $\mathcal{N}$. Let us point out now that the Ricci equation in our setting is expressed as

$$\begin{aligned}R^\perp(X,Y,\xi,\eta) &= f_2[g(X,J\xi)g(JY,\eta) - g(JX,\eta)g(Y,J\xi)]\\ &\quad - g([S_\xi, S_\eta]X,Y),\end{aligned} \quad (3)$$

for all vector fields X,Y tangent to $\mathcal{N}$ and ξ,η normal to $\mathcal{N}$.

If $\mathcal{N}$ is a submanifold of real dimension n in a generalized complex space form $\overline{\mathcal{N}}(f_1,f_2)$ of complex dimension m, then, for any $X \in T\mathcal{N}$, we have the decomposition $JX = PX + QX$, where P and Q denote the tangential component and the normal component of JX, respectively. We recall that, in the case $P = 0$, the submanifold $\mathcal{N}$ is called anti-invariant, while, in the case f $Q = 0$, the submanifold $\mathcal{N}$ is called invariant.

Now, let $\{e_1,\ldots,e_n\}$ be a tangent orthonormal frame on $\mathcal{N}$ and let $\{\xi_1,\ldots,\xi_{2m-n}\}$ be a normal orthonormal frame on $\mathcal{N}$. Then, the squared norm of P at $p \in \mathcal{N}$ is defined as

$$\|P\|^2 = \sum_{i,j=1}^n g^2(Pe_i,e_j), \quad (4)$$

while the mean curvature vector field is given by

$$\mathcal{H} = \frac{1}{n}\sum_{i=1}^n h(e_i,e_i). \quad (5)$$

We also set

$$h^r_{ij} = g(h(e_i,e_j),\xi_r), \quad i,j = 1,\ldots,n, \ r = 1,\ldots,2m-n. \quad (6)$$

and

$$\|h\|^2 = \sum_{i,j=1}^{n} g(h(e_i, e_j), h(e_i, e_j)). \tag{7}$$

3. Generalized Wintgen Inequality for Lagrangian Submanifolds in Generalized Complex Space Form

Let $\mathcal{N}$ be a submanifold of real dimension n in a generalized complex space form $\overline{\mathcal{N}}(f_1, f_2)$ of complex dimension m. In the following, let $\{e_1, \ldots, e_n\}$ and $\{\xi_1, \ldots, \xi_{2m-n}\}$ be tangent orthonormal frame and normal orthonormal frame on $\mathcal{N}$, respectively. If we denote by K the sectional curvature function and by τ the scalar curvature, then the normalized scalar curvature ρ of $\mathcal{N}$ can be expressed as [8]

$$\rho = \frac{2\tau}{n(n-1)} = \frac{2}{n(n-1)} \sum_{1 \leq i < j \leq n} K(e_i \wedge e_j). \tag{8}$$

On the other hand, the normalized normal scalar curvature of $\mathcal{N}$ is given by [8]

$$\rho^\perp = \frac{2\tau^\perp}{n(n-1)} = \frac{2}{n(n-1)} \sqrt{\sum_{1 \leq i < j \leq n} \sum_{1 \leq r < s \leq 2m-n} (R^\perp(e_i, e_j, \xi_r, \xi_s))^2}, \tag{9}$$

where $R^\perp$ denotes the normal curvature tensor on $\mathcal{N}$.

The scalar normal curvature of $\mathcal{N}$ can be defined following [31] as

$$\mathcal{K}_N = \frac{1}{4} \sum_{r,s=1}^{2m-n} (Trace[S_r, S_s])^2. \tag{10}$$

Now, the normalized scalar normal curvature can be defined with the help of $\mathcal{K}_N$ by [8]

$$\rho_N = \frac{2}{n(n-1)} \sqrt{\mathcal{K}_N}.$$

Obviously

$$\begin{aligned}
\mathcal{K}_N &= \frac{1}{2} \sum_{1 \leq r < s \leq 2m-n} (Trace[S_r, S_s])^2 \\
&= \sum_{1 \leq r < s \leq 2m-n} \sum_{1 \leq i < j \leq n} (g([S_r, S_s]e_i, e_j))^2.
\end{aligned} \tag{11}$$

It is easy to verify now that $\mathcal{K}_N$ can be expressed by

$$\mathcal{K}_N = \sum_{1 \leq r < s \leq 2m-n} \sum_{1 \leq i < j \leq n} \Big(\sum_{k=1}^{n} h^r_{jk} h^s_{ik} - h^s_{jk} h^r_{ik} \Big)^2. \tag{12}$$

Among the classes of submanifolds in complex geometry, we can distinguish two fundamental families depending on the behavior of J: holomorphic and totally real submanifolds. A submanifold $\mathcal{N}$ of a generalized complex space form $\overline{\mathcal{N}}(f_1, f_2)$ is said to be a holomorphic submanifold if each tangent space of $\mathcal{N}$ is carried into itself by J, i.e., $J(T_p\mathcal{N}) \subset T_p\mathcal{N}$, for all $p \in \mathcal{N}$. Similarly, the submanifold $\mathcal{N}$ is called a totally real submanifold if J maps each tangent space of $\mathcal{N}$ into the normal space, i.e., $J(T_p\mathcal{N}) \subset T_p^\perp\mathcal{N}$, for all $p \in \mathcal{N}$. In particular, if $n = m$, then $\mathcal{N}$ is said to be a Lagrangian submanifold.

Next, we prove the following lemma, which is required in the proof of the main result of this section.

Lemma 1. *Let $\mathcal{N}$ be a totally real submanifold of dimension n in a generalized complex space form $\overline{\mathcal{N}}(f_1, f_2)$ of complex dimension m. Then, we have*

$$\rho_\mathcal{N} \leq \|\mathcal{H}\|^2 - \rho + f_1, \tag{13}$$

and the equality holds at a point $p \in \mathcal{N}$ if and only if the shape operator S of $\mathcal{N}$ in $\overline{\mathcal{N}}(f_1, f_2)$ with respect to some suitable orthonormal bases $\{e_1, \ldots, e_n\}$ of $T_p\mathcal{N}$ and $\{\tilde{\zeta}_1, \ldots, \tilde{\zeta}_{2m-n}\}$ of $T_p^\perp \mathcal{N}$ takes the following forms

$$S_{\tilde{\zeta}_1} = \begin{pmatrix} \gamma_1 & \nu & 0 & \cdots & 0 \\ \nu & \gamma_1 & 0 & \cdots & 0 \\ 0 & 0 & \gamma_1 & \cdots & 0 \\ \vdots & \vdots & \vdots & \ddots & \vdots \\ 0 & 0 & 0 & \cdots & \gamma_1 \end{pmatrix},$$

$$S_{\tilde{\zeta}_2} = \begin{pmatrix} \gamma_2 + \nu & 0 & 0 & \cdots & 0 \\ 0 & \gamma_2 - \nu & 0 & \cdots & 0 \\ 0 & 0 & \gamma_2 & \cdots & 0 \\ \vdots & \vdots & \vdots & \ddots & \vdots \\ 0 & 0 & 0 & \cdots & \gamma_2 \end{pmatrix},$$

$$S_{\tilde{\zeta}_3} = \begin{pmatrix} \gamma_3 & 0 & 0 & \cdots & 0 \\ 0 & \gamma_3 & 0 & \cdots & 0 \\ 0 & 0 & \gamma_3 & \cdots & 0 \\ \vdots & \vdots & \vdots & \ddots & \vdots \\ 0 & 0 & 0 & \cdots & \gamma_3 \end{pmatrix}, \quad S_{\tilde{\zeta}_4} = \cdots = S_{\tilde{\zeta}_{2m-n}} = 0,$$

where $\gamma_1, \gamma_2, \gamma_3$, and ν are real functions on $\mathcal{N}$.

Proof. We know that

$$\begin{aligned}
n^2 \|\mathcal{H}\|^2 &= \sum_{r=1}^{2m-n} \left(\sum_{i=1}^{n} h_{ii}^r \right)^2 \\
&= \frac{1}{n-1} \sum_{r=1}^{2m-n} \sum_{1 \leq i < j \leq n} (h_{ii}^r - h_{jj}^r)^2 \\
&\quad + \frac{2n}{n-1} \sum_{r=1}^{2m-n} \sum_{1 \leq i < j \leq n} h_{ii}^r h_{jj}^r.
\end{aligned} \tag{14}$$

Further, from [6], we have

$$\sum_{r=1}^{2m-n} \sum_{1 \leq i < j \leq n} (h_{ii}^r - h_{jj}^r)^2 + 2n \sum_{r=1}^{2m-n} \sum_{1 \leq i < j \leq n} (h_{ij}^r)^2$$
$$\geq 2n \left[\sum_{1 \leq r < s \leq 2m-n} \sum_{1 \leq i < j \leq n} \left(\sum_{k=1}^{n} (h_{jk}^r h_{ik}^s - h_{ik}^r h_{jk}^s) \right)^2 \right]^{\frac{1}{2}}. \tag{15}$$

Now, combining Equations (12), (14) and (15), we find

$$n^2 \|\mathcal{H}\|^2 - n^2 \rho_N \geq \frac{2n}{n-1} \sum_{r=1}^{2m-n} \sum_{1 \leq i < j \leq n} [h_{ii}^r h_{jj}^r - (h_{ij}^r)^2]. \tag{16}$$

In addition, due to the fact that $\mathcal{N}$ is a totally real submanifold, we get from Equation (2):

$$\tau = \frac{n(n-1)}{2} f_1 + \sum_{r=1}^{2m-n} \sum_{1 \leq i < j \leq n} [h_{ii}^r h_{jj}^r - (h_{ij}^r)^2]. \tag{17}$$

Next, using Equations (8) and (17) in Equation (16), we obtain the inequality in Equation (13). Moreover, it follows easily that the equality case holds in Equation (13) if and only if the shape operator takes the above stated forms. □

Now, we prove the following.

Theorem 1. *Let $\mathcal{N}$ be a Lagrangian submanifold of a generalized complex space form $\overline{\mathcal{N}}(f_1, f_2)$ of complex dimension n. Then,*

$$\begin{aligned}
(\rho^\perp)^2 &\leq (\|\mathcal{H}\|^2 - \rho + f_1)^2 \\
&\quad + \frac{2}{n(n-1)} f_2^2 + \frac{4 f_2}{n(n-1)} (\rho - f_1)
\end{aligned} \tag{18}$$

and the equality in Equation (18) holds at a point $p \in \mathcal{N}$ if and only if the shape operator takes similar forms as in Lemma 1 with respect to some suitable tangent and normal orthonormal bases.

Proof. Let $\mathcal{N}$ be a Lagrangian submanifold of a generalized complex space form $\overline{\mathcal{N}}(f_1, f_2)$. We choose $\{e_1, \ldots, e_n\}$ and $\{\xi_1 = Je_1, \ldots, \xi_n = Je_n\}$ as orthonormal frame and orthonormal normal frame on $\mathcal{N}$, respectively. Putting $X = W = e_i$, $Y = Z = e_j$, $i \neq j$ in Equation (1), we obtain

$$\overline{R}(e_i, e_j, e_j, e_i) = f_1\{g(e_j, e_j)g(e_i, e_i) - g(e_i, e_j)g(e_j, e_i)\}. \tag{19}$$

Combining Equations (2) and (19), we derive

$$\begin{aligned}
R(e_i, e_j, e_j, e_i) &= f_1\{\delta_{ii}\delta_{jj} - \delta_{ij}^2\} - g(h(e_i, e_j), h(e_j, e_i)) \\
&\quad + g(h(e_i, e_i), h(e_j, e_j)).
\end{aligned} \tag{20}$$

By taking summation for $1 \leq i, j \leq n$ in Equation (20) and making use of Equations (5) and (7), we obtain

$$2\tau = n(n-1)f_1 + n^2 \|\mathcal{H}\|^2 - \|h\|^2. \tag{21}$$

Using Equation (8) in Equation (21), we get

$$\rho = f_1 + \frac{n}{n-1}\|\mathcal{H}\|^2 - \frac{1}{n(n-1)}\|h\|^2, \tag{22}$$

which implies

$$n^2\|\mathcal{H}\|^2 - \|h\|^2 = n(n-1)(\rho - f_1). \tag{23}$$

Further, Equation (3) gives

$$R^\perp(e_i, e_j, \zeta_r, \zeta_s) = f_2\{-(\delta_{ir}\delta_{js} - \delta_{jr}\delta_{is})\} - g([S_{\zeta_r}, S_{\zeta_s}]e_i, e_j), \tag{24}$$

for any indices $i, j, r, s \in \{1, \ldots, n\}$.

Next, by taking summation for $1 \leq r < s \leq n$ and $1 \leq i < j \leq n$ in Equation (24), we derive easily the following relation:

$$(\tau^\perp)^2 = \frac{n(n-1)}{2}f_2^2 + \frac{n^2(n-1)^2}{4}\rho_N^2 - f_2\|h\|^2 + f_2 n^2\|\mathcal{H}\|^2. \tag{25}$$

However, the above Equation (25) can be rewritten as

$$(\rho^\perp)^2 = \frac{2}{n(n-1)}f_2^2 + \rho_N^2 - \frac{4f_2}{n^2(n-1)^2}\|h\|^2 + \frac{4f_2}{(n-1)^2}\|\mathcal{H}\|^2. \tag{26}$$

Now, from Equations (23) and (26), we have

$$(\rho^\perp)^2 = \frac{2}{n(n-1)}f_2^2 + \rho_N^2 + \frac{4f_2}{n(n-1)}(\rho - f_1). \tag{27}$$

Combining now Equations (13) and (27), we obtain the required inequality and the equality case of the inequality is also clear from Lemma 1. □

Remark 1. *Theorem 2 generalizes the main result of [8], namely the generalized Wintgen inequality for the class of Lagrangian submanifolds in a complex space form. Indeed, if in the statement of Theorem 2 one particularizes the generalized complex space form by putting $f_1 = f_2 = c$, then $\overline{\mathcal{N}}$ reduces to a complex space form and one arrives at ([8] Theorem 2.3).*

4. Generalized Wintgen Inequality for bi-Slant Submanifolds in Generalized Complex Space Form

A submanifold $\mathcal{N}$ of an almost Hermitian manifold $(\overline{\mathcal{N}}, J, g)$ is said to be a slant submanifold if for any point $p \in \mathcal{N}$ and any non-zero vector $X \in T_p\mathcal{N}$, the angle θ between the vector JX and the tangent space $T_p\mathcal{N}$ is constant, i.e., this angle does not depend on the choice of $p \in \mathcal{N}$ and $X \in T_p\mathcal{N}$. Moreover, $\theta \in [0, \frac{\pi}{2}]$ is called the slant angle of $\mathcal{N}$ in $\overline{\mathcal{N}}$. Recall that both invariant and anti-invariant submanifolds are particular examples of slant submanifolds with slant angle $\theta = 0$ and $\theta = \frac{\pi}{2}$, respectively. Moreover, if $0 < \theta < \frac{\pi}{2}$, then $\mathcal{N}$ is said to be a θ-slant submanifold or a proper slant submanifold. It is known that any proper slant submanifold has even dimension. The concept of slant submanifold originally introduced by Chen [32,33] was later generalized as follows.

Definition 1. ([34]) *A submanifold $\mathcal{N}$ of an almost Hermitian manifold $\overline{\mathcal{N}}$ is said to be a bi-slant submanifold, if there exist two orthogonal distributions D_1 and D_2, such that:*

(i) *$T\mathcal{N}$ admits the orthogonal direct decomposition:*

$$T\mathcal{N} = D_1 \oplus D_2.$$

(ii) *$JD_1 \perp D_2$ and $JD_2 \perp D_1$.*

(iii) *For $i = 1, 2$, the distribution D_i is slant with slant angle θ_i.*

It is easy to see that the class of bi-slant submanifolds of almost Hermitian manifolds naturally englobes not only the class of slant submanifolds, but also the classes of semi-slant submanifolds [35], hemi-slant submanifolds [36], and CR-submanifolds [37], as synthesized in ([38] Table 1).

In the following, let us denote $d_1 = dimD_1$ and $d_2 = dimD_2$. We say that a bi-slant submanifold $\mathcal{N}$ of an almost Hermitian manifold $\overline{\mathcal{N}}$ with slant angles θ_1 and θ_2, respectively, is a proper bi-slant submanifold if $d_1 d_2 \neq 0$ and $0 < \theta_i < \frac{\pi}{2}$, for $i = 1, 2$. If $\mathcal{N}$ is a proper bi-slant submanifold in a generalized complex space form $\overline{\mathcal{N}}(f_1, f_2)$, then one can check that

$$\sum_{i,j=1}^{n} g^2(Je_i, e_j) = (d_1 cos^2\theta_1 + d_2 cos^2\theta_2). \tag{28}$$

Now, we state and prove the generalized Wintgen inequality for proper bi-slant submanifolds in generalized complex space forms.

Theorem 2. *Let $\mathcal{N}$ be a proper bi-slant submanifold of dimension n in a generalized complex space form $\overline{\mathcal{N}}(f_1, f_2)$ of complex dimension m, with slant angles θ_1, θ_2 and $d_i = dimD_i, i = 1, 2$. Then,*

$$\begin{aligned}\rho_{\mathcal{N}} &\leq \|\mathcal{H}\|^2 - \rho + f_1 \\ &+ \frac{3f_2}{n(n-1)}(d_1 cos^2\theta_1 + d_2 cos^2\theta_2).\end{aligned} \tag{29}$$

Proof. Let $\{e_1, \ldots, e_{n-1}, e_n\}$ be an orthonormal frame on $\mathcal{N}$ and $\{\xi_1, \ldots, \xi_{2m-n}\}$ be a normal orthonormal frame on $\mathcal{N}$.

Equation (2) can be re-written in view of Equation (1) as

$$\begin{aligned}R(X,Y,Z,W) &= f_1\{g(Y,Z)g(X,W) - g(X,Z)g(Y,W)\} \\ &+ f_2\{g(X,JZ)g(JY,W) - g(Y,JZ)g(JX,W) \\ &+ 2g(X,JY)g(JZ,W)\} \\ &- g(h(X,Z), h(Y,W)) + g(h(X,W), h(Y,Z))\end{aligned} \tag{30}$$

and this implies

$$\begin{aligned}
\tau &= \sum_{1 \leq i < j \leq n} R(e_i, e_j, e_j, e_i) \\
&= \frac{n(n-1)}{2} f_1 + \frac{3}{2} f_2 \sum_{1 \leq i < j \leq n} g^2(J e_j, e_i) \\
&\quad + \sum_{r=1}^{2m-n} \sum_{1 \leq i < j \leq n} [h_{ii}^r h_{jj}^r - (h_{ij}^r)^2] \\
&= \frac{n(n-1)}{2} f_1 + \frac{3}{2} f_2 (d_1 \cos^2 \theta_1 + d_2 \cos^2 \theta_2) \\
&\quad + \sum_{r=1}^{2m-n} \sum_{1 \leq i < j \leq n} [h_{ii}^r h_{jj}^r - (h_{ij}^r)^2].
\end{aligned} \qquad (31)$$

However, we know from the proof of Lemma 1 that

$$n^2 \|\mathcal{H}\|^2 - n^2 \rho_N \geq \frac{2n}{n-1} \sum_{r=1}^{2m-n} \sum_{1 \leq i < j \leq n} [h_{ii}^r h_{jj}^r - (h_{ij}^r)^2]. \qquad (32)$$

Combining Equations (31) and (32), we find

$$\begin{aligned}
\rho_N &\leq \|\mathcal{H}\|^2 - (\rho - f_1) \\
&\quad + \frac{3 f_2}{n(n-1)} (d_1 \cos^2 \theta_1 + d_2 \cos^2 \theta_2)
\end{aligned} \qquad (33)$$

and the proof is now complete. □

Remark 2. *If in the statement of the above theorem one takes $f_1 = f_2 = c$, then $\overline{\mathcal{N}}$ reduces to a complex space form and we can immediately see that Theorem 2 generalizes the generalized Wintgen inequality for the class of proper slant submanifolds in a complex space form, namely ([8] Theorem 3.1).*

5. Generalized Wintgen Inequalities for Submanifolds in Generalized Sasakian Space Form

Let $\overline{\mathcal{N}}$ be an almost contact metric manifold of dimension $(2m+1)$, equipped with the almost contact structure (ϕ, ξ, η, g). Then, it is known that the (1,1) tensor field ϕ, the structure vector field ξ, the 1-form η, and the Riemannian metric g on $\overline{\mathcal{N}}$ verify the compatibility relations

$$\phi^2 = -I + \eta \otimes \xi, \qquad \eta(\xi) = 1,$$
$$g(\phi X, \phi Y) = g(X, Y) - \eta(X) \eta(Y).$$

These conditions also imply that [39]

$$\phi \xi = 0, \qquad \eta(\phi X) = 0, \qquad \eta(X) = g(X, \xi)$$

and

$$g(\phi X, Y) + g(X, \phi Y) = 0,$$

for all vector fields X, Y on $\overline{\mathcal{N}}$.

Let $(\overline{\mathcal{N}}, \phi, \xi, \eta, g)$ be an almost contact metric manifold whose curvature tensor satisfies

$$\begin{aligned}\overline{R}(X,Y)Z &= f_1\{g(Y,Z)X - g(X,Z)Y\} \\ &+ f_2\{g(X,\phi Z)\phi Y - g(Y,\phi Z)\phi X + 2g(X,\phi Y)\phi Z\} \\ &+ f_3\{\eta(X)\eta(Z)Y - \eta(Y)\eta(Z)X + g(X,Z)\eta(Y)\xi \\ &- g(Y,Z)\eta(X)\xi\},\end{aligned} \qquad (34)$$

for all vector fields X, Y, Z on $\overline{\mathcal{N}}$, where f_1, f_2, f_3 are differentiable functions on $\overline{\mathcal{N}}$. Then, $\overline{\mathcal{N}}(f_1, f_2, f_3)$ is said to be a generalized Sasakian space form. It is important to outline that the generalized Sasakian space forms are an umbrella of the following well known spaces:

i. Sasakian space forms, i.e., Sasakian manifolds with constant ϕ-sectional curvature c. In this case, $f_1 = \frac{c+3}{4}, f_2 = f_3 = \frac{c-1}{4}$.
ii. Kenmotsu space forms, i.e., Kenmotsu manifolds of constant ϕ-sectional curvature c. In this case, $f_1 = \frac{c-3}{4}$ and $f_2 = f_3 = \frac{c+1}{4}$).
iii. cosymplectic space forms, i.e., cosymplectic manifolds of constant ϕ-sectional curvature c. In this case, $f_1 = f_2 = f_3 = \frac{c}{4}$.

For definitions, basic results, and examples of such spaces, the readers are referred to the monographs [39,40].

A Riemannian manifold $\mathcal{N}$ isometrically immersed in an almost contact metric manifold $(\overline{\mathcal{N}}, \phi, \xi, \eta, g))$ is called a C-totally real submanifold of $\overline{\mathcal{N}}$ if the structure vector field ξ is a normal vector field on $\mathcal{N}$. As an immediate consequence of the definition of a C-totally real submanifold, we deduce that ϕ maps any tangent space of $\mathcal{N}$ into the normal space. We recall that, if the dimension of the C-totally real submanifold $\mathcal{N}$ is $n = \frac{\dim \overline{\mathcal{N}} - 1}{2}$, then $\mathcal{N}$ is said to be a Legendrian submanifold. Notice that Legendrian submanifolds are the counterpart in odd dimension of Lagrangian submanifolds investigated in Section 3.

The first aim of this section is to obtain the generalized Wintgen inequality for Legendrian submanifolds in generalized Sasakian space forms. Similar to the case of Lemma 1, we can prove the following.

Lemma 2. *Let $\mathcal{N}$ be a C-totally real submanifold of dimension n in a generalized Sasakian space form $\overline{\mathcal{N}}(f_1, f_2, f_3)$ of dimension $(2m+1)$. Then, we have*

$$\rho_N \leq \|\mathcal{H}\|^2 - \rho + f_1, \qquad (35)$$

with the equality case holding at $p \in \mathcal{N}$ if and only if the shape operator S of $\mathcal{N}$ in $\overline{\mathcal{N}}(f_1, f_2, f_3)$ with respect to some suitable orthonormal bases $\{e_1, \ldots, e_n\}$ of $T_p\mathcal{N}$ and $\{\xi_1, \ldots, \xi_{2m-n+1}\}$ of $T_p^\perp\mathcal{N}$ takes the following forms

$$S_{\xi_1} = \begin{pmatrix} \gamma_1 & \nu & 0 & \cdots & 0 \\ \nu & \gamma_1 & 0 & \cdots & 0 \\ 0 & 0 & \gamma_1 & \cdots & 0 \\ \vdots & \vdots & \vdots & \ddots & \vdots \\ 0 & 0 & 0 & \cdots & \gamma_1 \end{pmatrix},$$

$$S_{\xi_2} = \begin{pmatrix} \gamma_2 + \nu & 0 & 0 & \cdots & 0 \\ 0 & \gamma_2 - \nu & 0 & \cdots & 0 \\ 0 & 0 & \gamma_2 & \cdots & 0 \\ \vdots & \vdots & \vdots & \ddots & \vdots \\ 0 & 0 & 0 & \cdots & \gamma_2 \end{pmatrix},$$

$$S_{\xi_3} = \begin{pmatrix} \gamma_3 & 0 & 0 & \cdots & 0 \\ 0 & \gamma_3 & 0 & \cdots & 0 \\ 0 & 0 & \gamma_3 & \cdots & 0 \\ \vdots & \vdots & \vdots & \ddots & \vdots \\ 0 & 0 & 0 & \cdots & \gamma_3 \end{pmatrix}, \quad S_{\xi_4} = \cdots = S_{\xi_{2m-n+1}} = 0,$$

where $\gamma_1, \gamma_2, \gamma_3$, and ν are real functions on $\mathcal{N}$.

Next, we can state a generalized Wintgen-type inequality for Legendrian submanifolds in a generalized Sasakian ambient.

Theorem 3. *If $\mathcal{N}$ is a Legendrian submanifold of a $(2n+1)$-dimensional generalized Sasakian space form $\overline{\mathcal{N}}(f_1, f_2, f_3)$, then*

$$\begin{aligned}(\rho^\perp)^2 &\leq (\|\mathcal{H}\|^2 - \rho + f_1)^2 + \frac{2}{n(n-1)} f_2^2 \\ &\quad + \frac{4 f_2}{n(n-1)} (\rho - f_1)\end{aligned} \tag{36}$$

and the equality holds at a point $p \in \mathcal{N}$ if and only if the shape operator takes the forms as in Lemma 2 with respect to some suitable tangent and normal orthonormal bases.

Proof. Let $\{e_1, \ldots, e_n\}$ be an orthonormal frame on $\mathcal{N}$. Due to the fact that $\mathcal{N}$ is a Legendrian submanifold of $\overline{\mathcal{N}}$, it follows that $\{\xi_1 = \phi e_1, \ldots, \xi_n = \phi e_n, \xi_{n+1} = \xi\}$ is an orthonormal frame in the normal bundle of $\mathcal{N}$. Next, the proof is similar to the one of Theorem 2, being based on Lemma 2 instead of Lemma 1, so we omit it. □

Remark 3. *We note that function f_3 does not appear in the generalized Wintgen inequality in Equation (36) for a Legendrian submanifold $\mathcal{N}$ in a generalized Sasakian space form $\overline{\mathcal{N}}(f_1, f_2, f_3)$. This is a consequence of the fact that ξ is normal to $\mathcal{N}$. However, for a submanifold tangent to the structure vector field ξ, the corresponding generalized Wintgen inequality will depend on f_3, as we can see in the second part of this section.*

Remark 4. *Theorem 3 generalizes the main result of [9], namely the generalized Wintgen inequality for the class of Legendrian submanifolds in a Sasakian space form. Actually, if in the statement of Theorem 3, one considers $f_1 = \frac{c+3}{4}$ and $f_2 = f_3 = \frac{c-1}{4}$, then $\overline{\mathcal{N}}$ reduces to a Sasakian space form and Theorem 3 becomes nothing but ([9] Theorem 3.2).*

Corollary 1. Let $\mathcal{N}$ be a Legendrian submanifold of a $(2n+1)$-dimensional Kenmotsu space form $\overline{\mathcal{N}}(c)$. Then

$$(\rho^\perp)^2 \leq \left(\|\mathcal{H}\|^2 - \rho + \frac{c-3}{4}\right)^2 + \frac{(c+1)^2}{8n(n-1)}$$
$$+ \frac{c+1}{n(n-1)}\left(\rho - \frac{c-3}{4}\right) \tag{37}$$

and the equality holds at a point $p \in \mathcal{N}$ if and only if the shape operator takes the forms as in Lemma 2 with respect to some suitable tangent and normal orthonormal bases.

Proof. The proof follows immediately from Theorem 3 by replacing $f_1 = \frac{c-3}{4}$ and $f_2 = f_3 = \frac{c+1}{4}$. □

Corollary 2. Let $\mathcal{N}$ be a Legendrian submanifold of a $(2n+1)$-dimensional cosymplectic space form $\overline{\mathcal{N}}(c)$. Then,

$$(\rho^\perp)^2 \leq \left(\|\mathcal{H}\|^2 - \rho + \frac{c}{4}\right)^2 + \frac{c^2}{8n(n-1)}$$
$$+ \frac{c}{n(n-1)}\left(\rho - \frac{c}{4}\right) \tag{38}$$

and the equality holds at a point $p \in \mathcal{N}$ if and only if the shape operator takes the forms as in Lemma 2 with respect to some suitable tangent and normal orthonormal bases.

Proof. The proof follows immediately from Theorem 3 by putting $f_1 = f_2 = f_3 = \frac{c}{4}$. □

Remark 5. We note that the proof of Theorem 3.3 of [41] contains an error. Consequently, Theorem 3.3 of [41] must be replaced by Corollary 1 of the present article.

In 1996, Lotta [42] introduced the notion of slant submanifold in almost contact geometry as follows. A submanifold $\mathcal{N}$ of an almost contact metric manifold $(\overline{\mathcal{N}}, \phi, \xi, \eta, g)$ tangent to the structure vector field ξ is said to be a contact slant submanifold if, for any point $p \in \mathcal{N}$ and any vector $X \in T_p\mathcal{N}$ linearly independent on ξ_p, the angle between the vector ϕX and the tangent space $T_p\mathcal{N}$ is constant. This constant, usually denoted by θ, is said to be the slant angle of $\mathcal{N}$. We recall that invariant and anti-invariant submanifolds are particular examples of slant submanifolds with slant angle $\theta = 0$ and $\theta = \frac{\pi}{2}$, respectively. A contact slant submanifold is said to be θ-slant or proper if $0 < \theta < \frac{\pi}{2}$. Notice that ([42] Theorem 3.3) implies the dimension of a contact slant submanifold tangent to the structure vector field ξ and with slant angle $\theta \neq \frac{\pi}{2}$ is odd. The concept of contact slant submanifold is further generalized as follows.

Definition 2. [43] A submanifold $\mathcal{N}$ of an almost contact metric manifold $\overline{\mathcal{N}}$ is said to be a bi-slant submanifold, if there exist two orthogonal distributions D_1 and D_2 on $\mathcal{N}$, such that:

(i) $T\mathcal{N}$ admits the orthogonal direct decomposition $T\mathcal{N} = D_1 \oplus D_2 \oplus \xi$.
(ii) $JD_1 \perp D_2$ and $JD_2 \perp D_1$.
(iii) For i=1,2, the distribution D_i is slant with slant angle θ_i.

In the following, we denote by d_i the dimension of the distribution D_i, $i = 1, 2$. It is easy to check that, similar to in the case of complex geometry, the class of bi-slant submanifolds of almost contact metric manifolds naturally includes not only the class of slant submanifolds, but also the classes of semi-slant submanifolds [44], hemi-slant submanifolds (also named pseudo-slant submanifolds) [45], and contact CR-submanifolds (also known as semi-invariant submanifolds) [46]. For definitions and basic properties of the above classes of submanifolds, see also [47]. We only recall here that a bi-slant submanifold is called proper if $d_1 d_2 \neq 0$ and the slant angles $\theta_1, \theta_2 \neq 0, \frac{\pi}{2}$. Notice that various examples of proper bi-slant submanifolds in almost contact metric manifolds can be found in [43,44,48].

Next, we focus on the second aim of this section, that is to derive a generalized Wintgen-type inequality for bi-slant submanifolds in generalized Sasakian space form.

Theorem 4. *Let $\mathcal{N}$ be a proper bi-slant submanifold of dimension n in a generalized Sasakian space form $\overline{\mathcal{N}}(f_1, f_2, f_3)$ of dimension $(2m+1)$, with slant angles θ_1, θ_2 and $\dim D_i = d_i$, $i = 1, 2$. Then,*

$$\rho_{\mathcal{N}} \leq \|\mathcal{H}\|^2 - \rho + f_1 \\ + \frac{3 f_2}{n(n-1)}(d_1 \cos^2 \theta_1 + d_2 \cos^2 \theta_2) - \frac{2}{n} f_3. \tag{39}$$

Proof. First, we remark that the definition of a bi-slant submanifold implies that $d_1 + d_2 + 1 = n$. Next, let $\{e_1, \ldots, e_{d_1}, e_{d_1+1}, \ldots, e_{d_1+d_2}, e_n = \xi\}$ be an orthonormal frame on $\mathcal{N}$ and $\{\xi_1, \ldots, \xi_{2m-n+1}\}$ be a normal orthonormal frame on $\mathcal{N}$.

Using Equations (2) and (34), we obtain

$$\tau = \sum_{1 \leq i < j \leq n} R(e_i, e_j, e_j, e_i) \\ = \frac{n(n-1)}{2} f_1 + \frac{3}{2} f_2 (d_1 \cos^2 \theta_1 + d_2 \cos^2 \theta_2) \\ + (1-n) f_3 + \sum_{r=1}^{2m-n+1} \sum_{1 \leq i < j \leq n} [h_{ii}^r h_{jj}^r - (h_{ij}^r)^2]. \tag{40}$$

However, as in the proof of Lemma 1, we get

$$n^2 \|\mathcal{H}\|^2 - n^2 \rho_{\mathcal{N}} \geq \frac{2n}{n-1} \sum_{r=1}^{2m-n+1} \sum_{1 \leq i < j \leq n} [h_{ii}^r h_{jj}^r - (h_{ij}^r)^2]. \tag{41}$$

Combining now Equations (40) and (41), we obtain Equation (44) and the conclusion follows. □

As immediate consequences of Theorem 4, we derive the following results.

Corollary 3. *Let $\mathcal{N}$ be a proper bi-slant submanifold of dimension n in a Sasakian space form $\overline{\mathcal{N}}(c)$ of dimension $(2m+1)$, with slant angles θ_1, θ_2 and $\dim D_i = d_i$, $i = 1, 2$. Then,*

$$\rho_{\mathcal{N}} \leq \|\mathcal{H}\|^2 - \rho + \frac{c+3}{4} \\ + \frac{3(c-1)}{4n(n-1)}(d_1 \cos^2 \theta_1 + d_2 \cos^2 \theta_2) - \frac{c-1}{2n}. \tag{42}$$

Corollary 4. *Let $\mathcal{N}$ be a proper bi-slant submanifold of dimension n in a Kenmotsu space form $\overline{\mathcal{N}}(c)$ of dimension $(2m+1)$, with slant angles θ_1, θ_2 and $\dim D_i = d_i$, $i = 1, 2$. Then,*

$$\rho_{\mathcal{N}} \leq \|\mathcal{H}\|^2 - \rho + \frac{c-3}{4} \\ + \frac{3(c+1)}{4n(n-1)}(d_1 \cos^2 \theta_1 + d_2 \cos^2 \theta_2) - \frac{c+1}{2n}. \tag{43}$$

Corollary 5. Let $\mathcal{N}$ be a proper bi-slant submanifold of dimension n in a cosymplectic space form $\overline{\mathcal{N}}(c)$ of dimension $(2m+1)$, with slant angles θ_1, θ_2 and $\dim D_i = d_i$, $i = 1, 2$. Then,

$$\rho_N \leq \|\mathcal{H}\|^2 - \rho + \frac{c}{4}$$
$$+ \frac{3c}{4n(n-1)}(d_1\cos^2\theta_1 + d_2\cos^2\theta_2) - \frac{c}{2n}. \qquad (44)$$

Remark 6. Corollary 3 generalizes Theorem 4.1 of [9].

Remark 7. We note that the authors of [8,9] provided non-trivial examples of Lagrangian and Legendrian submanifolds satisfying the equality case of the corresponding Wintgen-type inequalities stated in this paper, because the shape operators have the appropriate form (see also [49]).

6. The First Fundamental Equation of Generalized Space Forms

For a given Riemannian manifold $(\overline{\mathcal{N}}, g)$, let us denote by $\overline{\nabla}$ the Levi–Civita connection of the metric g and by $\overline{R}$ the curvature tensor of $\overline{\nabla}$. We consider the differential operator $D_{\overline{\nabla}}$ defined in the tangent vector bundle $T\overline{\mathcal{N}}$ with values belonging to the vector bundle $hom(\otimes^2 T\overline{\mathcal{N}}, T\overline{\mathcal{N}})$. Hence, for a given vector field X on $\overline{\mathcal{N}}$, we have that $D_{\overline{\nabla}}(X)$ is a section of the vector bundle $T\overline{\mathcal{N}} \otimes T^{*\otimes 2}\overline{\mathcal{N}}$ defined by

$$D_{\overline{\nabla}}(X) = \overline{\nabla}^2 X.$$

Obviously, the complete expression is

$$D_{\overline{\nabla}}(X)(Y, Z) = \overline{\nabla}_Y \overline{\nabla}_Z X - \overline{\nabla}_{\overline{\nabla}_Y Z} X, \quad \forall Y, Z \in \mathcal{X}(M).$$

We recall now that the first fundamental equation of $(\overline{\mathcal{N}}, \overline{\nabla})$ is the second-order differential equation [50]

$$D_{\overline{\nabla}}(X) = 0. \qquad (45)$$

In the following, we denote by $\mathcal{J}_{\overline{\nabla}}$ the sheaf of germs of solutions to Equation (45) and by $J_{\overline{\nabla}}$ the vector space of sections of $\mathcal{J}_{\overline{\nabla}}$.

We would like to investigate next the consequences of the condition $\dim \mathcal{J}_{\overline{\nabla}} > 0$, i.e., the first fundamental in Equation (45) admits non-null solutions, on the geometry and topology of generalized complex space forms and generalized Sasakian space forms. Before answering the above question, we need the following.

Proposition 1. Let $\overline{\nabla}$ be the Levi–Civita connection of a Riemannian metric g on a manifold $\overline{\mathcal{N}}$. If Z is a solution to the first fundamental equation of $(\overline{\mathcal{N}}, \overline{\nabla})$, then one has

(i) $\overline{R}(X, Y)Z = 0$,
(ii) $\overline{R}(X, Z)Y = 0$,

for all vector fields X, Y on M.

Proof. (i) If Z is a solution to the first fundamental equation of $(\overline{\mathcal{N}}, \overline{\nabla})$, then

$$D_{\overline{\nabla}}(Z)(X, Y) = \overline{\nabla}_X \overline{\nabla}_Y Z - \overline{\nabla}_{\overline{\nabla}_X Y} Z = 0, \qquad (46)$$

for all vector fields X, Y on M.

However, since the connection $\overline{\nabla}$ is torsion-free, we can express its Riemann curvature tensor $\overline{R}$ by

$$\overline{R}(X,Y)Z = D_{\overline{\nabla}}(Z)(X,Y) - D_{\overline{\nabla}}(Z)(Y,X). \tag{47}$$

Consequently, from Equations (46) and (47), we derive $\overline{R}(X,Y)Z = 0$.
(ii) Using (i) and the Bianchi identity, one has

$$\overline{R}(Z,X)Y = \overline{R}(Z,Y)X. \tag{48}$$

Then, we have

$$\begin{aligned}
g(\overline{R}(Z,X)Y,W) &= -g(Y,\overline{R}(Z,X)W) \\
&= -g(Y,\overline{R}(Z,W)X) \\
&= g(\overline{R}(Z,W)Y,X) \\
&= g(\overline{R}(Z,Y)W,X) \\
&= -g(W,\overline{R}(Z,Y)X) \\
&= -g(W,\overline{R}(Z,X)Y) \\
&= -g(\overline{R}(Z,X)Y,W),
\end{aligned} \tag{49}$$

which implies

$$g(\overline{R}(Z,X)Y,W) = 0$$

and the conclusion is now clear. □

Theorem 5. *Let $\overline{\mathcal{N}}(f_1,f_2)$ be a generalized complex space form of real dimension $2m > 2$. If $\dim \mathcal{J}_{\overline{\nabla}} > 0$, then $\overline{\mathcal{N}}$ is flat. Moreover, $\overline{\mathcal{N}}$ admits a normal Riemannian covering by a flat $2m$-dimensional torus, provided that the manifold is compact and connected.*

Proof. Let Z be a non-null solution of the first fundamental equation of $(\overline{\mathcal{N}}(f_1,f_2),\overline{\nabla})$, where $\overline{\nabla}$ is the Levi–Civita connection on $\overline{\mathcal{N}}(f_1,f_2)$. Then, using Equation (1) and Proposition 1 (i), we get

$$\begin{aligned}
f_1\{g(Z,Y)X - g(Z,X)Y\} & \\
+ f_2\{g(JZ,X)JY - g(JZ,Y)JX\} &= -2f_2 g(X,JY)JZ,
\end{aligned} \tag{50}$$

for all vector fields X,Y on $\overline{\mathcal{N}}$.

In addition, using Equation (1) and Proposition 1 (ii), we obtain

$$\begin{aligned}
f_1 g(Z,Y)X + f_2\{2g(X,JZ)JY - g(Z,JY)JX\} \\
= f_1 g(X,Y)Z - f_2 g(X,JY)JZ.
\end{aligned} \tag{51}$$

Replacing now $X = Z$ and $Y = JZ$ in Equation (50), we derive:

$$(f_1 + 3f_2)g(Z,Z)JZ = 0$$

and therefore we obtain

$$3f_2 + f_1 = 0. \tag{52}$$

Combining Equations (51) and (52), we get

$$f_2\{-3g(X,Y)Z - g(X,JY)JZ\}$$
$$+f_2\{3g(Z,Y)X - 2g(X,JZ)JY + g(Z,JY)JX\} = 0 \qquad (53)$$

and choosing $Y = X$ in Equation (53) we derive

$$f_2\{-3g(X,X)Z + 3g(Z,X)X - 3g(X,JZ)JX\} = 0. \qquad (54)$$

Now, because $m > 1$, we can choose a vector field X on $\overline{\mathcal{N}}$ subjected to

1. $g(X,JZ) = 0$,
2. $g(X,Z) = 0$,

and therefore Equation (54) yields

$$3f_2 g(X,X)Z = 0. \qquad (55)$$

Thereby,

$$f_2 = 0$$

and, from Equation (52), we also derive

$$f_1 = 0.$$

Thus, Equation (1) implies that $\overline{\mathcal{N}}$ is flat and the conclusion follows immediately (see ([51] Theorem 3.3.1)). □

Theorem 6. Let $\overline{\mathcal{N}}(f_1, f_2, f_3)$ be a generalized Sasakian space form of dimension $2m + 1 > 3$. If the first fundamental equation admits solutions linearly independent on the structure vector field ξ, then $\overline{\mathcal{N}}$ is flat. Moreover, $\overline{\mathcal{N}}$ admits a normal Riemannian covering by a flat $(2m + 1)$-dimensional torus, provided that the manifold is compact and connected.

Proof. Let Z be a solution to the first fundamental equation of $(\overline{\mathcal{N}}(f_1, f_2, f_3), \overline{\nabla})$ linearly independent on the structure vector field ξ, where $\overline{\nabla}$ is the Levi–Civita connection on $\overline{\mathcal{N}}(f_1, f_2, f_3)$. Then, using Equation (34) and Proposition 1, we get the following identities:

$$f_1\{g(Y,Z)X - g(X,Z)Y\} + f_2\{g(X,\phi Z)\phi Y - g(Y,\phi Z)\phi X\}$$
$$+ f_3\{\eta(X)\eta(Z)Y - \eta(Y)\eta(Z)X\}$$
$$= -2f_2 g(X,\phi Y)\phi Z - f_3\{g(X,Z)\eta(Y) - g(Y,Z)\eta(X)\}\xi, \qquad (56)$$
$$f_1 g(Z,Y)X + f_2\{2g(X,\phi Z)\phi Y - g(Z,\phi Y)\phi X\}$$
$$- f_3 \eta(Z)\eta(Y)X = f_1 g(X,Y)Z - f_2 g(X,\phi Y)\phi Z$$
$$- f_3\{\eta(X)\eta(Y)Z + g(X,Y)\eta(Z)\xi - g(Z,Y)\eta(X)\xi\} \qquad (57)$$

for all vector fields X, Y on $\overline{\mathcal{N}}$.

Choosing now in Equation (56) the vector field X to be orthogonal to Z, ϕZ, and ξ, we derive

$$\{f_1 g(Y,Z) - f_3 \eta(Y)\eta(Z)\}X - f_2 g(Y,\phi Z)\phi X + 2f_2 g(X,\phi Y)\phi Z = 0 \qquad (58)$$

and, particularizing $Y = \phi Z$ in Equation (58), one immediately gets

$$f_2 = 0. \tag{59}$$

Therefore, Equations (56) and (57) become

$$f_1\{g(Y,Z)X - g(X,Z)Y\} + f_3\{\eta(X)\eta(Z)Y - \eta(Y)\eta(Z)X\}$$
$$= -f_3\{g(X,Z)\eta(Y) - g(Y,Z)\eta(X)\}\xi, \tag{60}$$
$$f_1 g(Z,Y)X - f_3\eta(Z)\eta(Y)X = f_1 g(X,Y)Z$$
$$- f_3\{\eta(X)\eta(Y)Z + g(X,Y)\eta(Z)\xi - g(Z,Y)\eta(X)\xi\}, \tag{61}$$

for all vector fields X, Y on $\overline{\mathcal{N}}$.

Similarly, considering in Equation (61) the vector field X to be orthogonal to $Z, \phi Z$, and ξ, we deduce

$$f_1 g(Z,Y)X + f_3 \eta(Z)\eta(Y)X = f_1 g(X,Y)Z - f_3 g(X,Y)\eta(Z)\xi \tag{62}$$

and, particularizing $Y = X$ in Equation (58), one obtains

$$f_1 g(X,X)Z - f_3 g(X,X)\eta(Z)\xi = 0. \tag{63}$$

As Z and ξ are linearly independent, Equation (63) implies

$$f_1 = 0 \tag{64}$$

and

$$f_3 \eta(Z) = 0. \tag{65}$$

Now, we have to distinguish two cases.

Case I: $\eta(Z) \neq 0$. Then, it follows from Equation (65) that

$$f_3 = 0 \tag{66}$$

and replacing Equations (59), (64), and (66) in Equation (1), we conclude that $\overline{\mathcal{N}}$ is flat.

Case II: $\eta(Z) = 0$. Then, taking account of Equation (64), we obtain from Equation (60) that

$$f_3\{g(X,Z)\eta(Y) - g(Y,Z)\eta(X)\}\xi = 0. \tag{67}$$

Particularizing now $X = Z$ and $Y = \xi$ in Equation (67), one obtains also Equation (66) and therefore we reach again the required conclusion. □

Remark 8. *Theorems 5 and 6 provide obstructions to the existence of non-flat generalized space forms. Therefore, the existence of non-null solutions for the first fundamental equation of a generalized complex space form $\overline{\mathcal{N}}(f_1, f_2)$ implies the flatness of this space. On the other hand, the existence of solutions linearly independent on the structure vector field for the first fundamental equation of a generalized Sasakian space form $\overline{\mathcal{N}}(f_1, f_2, f_3)$ also implies that its Riemannian curvature tensor vanishes identically.*

Author Contributions: Conceptualization, methodology and writing-original manuscript, M.A., M.N.B., M.H.S. and G.-E.V.; project management, supervision and writing-review, M.A., M.N.B., M.H.S. and G.-E.V. All authors contributed equally in this work.

Funding: This research received no external funding.

Conflicts of Interest: The authors declare no conflict of interest.

References

1. Wintgen, P. Sur l'inegalite de Chen-Wilmore. *C. R. Acad. Sci. Paris Ser. A-B* **1979**, *288*, A993–A995.
2. Rouxel, B. *Sur une Famille des A-Surfaces d'un Espace Euclidien* $\mathbb{E}^4$; Österreischer Mathematiker Kongress: Insbruck, Austria, 1981.
3. Guadalupe, I.V.; Rodriguez, L. Normal curvature of surfaces in space forms. *Pac. J. Math.* **1983**, *106*, 95–103.
4. De Smet, P.J.; Dillen, F.; Verstraelen, L.; Vrancken, L. A pointwise inequality in submanifold theory. *Arch. Math.* **1999**, *35*, 115–128.
5. Ge, J.; Tang, Z.Z. A proof of the DDVV conjecture and its equality case. *Pac. J. Math.* **2008**, *237*, 87–95.
6. Lu, Z. Normal scalar curvature conjecture and its applications. *J. Funct. Anal.* **2011**, *261*, 1284–1308.
7. Haesen, S.; Verstraelen, L. Natural Intrinsic Geometrical Symmetries. *Symmetry Integr. Geom. Methods Appl.* **2009**, *5*, 15.
8. Mihai, I. On the generalized Wintgen inequality for Lagrangian submanifolds in complex space forms. *Nonlinear Anal.* **2014**, *95*, 714–720.
9. Mihai, I. On the generalized Wintgen inequality for Legendrian submanifolds in Sasakian space forms. *Tohoku Math. J.* **2017**, *69*, 43–53.
10. Macsim, G.; Ghişoiu, V. Generalized Wintgen inequality for Lagrangian submanifolds in quaternionic space forms. *Math. Inequal. Appl.* **2019**, *22*, 803–813.
11. Roth, J. A DDVV inequality for submanifolds of warped products. *Bull. Aust. Math. Soc.* **2017**, *95*, 495–499.
12. Bansal, P.; Uddin, S.; Shahid, M.-H. On the normal scalar curvature conjecture in Kenmotsu statistical manifolds. *J. Geom. Phys.* **2019**, *142*, 37–46.
13. Tricerri, F.; Vanhecke, L. Curvature tensors on almost Hermitian manifolds. *Trans. Am. Math. Soc.* **1981**, *267*, 365–397.
14. Olszak, Z. On the existence of generalized complex space forms. *Israel J. Math.* **1989**, *65*, 214–218.
15. Gray, A. Nearly Kähler manifolds. *J. Differ. Geom.* **1970**, *4*, 283–309.
16. Alegre, P.; Blair, D.E.; Carriazo, A. Generalized Sasakian space forms. *Israel J. Math.* **2004**, *141*, 157–183.
17. Alegre, P.; Carriazo, A. Structures on generalized Sasakian-space-forms. *Differ. Geom. Appl.* **2008**, *26*, 656–666.
18. Alegre, P.; Carriazo, A. Submanifolds of generalized Sasakian space forms. *Taiwan. J. Math.* **2009**, *13*, 923–941.
19. Alegre, P.; Carriazo, A. Generalized Sasakian space forms and conformal changes of the metric. *Results Math.* **2011**, *59*, 485–493.
20. Falcitelli, M. Locally conformal C_6-manifolds and generalized Sasakian-space-forms. *Mediterr. J. Math.* **2010**, *7*, 19–36.
21. Al-Ghefari, R.; Al-Solamy, F.R.; Shahid, M.H. CR-submanifolds of generalized Sasakian-space-forms. *JP J. Geom. Topol.* **2006**, *6*, 151–166.
22. Hui, S.K.; Uddin, S.; Alkhaldi, A.; Mandal, P. Invariant submanifolds of generalized Sasakian-space-forms. *Int. J. Geom. Methods Mod. Phys.* **2018**, *15*, 21.
23. Malek, F.; Nejadakbary, V. A lower bound for the Ricci curvature of submanifolds in generalized Sasakian space forms. *Adv. Geom.* **2013**, *13*, 695–711.
24. Kim, U.K. On 4-dimensional generalized complex space forms. *J. Austral. Math. Soc. Ser. A* **1999**, *66*, 379–387.
25. Kim, U.K. Conformally flat generalized Sasakian-space-forms and locally symmetric generalized Sasakian-space-forms. *Note Mat.* **2006**, *26*, 55–67.
26. Özgür, C.; Güvenç, Ş. On some classes of biharmonic Legendre curves in generalized Sasakian space forms. *Collect. Math.* **2014**, *65*, 203–218.
27. Prakasha, D.G.; Chavan, V. E-Bochner curvature tensor on generalized Sasakian space forms. *C. R. Math. Acad. Sci. Paris* **2016**, *354*, 835–841.

28. Bejan, C.-L.; Güler, S. Kähler manifolds of quasi-constant holomorphic sectional curvature and generalized Sasakian space forms. *Rev. R. Acad. Cienc. Exactas Fis. Nat. Ser. A Mat. RACSAM* **2019**, *113*, 1173–1189.
29. Chen, B.-Y. *Pseudo-Riemannian Geometry, δ-Invariants and Applications*; World Scientific: Hackensack, NJ, USA, 2011.
30. Chen, B.-Y. *Differential Geometry of Warped Product Manifolds and Submanifolds*; World Scientific: Hackensack, NJ, USA, 2017.
31. Yano, K.; Kon, M. *Anti-Invariant Submanifolds*; M. Dekker: New York, NY, USA, 1976.
32. Chen, B.-Y. Slant Immersions. *Bull. Australian Math. Soc.* **1990**, *41*, 135–147.
33. Chen, B.-Y. *Geometry of Slant Submanifolds*; Publishing House of "Katholieke Universiteit Leuven": Louvain, Belgium, 1990.
34. Uddin, S.; Chen, B.-Y.; Al-Solamy, F.R. Warped product bi-slant immersions in Kaehler manifolds. *Mediterr. J. Math.* **2017**, *14*, 11.
35. Papaghiuc, N. Semi-slant submanifold of Kaehlerian manifold. *An. Ştiinţ. Univ. "Alexandru Ioan Cuza" Iaşi Secţ. I Mat.* **1994**, *40*, 55–61.
36. Şahin, B. Warped product submanifolds of Kaehler manifolds with a slant factor. *Ann. Polon. Math.* **2009**, *95*, 207–226.
37. Bejancu, A. CR submanifolds of a Kaehler manifold. I. *Proc. Am. Math. Soc.* **1978**, *69*, 135–142.
38. Aquib, M.; Shahid, M.H.; Jamali, M. Lower extremities for generalized normalized δ-Casorati curvatures of bi-slant submanifolds in generalized complex space forms. *Kragujevac J. Math.* **2018**, *42*, 591–605.
39. Blair, D.E. *Contact Manifolds in Riemannian Geometry*; Springer: Berlin, Germany, 1976.
40. Pitiş, G. *Geometry of Kenmotsu Manifolds*; Publishing House of "Transilvania" University of Braşov: Braşov, Romania, 2007.
41. Aquib, M.; Shahid, M.H. Generalized Wintgen inequality for submanifolds in Kenmotsu space forms. *Tamkang J. Math.* **2019**, *50*, 155–164.
42. Lotta, A. Slant submanifolds in contact geometry. *Bull. Math. Soc. Roum.* **1996**, *39*, 183–198.
43. Alqahtani, L.S.; Stanković, M.S.; Uddin, S. Warped product bi-slant submanifolds of cosymplectic manifolds. *Filomat* **2017**, *31*, 5065–5071.
44. Cabrerizo, J.L.; Carriazo, A.; Fernández, L.M.; Fernández, M. Semi-slant submanifolds of a Sasakian manifold. *Geom. Dedicata* **1999**, *78*, 183–199.
45. Carriazo, A. *New Developments in Slant Submanifolds Theory*; Narosa Publishing House: New Delhi, India, 2002.
46. Bejancu, A.; Papaghiuc, N. Semi-invariant submanifolds of a Sasakian manifold. *An. Ştiinţ. Univ. "Alexandru Ioan Cuza" Iaşi Secţ. I Mat.* **1981**, *27*, 163–170.
47. Yano, K.; Kon, M. *CR Submanifolds of Kaehlerian and Sasakian Manifolds*; Birkhäuser: Basel, Switzerland, 1983.
48. Gupta, R.S. B.Y. Chen's inequalities for bi-slant submanifolds in cosymplectic space forms. *Sarajevo J. Math.* **2013**, *9*, 1, 117–128.
49. Chen, B.-Y.; Dillen, F.; Verstraelen, L.; Vrancken, L. Totally real submanifolds of CP^n satisfying a basic equality. *Arch. Math.* **1994**, *63*, 553–564.
50. Boyom, M.N. Numerical properties of Koszul connections. *arXiv* **2017**, arXiv:1708.01106.
51. Wolf, J.A. *Spaces of Constant Curvature*, 6th ed.; American Mathematical Society: Providence, RI, USA, 2010.

© 2019 by the authors. Licensee MDPI, Basel, Switzerland. This article is an open access article distributed under the terms and conditions of the Creative Commons Attribution (CC BY) license (http://creativecommons.org/licenses/by/4.0/).

Article

Almost Hermitian Identities

Joana Cirici [1] and Scott O. Wilson [2],*

1. Department of Mathematics and Computer Science, Universitat de Barcelona, Gran Via 585, 08007 Barcelona, Spain; jcirici@ub.edu
2. Department of Mathematics, Queens College, City University of New York, 65-30 Kissena Blvd., Flushing, NY 11367, USA
* Correspondence: scott.wilson@qc.cuny.edu

Received: 30 July 2020; Accepted: 11 August 2020; Published: 13 August 2020

Abstract: We study the local commutation relation between the Lefschetz operator and the exterior differential on an almost complex manifold with a compatible metric. The identity that we obtain generalizes the backbone of the local Kähler identities to the setting of almost Hermitian manifolds, allowing for new global results for such manifolds.

Keywords: almost Hermitian manifolds; Kähler identities; Lefschetz operator

1. Introduction

On a Kähler manifold (M, J, ω), the most fundamental local identity is perhaps the commutation relation between the exterior differential d and the adjoint Λ to the Lefschetz operator,

$$[\Lambda, d] = \star \, \mathbb{I}^{-1} \, d \, \mathbb{I} \, \star, \qquad (1)$$

where $\star$ denotes the Hodge star operator and $\mathbb{I}$ denotes the extension of J to all forms.

This identity, due to A. Weil [1], strongly depends on the Kähler condition, $d\omega = 0$, and in fact is true when removing the integrability condition $N_J \equiv 0$. So, it is valid for almost Kähler and also symplectic manifolds as well [2–4]. On the other hand, there is also a generalization of the Kähler identities in the Hermitian setting (see [5,6]), which strongly uses integrability.

When the manifold is only almost Hermitian, then the above local identity does not hold in general, as noticed implicitly in [7]. The purpose of this short note is to show precisely how the above Kähler identity (1) becomes modified when the form ω is not closed.

The main result is given in Theorem 1 below, which has several applications including the uniqueness of the Dirichlet problem

$$\partial \bar{\partial} u = g \quad \text{with} \quad u|_{\partial \Omega} = \phi,$$

on any compact domain Ω in an almost complex manifold. This in turn implies that the Dolbeault cohomology introduced in [8], for all almost complex manifolds, satisfies $H^{0,0}_{\mathrm{Dol}}(M) \cong \mathbb{C}$ for a compact connected almost complex manifold.

Another application of the almost Hermitian identities of Theorem 1 appears in forthcoming work by Feehan and Leness [9]. There the fundamental relation of Proposition 1 is used to show that the moduli spaces of unitary anti-self-dual connections over any almost Hermitian 4-manifold is almost Hermitian, whenever the Nijenhuis tensor has sufficiently small C^0-norm. This generalizes a well known result for Kähler manifolds that was exploited in Donaldson's work in the 1980s, and is expected to have consequences for the topology of almost complex 4-manifolds which are of so-called Seiberg–Witten simple type.

When M is compact, local identities lead to consequences in cohomology, often governed by geometric-topological inequalities. Indeed, the exterior differential inherits a bidegree decomposition into four components $d = \bar{\mu} + \bar{\partial} + \partial + \mu$ and the Hermitian metric allows one to consider the Laplacian operators associated to each of these components. In the compact case, the numbers

$$\ell^{p,q} := \dim \operatorname{Ker}(\Delta_{\bar{\partial}} + \Delta_{\mu})|_{(p,q)}$$

given by the kernel of $\Delta_{\bar{\partial}} + \Delta_{\mu}$ in bidegree (p,q) are finite by elliptic operator theory. When J is integrable (and so M is a complex manifold) the operator Δ_{μ} vanishes and these are just the Hodge numbers $\ell^{p,q} = h^{p,q}$. In this case, the Hodge-to-de Rham spectral sequence gives inequalities

$$\sum_{p+q=k} \ell^{p,q} \geq b^k,$$

where b^k denotes the k-th Betti number. On the other hand, as shown in [4], one main consequence of the local identity (1) in the almost Kähler case $d\omega = 0$ is the converse inequality

$$\sum_{p+q=k} \ell^{p,q} \leq b^k.$$

Of course, in the integrable Kähler case both inequalities are true and so one recovers the well-known consequence of the Hodge decomposition

$$\sum_{p+q=k} \ell^{p,q} = b^k.$$

The local identities of [5,6] for complex non-Kähler manifolds include other algebra terms which lead to further Laplacian operators, leading also to various inequalities relating the geometry with the topology of the manifold.

With this note, we aim to further understand the origin of these inequalities by means of the correct version of (1) for almost Hermitian manifolds for which, a priori, the only geometric-topological inequality in the compact case is given by

$$\sum_{p+q=k} \dim \operatorname{Ker}(\Delta_{\bar{\mu}} + \Delta_{\bar{\partial}} + \Delta_{\partial} + \Delta_{\mu})|_{(p,q)} \leq b^k.$$

2. Preliminaries

Let $(\mathcal{A}, d)$ denote the complex valued differential forms of an almost complex manifold (M, J). For any Hermitian metric, define the associated Hodge-star operator

$$\star : \mathcal{A}_x^{p,q} \to \mathcal{A}_x^{n-q, n-p} \quad \text{by} \quad \omega \wedge \star \bar{\eta} = \langle \omega, \eta \rangle \operatorname{vol},$$

where ω is the fundamental $(1,1)$-form, and $\operatorname{vol} = \frac{1}{n!}\omega^n \in \mathcal{A}^{n,n}$ is the volume form determined by the Hermitian metric. Note $\star^2 = (-1)^k$ on $\mathcal{A}^k$.

Define $d^* = -\star d\star$, so that $d^*\star = (-1)^{k+1} \star d$ on $\mathcal{A}^k$. Similarly, consider the bidegree decomposition of the exterior differential

$$d = \bar{\mu} + \bar{\partial} + \partial + \mu,$$

where the bidegree of each component is given by

$$|\bar{\mu}| = (-1, 2), |\bar{\partial}| = (0, 1), |\partial| = (1, 0) \text{ and } |\mu| = (2, -1).$$

We then let $\delta^* = - \star \delta \star$ for $\delta = \bar{\mu}, \bar{\partial}, \partial, \mu$ and we have the bidegree decomposition

$$d^* = \bar{\mu}^* + \bar{\partial}^* + \partial^* + \mu^*.$$

where

$$|\bar{\mu}^*| = (1,-2), |\bar{\partial}^*| = (0,-1), |\partial^*| = (-1,0) \text{ and } |\mu^*| = (-2,1).$$

Let $L : \mathcal{A}^{p,q} \to \mathcal{A}^{p+1,q+1}$ be the real $(1,1)$-operator given by $L(\eta) = \omega \wedge \eta$. Let $\Lambda = L^* = \star^{-1} L \star$. Then $\star \Lambda = L \star$ and $\star L = \Lambda \star$. Let $P^k = \operatorname{Ker} \Lambda \cap \mathcal{A}^k$ denote the primitive forms of total degree k.

It is well known that $\{L, \Lambda, [L, \Lambda]\}$ defines a representation of $sl(2, \mathbb{C})$ and induces the Lefschetz decomposition on forms:

Lemma 1. *We have*

$$\mathcal{A}^k = \bigoplus_{r=0}^{k/2} L^r P^{k-2r},$$

and this direct sum decomposition respects the (p,q) bigrading.

Let $[A, B] = AB - (-1)^{|A||B|} BA$ be the graded commutator, where $|A|$ denotes the total degree of A. This defines a graded Poisson algebra

$$[A, BC] = [A, B]C + (-1)^{|A||B|} B[A, C].$$

The following is well known (e.g., [10] Corollary 1.2.28):

Lemma 2. *For all $j \geq 0$ and $\alpha \in \mathcal{A}^k$*

$$[L^j, \Lambda]\alpha = j(k - n + j - 1)L^{j-1}\alpha.$$

By induction, and the fact that $[d, L]$ and L commute, we have:

Lemma 3. *For all $n \geq 1$*

$$[d, L^n] = n[d, L]L^{n-1},$$

and

$$\star[d, L]\alpha = (-1)^{k+1}[d^*, \Lambda] \star \alpha \quad \text{for } \alpha \in \mathcal{A}^k.$$

Let $\mathbb{I}$ be the extension of J to all forms as an algebra map with respect to wedge product, so that $\mathbb{I}_{p,q}$ acts on $\mathcal{A}^{p,q}$ by multiplication by i^{p-q}. Then $\mathbb{I}_{p,q}^2 = (-1)^{p+q}$ so that $\mathbb{I}_{p,q}^{-1} = (-1)^{p+q} \mathbb{I}_{p,q}$. Note that $\mathbb{I}$ and $\star$ commute, and $\mathbb{I}$ and L^n commute for all $n \geq 0$. The following is a direct calculation.

Lemma 4. *If an operator $T_{r,s} : \mathcal{A}^{p,q} \to \mathcal{A}^{p+r,q+s}$ has bidegree (r,s), then*

$$\mathbb{I}_{r+p,s+q}^{-1} \circ T_{r,s} \circ \mathbb{I}_{p,q} = (-i)^{r-s} T_{r,s}.$$

The above result readily implies that

$$\mathbb{I}^{-1} \circ d \circ \mathbb{I} = -i(\bar{\mu} - \bar{\partial} + \partial - \mu).$$

Finally, the following is well known (e.g., [10] Proposition 1.2.31):

Lemma 5. *If M is an almost Hermitian manifold of dimension $2n$, then for all $j \geq 0$ and all $\alpha \in P^k$,*

$$\star L^j \alpha = (-1)^{\frac{k(k+1)}{2}} \frac{j!}{(n-k-j)!} L^{n-k-j} \mathbb{I} \alpha.$$

3. Almost Hermitian Identities

By the previous section, any differential form η can be written as $\eta = L^j \alpha$ for unique $j, k \geq 0$ and $\alpha \in P^k$. We now state the main result:

Theorem 1. *For any almost Hermitian manifold of dimension $2n$, let $\alpha \in P^k$, with $d\alpha$ written as*

$$d\alpha = \alpha_0 + L\alpha_1 + L^2 \alpha_2 + \cdots, \qquad (2)$$

for unique $\alpha_r \in P^{k+1-2r}$. Then, for all $j \geq 0$,

$$[\Lambda, d]L^j \alpha - \star \mathbb{I}^{-1} d\mathbb{I} \star L^j \alpha = \frac{1}{j+1} \mathbb{I}^{-1} [d^*, \Lambda] \mathbb{I} L^{j+1} \alpha$$
$$+ j\Lambda [d, L] L^{j-1} \alpha + j(j-1)(k-n+j-1)[d, L]L^{j-2}\alpha$$
$$+ \sum_{r=2}^{\infty} f_{n,j,k}(r) L^{j+r-1} \alpha_r,$$

where

$$f_{n,k,j}(r) = (r(n-k+r) - j) + (-1)^r \frac{j!(n-k-j+r)!}{(j+r-1)!(n-k-j)!}.$$

Remark 1. *In the almost Kähler case we have $[d^*, \Lambda] = [d, L] = 0$, and $d\alpha = \alpha_0 + L\alpha_1$, so we recover the identity*

$$[\Lambda, d] = \star \mathbb{I}^{-1} d\mathbb{I} \star,$$

as expected.

Proof. The proof consists of several calculations using the lemmas in the previous section. Using $[\mathbb{I}, L] = 0$, and $\mathbb{I}^2 = (-1)^k$ on $\mathcal{A}^k$, we have

$$\star \mathbb{I}^{-1} d\mathbb{I} \star \eta = \star \mathbb{I}^{-1} d\mathbb{I} \left((-1)^{\frac{k(k+1)}{2}} \frac{j!}{(n-k-j)!} L^{n-k-j} \mathbb{I} \alpha \right)$$
$$= (-1)^{\frac{k(k+1)}{2}+k} \frac{j!}{(n-k-j)!} \star \mathbb{I}^{-1} d L^{n-k-j} \alpha.$$

By Lemma 3 this is equal to

$$(-1)^{\frac{k(k+1)}{2}+k} \frac{j!}{(n-k-j)!} \star \mathbb{I}^{-1} L^{n-k-j} d\alpha + (-1)^{\frac{k(k+1)}{2}+k} \frac{j!}{(n-k-j-1)!} \star \mathbb{I}^{-1} [d, L] L^{n-k-j-1} \alpha. \qquad (3)$$

We first simplify each of these last two summands. By Equation (2), the fact that $\star$ commutes with $\mathbb{I}$, and Lemma 5 applied to $\alpha_r \in P^{k+1-2r}$, the first summand of Equation (3) is equal to:

$$(-1)^{\frac{k(k+1)}{2}+k} \frac{j!}{(n-k-j)!} \star \mathbb{I}^{-1} \left(\sum_{r=0}^{\infty} L^{n-k-j+r} \alpha_r \right)$$
$$= (-1)^{\frac{k(k+1)}{2}+k} \frac{j!}{(n-k-j)!} \mathbb{I}^{-1} \left(\sum_{r=0}^{\infty} (-1)^{\frac{(k+1-2r)(k-2r+2)}{2}} \frac{(n-k-j+r)!}{(j+r-1)!} L^{j+r-1} \mathbb{I} \alpha_r \right)$$
$$= \sum_{r=0}^{\infty} (-1)^{r+1} \frac{j!(n-k-j+r)!}{(j+r-1)!(n-k-j)!} L^{j+r-1} \alpha_r.$$

For the second summand, we use the fact that for all $m \geq 0$ and $\beta \in \mathcal{A}^k$,

$$\star L^m [d, L] \beta = \star [d, L] L^m \beta = (-1)^{k+1} [d^*, \Lambda] \star L^m \beta.$$

So, the second summand in Equation (3) is equal to

$$(-1)^{\frac{k(k+1)}{2}+k}\frac{j!}{(n-k-j-1)!}\star\mathbb{I}^{-1}[d,L]L^{n-k-j-1}\alpha$$

$$=(-1)^{\frac{k(k+1)}{2}+1}\frac{j!}{(n-k-j-1)!}\mathbb{I}^{-1}[d^*,\Lambda]\star L^{n-k-j-1}\alpha$$

$$=(-1)^{\frac{k(k+1)}{2}+1}\frac{j!}{(n-k-j-1)!}\mathbb{I}^{-1}[d^*,\Lambda](-1)^{\frac{k(k+1)}{2}}\frac{(n-k-j-1)!}{(j+1)!}L^{j+1}\mathbb{I}\alpha$$

$$=\frac{-1}{j+1}\mathbb{I}^{-1}[d^*,\Lambda]\,\mathbb{I}\,L^{j+1}\alpha,$$

where in the second to last step we used Lemma 5.

In summary, we have

$$\star\,\mathbb{I}^{-1}\,d\,\mathbb{I}\star\eta = \sum_{r=0}^{\infty}(-1)^{r+1}\frac{j!(n-k-j+r)!}{(j+r-1)!(n-k-j)!}L^{j+r-1}\alpha_r - \frac{1}{j+1}\mathbb{I}^{-1}[d^*,\Lambda]\,\mathbb{I}\,L^{j+1}\alpha. \qquad (4)$$

We now compute $[\Lambda,d]\eta$, by first computing $\Lambda dL^j\alpha$, using that all α_r are primitive. By Equation (2), Lemma 2, and Lemma 3, we have:

$$\Lambda dL^j\alpha = \Lambda L^j d\alpha + \Lambda[d,L^j]\alpha$$

$$= \Lambda L^j\left(\sum_{r=0}^{\infty}L^r\alpha_r\right) + j\Lambda[d,L]L^{j-1}\alpha$$

$$= \sum_{r=0}^{\infty}\Lambda L^{j+r}\alpha_r + j\Lambda[d,L]L^{j-1}\alpha$$

$$= -\sum_{r=0}^{\infty}(j+r)(k+1-2r-n+j+r-1)L^{j+r-1}\alpha_r + j\Lambda[d,L]L^{j-1}\alpha.$$

Next using, α is primitive, and Lemma 2 again, we have

$$d\Lambda L^j\alpha = -j(k-n+j-1)dL^{j-1}\alpha$$

$$= -j(k-n+j-1)L^{j-1}d\alpha - j(k-n+j-1)(j-1)[d,L]L^{j-2}\alpha$$

$$= -j(k-n+j-1)\left(\sum_{r=0}^{\infty}L^{j+r-1}\alpha_r\right) - j(k-n+j-1)(j-1)[d,L]L^{j-2}\alpha.$$

So,

$$[\Lambda,d]\eta = \sum_{r=0}^{\infty}(r(n-k+r)-j)L^{j+r-1}\alpha_r + j\Lambda[d,L]L^{j-1}\alpha + j(j-1)(k-n+j-1)[d,L]L^{j-2}\alpha.$$

Using this last equation and combining with Equation (4) we obtain the desired result:

$$[\Lambda,d]\eta - \star\,\mathbb{I}^{-1}\,d\,\mathbb{I}\star\eta = \frac{1}{j+1}\mathbb{I}^{-1}[d^*,\Lambda]\,\mathbb{I}\,L^{j+1}\alpha$$

$$+ j\Lambda[d,L]L^{j-1}\alpha + j(j-1)(k-n+j-1)[d,L]L^{j-2}\alpha$$

$$+ \sum_{r=0}^{\infty}f_{n,k,j}(r)L^{j+r-1}\alpha_r,$$

where

$$f_{n,k,j}(r) = (r(n-k+r)-j) + (-1)^r\frac{j!(n-k-j+r)!}{(j+r-1)!(n-k-j)!}.$$

It is a curious fact that $f(0) = f(1) = 0$, whereas for $r \geq 2$, $f(r)$ is in general non-zero. □

4. Applications

On an almost Kähler manifold, using the bidegree decompositions of d and d^*, one may derive from (1) the relation
$$[\Lambda, \partial] = i\bar{\partial}^*,$$
involving Λ, ∂ and the adjoint of $\bar{\partial}$. For a non-Kähler Hermitian manifold there is an additional term
$$[\Lambda, \partial] = i(\bar{\partial}^* + \bar{\tau}^*)$$
where $\bar{\tau} = [\Lambda, [\bar{\partial}, L]]$ is the zero-order *torsion operator* (see [5,6]). In the case of $(0,q)$-forms this gives
$$\Lambda \partial \alpha = i\bar{\partial}^*\alpha + i[\Lambda, \bar{\partial}^*]L\alpha.$$

Next we use Theorem 1 to derive this local identity also in the non-integrable case.

Proposition 1. *For all $\alpha \in \mathcal{A}^{0,q}$ in an almost Hermitian manifold we have*
$$\Lambda \partial \alpha = i\bar{\partial}^*\alpha + i[\Lambda, \bar{\partial}^*]L\alpha.$$

Proof. By bidegree reasons α is a primitive form and we have $d\alpha = \alpha_0 + L\alpha_1 + L^2\alpha_2$ where α_i are primitive. By expanding each term in the equality of Theorem 1 with respect to the bidegree decomposition $d = \bar{\mu} + \bar{\partial} + \partial + \mu$, in the case $j = 0$, we obtain:
$$[\Lambda, d]\alpha = \Lambda d\alpha = \Lambda(\partial + \mu)\alpha,$$
$$\star \mathbb{I}^{-1} d\,\mathbb{I} \star \alpha = i(\bar{\partial}^* - \bar{\mu}^*)\alpha,$$
and
$$\mathbb{I}^{-1}[d^*, \Lambda]\,\mathbb{I}\,L\alpha = i[\Lambda, \bar{\partial}^* - \bar{\mu}^*]L\alpha.$$

In particular, all terms decompose into sums of pure bidegrees $(0, q-1)$ and $(1, q-2)$. Note as well that the remaining term
$$f_{n,q,0}(2)L\alpha_2$$
given in Theorem 1 has pure bidegree $(1, q-2)$, since α_2 must have bidegree $(0, q-3)$. By putting together all terms of bidegree $(0, q-1)$ we obtain the desired identity. □

Remark 2. *The proof of Proposition 1 gives a second identity relating the operators Λ, μ and $\bar{\mu}$ and their adjoints, which also contains the term $f_{n,q,0}(2)L\alpha_2$. For forms in $\mathcal{A}^{0,2}$, this extra term vanishes by bidegree reasons, since $\alpha_2 = 0$. Then the second identity reads*
$$\Lambda \mu \alpha = -i\bar{\mu}^*\alpha - i[\Lambda, \bar{\mu}^*]L\alpha.$$

This corrects the identity
$$[\Lambda, \mu] = -i\bar{\mu}^*$$
known in the almost Kähler case for arbitrary forms (see [4]).

The previous proposition can be used to give a uniqueness result for the Dirichlet problem on compact domains with a boundary.

Corollary 1. *Let Ω be a compact domain in an almost complex manifold (M, J), with smooth boundary, and let $g : \Omega \to \mathbb{C}$, and $\phi : \partial\Omega \to \mathbb{C}$ be smooth. Then the Dirichlet problem,*

$$\partial\bar{\partial} u = g \quad \text{with} \quad u|_{\partial\Omega} = \phi,$$

has at most one solution $u : \Omega \to \mathbb{C}$.

In particular, if (M, J) is a compact connected almost complex manifold, and $f : M \to \mathbb{C}$ is a smooth map of almost complex manifolds, then f is constant.

Proof. It suffices to show the only solution to the homogenous equation with $g = 0$ is a constant function.

In any coordinate chart $\psi : V \to \mathbb{R}^{2n}$ containing any maximum point, we pullback J to $\psi(V)$ and consider the J-preserving map $u \circ \psi^{-1} : \psi(V) \to \mathbb{C}$. The components of d are natural with respect to this J-preserving map and we use a compatible metric on $\psi(V)$ to define Λ and $\bar{\partial}^*$. Then by Proposition 1 with $q = 1$ we obtain

$$-i\Lambda\partial\bar{\partial}u = \bar{\partial}^*\bar{\partial}u + [\Lambda, \bar{\partial}^*]L\bar{\partial}u$$

on $\psi(V)$. Note $\bar{\partial}^*\bar{\partial}$ is quadratic, self-adjoint, and positive, and $[\Lambda, \bar{\partial}^*]L\bar{\partial}$ is first order since $[\Lambda, \bar{\partial}^*] = [d, L]^*$ is zeroth order, because $[d, L]\eta = d\omega \wedge \eta$. Then the right hand side is zero, so the maximum principle due to E. Hopf applies [11], showing u is constant in a neighborhood of the maximum point and therefore, by connectedness, u is constant.

The final claim follows taking $\Omega = M$, with empty boundary, $g = 0$, and noting the condition that f is a map of almost complex manifolds implies $\bar{\partial}f = 0$. □

Remark 3. *In [8], we introduce a Dolbeault cohomology theory that is valid for all almost complex manifolds. The above corollary is key in showing that, for a compact connected almost complex manifold, this cohomology is well-behaved in lowest bidegree, in the sense that $H^{0,0}_{\text{Dol}}(M) \cong \mathbb{C}$.*

Finally, we refer the reader to the work of Feehan and Leness [9], where the relation of Proposition 1, for $q = 1$, is used to show that the moduli spaces of unitary anti-self-dual connections over any almost Hermitian 4-manifold is almost Hermitian, whenever the Nijenhuis tensor has sufficiently small C^0-norm.

Author Contributions: Conceptualization, J.C. and S.O.W.; Writing—original draft, J.C. and S.O.W.; Writing—review & editing, J.C. and S.O.W. Both authors have read and agreed to the published version of the manuscript.

Funding: J.C. would like to acknowledge partial support from the AEI/FEDER, UE (MTM2016-76453-C2-2-P) and the Serra Húnter Program. S.O.W. acknowledges the support provided by a PSC-CUNY Award, jointly funded by The Professional Staff Congress and The City University of New York.

Acknowledgments: The authors would like to thank Paul Feehan for encouraging us to develop some previous notes into the present paper.

Conflicts of Interest: The authors declare no conflict of interest.

References

1. Weil, A. *Introduction à l'étude des Variétés Kählériennes*; l'Institut de Mathématique de l'Université de Nancago: Hermann, Paris, 1958; Volume VI, p. 175.
2. De Bartolomeis, P.; Tomassini, A. On formality of some symplectic manifolds. *Intern. Math. Res. Not.* **2001**, 1287–1314. [CrossRef]
3. Tardini, N.; Tomassini, A. Differential operators on almost-Hermitian manifolds and harmonic forms. *Complex Manifolds* **2020**, 7, 106–128. [CrossRef]

4. Cirici, J.; Wilson, S.O. Topological and geometric aspects of almost Kähler manifolds via harmonic theory. *Sel. Math.* **2020**, *26*, 35. [CrossRef]
5. Demailly, J.P. Sur l'identité de Bochner-Kodaira-Nakano en géométrie hermitienne. In *Séminaire d'analyse P. Lelong-P. Dolbeault-H. Skoda, Années 1983/1984*; Springer: Berlin, Germany, 1986; Volume 1198, pp. 88–97.
6. Wilson, S.O. Harmonic symmetries for Hermitian manifolds. *Proc. Am. Math. Soc.* **2020**, *148*, 3039–3045. [CrossRef]
7. Ohsawa, T. Isomorphism theorems for cohomology groups of weakly 1-complete manifolds. *Publ. Res. Inst. Math. Sci.* **1982**, *18*, 191–232. [CrossRef]
8. Cirici, J.; Wilson, S. Dolbeault cohomology for almost complex manifolds. *arXiv* **2018**, arXiv:1809.01416.
9. Feehan, P.M.N.; Leness, T.G. Virtual Morse theory, almost Hermitian manifolds, SO(3) monopoles, and applications to four-manifold topology. In preparation.
10. Huybrechts, D. *Complex Geometry*; An Introduction; Springer: Berlin, Germany, 2005; pp. xii, 309.
11. Hopf, E. *Selected Works of Eberhard Hopf with Commentaries*; Cathleen S., Serrin, M.J.B., Sinai, Y.G., Eds.; American Mathematical Society: Providence, RI, USA, 2002; pp. xxiv, 386.

© 2020 by the authors. Licensee MDPI, Basel, Switzerland. This article is an open access article distributed under the terms and conditions of the Creative Commons Attribution (CC BY) license (http://creativecommons.org/licenses/by/4.0/).

Article

Inequalities for the Casorati Curvature of Statistical Manifolds in Holomorphic Statistical Manifolds of Constant Holomorphic Curvature

Simona Decu [1,2,*,†], Stefan Haesen [3,4,†] and Leopold Verstraelen [5,†]

1. Department of Applied Mathematics, Bucharest University of Economic Studies, 6 Piața Romană, 010374 Bucharest, Romania
2. Romanian Academy, *Costin C. Kirițescu* National Institute of Economic Research—Centre of Mountain Economy (CE-MONT), 13 Calea 13 Septembrie, 030508 Bucharest, Romania
3. Department of Mathematics, University of Hasselt, BE 3590 Diepenbeek, Belgium; stefan.haesen@uhasselt.be
4. Department of Teacher Education, Thomas More University College, 2290 Vorselaar, Belgium
5. Department of Mathematics, Katholieke Universiteit Leuven, Celestijnenlaan 200B, Box 2400, BE-3001 Leuven, Belgium; leopold.verstraelen@wis.kuleuven.be
* Correspondence: simona.decu@gmail.com or simona.marinescu@csie.ase.ro
† These authors contributed equally to this work.

Received: 16 December 2019; Accepted: 9 February 2020; Published: 14 February 2020

Abstract: In this paper, we prove some inequalities in terms of the normalized δ-Casorati curvatures (extrinsic invariants) and the scalar curvature (intrinsic invariant) of statistical submanifolds in holomorphic statistical manifolds with constant holomorphic sectional curvature. Moreover, we study the equality cases of such inequalities. An example on these submanifolds is presented.

Keywords: Casorati curvature; statistical submanifold; holomorphic statistical manifold

1. Introduction

The problem of discovering simple relationships between the main *intrinsic invariants* and the main *extrinsic invariants* of submanifolds is a basic problem in submanifold theory [1]. In this respect, beautiful results focus on certain types of geometric inequalities. Moreover, another basic problem in this field is to study the *ideal submanifolds* in a space form, namely to investigate the submanifolds which satisfy the equality case of such inequalities [2].

The method of looking for *Chen invariants* answers the problems posed above. First, Chen demonstrated in [3] an optimal inequality for a submanifold on a real space form between the intrinsically defined δ-curvature and the extrinsically defined squared mean curvature. This approach initiated a new line of research and was extended to various types of submanifolds in several types of ambient spaces, e.g., submanifolds in complex space forms of constant holomorphic sectional curvature (see [4–7]). The submanifolds attaining the equality of these inequalities (called Chen ideal submanifolds) were also investigated. Recently, Chen et al. classified $\delta(2, n-2)$-ideal Lagrangian submanifolds in complex space forms in [8].

Moreover, new solutions to the above problems are given by the inequalities involving δ-*Casorati curvatures*, initiated in [9,10]. In the search for a true measure of curvature, Casorati in 1890 proposed the curvature which nowadays bears his name because it better corresponds with our *common intuition of curvature* than Gauss and mean curvature [11]. However, this notion of curvature was soon forgotten and was rediscovered by Koenderink working in the field of computer vision [12]. Verstraelen developed some geometrical models for early vision, presenting perception via the Casorati curvature of sensation [13]. A geometrical interpretation of this type of curvature for submanifolds in Riemannian spaces was given in [14]. In [15], the isotropical Casorati curvature of production surfaces was studied. The Casorati curvature was used to obtain optimal inequalities between intrinsic and extrinsic curvatures of submanifolds in real space forms in [9,10]. Later, this knowledge was extended (e.g., see [16–21]). Submanifolds which satisfy these equalities are named *Casorati ideal submanifolds*. Recently, Vîlcu established an optimal inequality for Lagrangian submanifolds in complex space forms involving Casorati curvature [22]. Aquib et al. obtained a classification of Casorati ideal Lagrangian submanifolds in complex space forms [23]. Very recently, Suceavă and Vajiac studied inequalities involving some Chen invariants, mean curvature, and Casorati curvature for strictly convex Euclidean hypersurfaces [24]. Brubaker and Suceavă investigated a geometric interpretation of Cauchy–Schwarz inequality in terms of Casorati curvature [25].

The concept of *statistical manifold* was defined by Amari in 1985, in the basic study on information geometry [26]. Currently, interest in the field of statistical manifolds is increasing, being focused on applications in differential geometry, information geometry, statistics, machine learning, etc. (see, e.g., [27–29]). Cuingnet et al. introduced a continuous framework to spatially regularize support vector machines (SVM) for brain image analysis, considering the images as elements of a statistical manifold, in order to classify patients with Alzheimer's disease [30]. The study of curvature invariants of submanifolds in statistical manifolds gives other solutions to the above research problems. Aydin et al. established some inequalities (Chen–Ricci and Wintgen) for submanifolds in statistical manifolds of constant curvature in [31,32]. Lee et al. obtained inequalities on Sasakian statistical manifolds in terms of Casorati curvatures [33]. Aquib and Shahid [34] proved some inequalities involving Casorati curvatures on statistical submanifolds in quaternion Kähler-like statistical space forms. The quaternionic theory of statistical manifolds is investigated in [35]. Very recently, new results have been published. Aytimur et al. established some Chen inequalities for submanifolds in Kähler-like statistical manifolds [36]. Aquib et al. achieved generalized Wintgen-type inequalities for submanifolds in generalized space forms [37]. Chen et al. established a Chen first inequality for statistical submanifolds in Hessian manifolds of constant Hessian curvature [38]. Moreover, Siddiqui et al. studied a Chen inequality for statistical warped products statistically immersed in a statistical manifold of constant curvature [39].

Recently, Furuhata et al. [40] defined the notion of a *holomorphic statistical manifold*, which can be considered as a generalization of a special Kähler manifold. The authors establish the basics for statistical submanifolds in holomorphic statistical manifolds.

In order to find out new solutions for the problems under debate, we obtain inequalities for statistical submanifolds in holomorphic statistical manifolds. The invariants involved in such inequalities are the extrinsic normalized δ-Casorati curvatures and the intrinsic scalar curvature. The method is focused on a constrained extremum problem. Moreover, the equality cases are investigated. This study revealed that the equality at all points characterizes submanifolds that are totally geodesic with respect to the Levi–Civita connection.

2. Preliminaries

Let $(\tilde{M}, \tilde{g})$ be a $2n$-dimensional manifold, $\tilde{\nabla}$ an affine connection on $\tilde{M}$, and $\tilde{g}$ a Riemannian metric on $\tilde{M}$. Consider $\tilde{T} \in \Gamma(T\tilde{M}^{(1,2)})$ the torsion tensor field of $\tilde{\nabla}$.

A pair $(\tilde{\nabla}, \tilde{g})$ is called a *statistical structure* on $\tilde{M}$ if the torsion tensor field $\tilde{T}$ vanishes and $\tilde{\nabla}\tilde{g} \in \Gamma(T\tilde{M}^{(0,3)})$ is symmetric.

A Riemannian manifold $(\tilde{M}, \tilde{g})$ is called a *statistical manifold* if it is endowed with a pair of torsion-free affine connections $\tilde{\nabla}$ and $\tilde{\nabla}^*$ satisfying

$$Z\tilde{g}(X,Y) = \tilde{g}(\tilde{\nabla}_Z X, Y) + \tilde{g}(X, \tilde{\nabla}^*_Z Y),$$

for any $X, Y, Z \in \Gamma(T\tilde{M})$. Denote $(\tilde{M}, \tilde{g}, \tilde{\nabla})$ as the statistical manifold. The connections $\tilde{\nabla}$ and $\tilde{\nabla}^*$ are named *dual connections* or *conjugate connections*.

Remark 1. *If $(\tilde{M}, \tilde{g}, \tilde{\nabla})$ is a statistical manifold, then we remark that*

1. $(\tilde{\nabla}^*)^* = \tilde{\nabla}$;
2. $(\tilde{M}, \tilde{g}, \tilde{\nabla}^*)$ is also a statistical manifold;
3. $\tilde{\nabla}$ always has a dual connection $\tilde{\nabla}^*$ satisfying

$$\tilde{\nabla} + \tilde{\nabla}^* = 2\tilde{\nabla}^0, \tag{1}$$

where $\tilde{\nabla}^0$ is the Levi–Civita connection on $\tilde{M}$.

Let M be an m-dimensional submanifold of a $2n$-dimensional statistical manifold $(\tilde{M}, \tilde{g})$ and g the induced metric on M. The Gauss formulas are given by

$$\tilde{\nabla}_X Y = \nabla_X Y + h(X, Y),$$
$$\tilde{\nabla}^*_X Y = \nabla^*_X Y + h^*(X, Y),$$

for any $X, Y \in \Gamma(TM)$, where h and h^* are symmetric and bilinear $(0,2)$-tensors, called the *imbedding curvature tensor* of M in $\tilde{M}$ for $\tilde{\nabla}$ and $\tilde{\nabla}^*$, respectively.

Denote the curvature tensor fields of ∇ and $\tilde{\nabla}$ by R and $\tilde{R}$, respectively. Then, the *Gauss equation* concerning the connection $\tilde{\nabla}$ is ([41])

$$\tilde{g}(\tilde{R}(X,Y)Z, W) = g(R(X,Y)Z, W) + \tilde{g}(h(X,Z), h^*(Y,W)) - \tilde{g}(h^*(X,W), h(Y,Z)), \tag{2}$$

for any $X, Y, Z, W \in \Gamma(TM)$.

In addition, denote the curvature tensor fields of the connections ∇^* and $\tilde{\nabla}^*$ by R^* and $\tilde{R}^*$, respectively. Then the *Gauss equation* concerning the connection $\tilde{\nabla}^*$ is ([41])

$$\tilde{g}(\tilde{R}^*(X,Y)Z, W) = g(R^*(X,Y)Z, W) + \tilde{g}(h^*(X,Z), h(Y,W)) - \tilde{g}(h(X,W), h^*(Y,Z)), \tag{3}$$

for any $X, Y, Z, W \in \Gamma(TM)$.

If M is a submanifold of a statistical manifold $(\tilde{M}, \tilde{g}, \tilde{\nabla})$, then (M, g, ∇) is also a statistical manifold with the induced metric g and the induced connection ∇.

Let S be the *statistical curvature tensor field of a statistical manifold* (M, g, ∇), where $S \in \Gamma(TM^{(1,3)})$ is defined by [40]

$$S(X,Y)Z = \frac{1}{2}\{R(X,Y)Z + R^*(X,Y)Z\}, \tag{4}$$

for $X, Y, Z \in \Gamma(TM)$.

If $\pi = \mathrm{span}_{\mathbb{R}}\{u_1, u_2\}$ is a 2-dimensional subspace of $T_p M$, for $p \in M$, then the *sectional curvature* of M is defined by [40]:

$$\sigma(\pi) = \frac{g(S(u_1, u_2)u_2, u_1)}{g(u_1, u_1)g(u_2, u_2) - g^2(u_1, u_2)}. \tag{5}$$

Let $\{e_1, ..., e_m\}$ be an orthonormal basis of the tangent space $T_p M$, for $p \in M$, and let $\{e_{m+1}, ..., e_{2n}\}$ be an orthonormal basis of the normal space $T_p^{\perp} M$. The *scalar curvature* τ at p is given by

$$\tau(p) = \sum_{1 \leq i < j \leq m} \sigma(e_i \wedge e_j) = \sum_{1 \leq i < j \leq m} g(S(e_i, e_j)e_j, e_i), \tag{6}$$

and the *normalized scalar curvature* ρ of M is defined as

$$\rho = \frac{2\tau}{m(m-1)}. \tag{7}$$

The *mean curvature* vector fields of M, denoted by H and H^*, are given by

$$H = \frac{1}{m}\sum_{i=1}^{m} h(e_i, e_i), \quad H^* = \frac{1}{m}\sum_{i=1}^{m} h^*(e_i, e_i).$$

From Equation (1), we get $2h^0 = h + h^*$ and $2H^0 = H + H^*$, where h^0 and H^0 are the second fundamental form and the mean curvature field of M, respectively, with respect to the Levi–Civita connection ∇^0 on M.

The *squared mean curvatures* of the submanifold M in $\tilde{M}$ have the expressions

$$\|H\|^2 = \frac{1}{m^2}\sum_{\alpha=m+1}^{2n}\left(\sum_{i=1}^{m} h_{ii}^{\alpha}\right)^2, \quad \|H^*\|^2 = \frac{1}{m^2}\sum_{\alpha=m+1}^{2n}\left(\sum_{i=1}^{m} h_{ii}^{*\alpha}\right)^2,$$

where $h_{ij}^{\alpha} = \tilde{g}(h(e_i, e_j), e_{\alpha})$ and $h_{ij}^{*\alpha} = \tilde{g}(h^*(e_i, e_j), e_{\alpha})$, for $i, j \in \{1, ..., m\}$, $\alpha \in \{m+1, ..., 2n\}$.

Denote by $\mathcal{C}$ and $\mathcal{C}^*$ the *Casorati curvatures* of the submanifold M, defined by the squared norms of h and h^*, respectively, over the dimension m, as follows:

$$\mathcal{C} = \frac{1}{m}\|h\|^2 = \frac{1}{m}\sum_{\alpha=m+1}^{2n}\sum_{i,j=1}^{m}\left(h_{ij}^{\alpha}\right)^2,$$

$$\mathcal{C}^* = \frac{1}{m}\|h^*\|^2 = \frac{1}{m}\sum_{\alpha=m+1}^{2n}\sum_{i,j=1}^{m}\left(h_{ij}^{*\alpha}\right)^2.$$

Let L be an s-dimensional subspace of T_pM, $s \geq 2$ and let $\{e_1, \ldots, e_s\}$ be an orthonormal basis of L. Hence, the Casorati curvatures $\mathcal{C}(L)$ and $\mathcal{C}^*(L)$ of L are given by

$$\mathcal{C}(L) = \frac{1}{s}\sum_{\alpha=m+1}^{2n}\sum_{i,j=1}^{s}\left(h_{ij}^{\alpha}\right)^2, \quad \mathcal{C}^*(L) = \frac{1}{s}\sum_{\alpha=m+1}^{2n}\sum_{i,j=1}^{s}\left(h_{ij}^{*\alpha}\right)^2.$$

The *normalized δ-Casorati curvatures* $\delta_{\mathcal{C}}(m-1)$ and $\hat{\delta}_{\mathcal{C}}(m-1)$ of the submanifold M^n are given by

$$\delta_{\mathcal{C}}(m-1)|_p = \frac{1}{2}\mathcal{C}|_p + \frac{m+1}{2m}\inf\{\mathcal{C}(L) | L \text{ a hyperplane of } T_pM\}$$

and

$$\hat{\delta}_{\mathcal{C}}(m-1)|_p = 2\mathcal{C}|_p - \frac{2m-1}{2m}\sup\{\mathcal{C}(L) | L \text{ a hyperplane of } T_pM\}.$$

Moreover, the *dual normalized δ^*-Casorati curvatures* $\delta_{\mathcal{C}}^*(m-1)$ and $\hat{\delta}_{\mathcal{C}}^*(m-1)$ of the submanifold M in $\tilde{M}$ are defined as

$$\delta_{\mathcal{C}}^*(m-1)|_p = \frac{1}{2}\mathcal{C}^*|_p + \frac{m+1}{2m}\inf\{\mathcal{C}^*(L) | L \text{ a hyperplane of } T_pM\}$$

and

$$\hat{\delta}_{\mathcal{C}}^*(m-1)|_p = 2\mathcal{C}^*|_p - \frac{2m-1}{2m}\sup\{\mathcal{C}^*(L) | L \text{ a hyperplane of } T_pM\}.$$

Denote by $\delta_{\mathcal{C}}(r; m-1)$ and $\hat{\delta}_{\mathcal{C}}(r; m-1)$, the *generalized normalized δ-Casorati curvatures* of M, defined in [10] as

$$\delta_{\mathcal{C}}(r; m-1)|_p = r\,\mathcal{C}|_p + a(r)\inf\{\mathcal{C}(L) \mid L \text{ a hyperplane of } T_pM\},$$

if $0 < r < m(m-1)$, and

$$\hat{\delta}_{\mathcal{C}}(r; m-1)|_p = r\,\mathcal{C}|_p + a(r)\sup\{\mathcal{C}(L) \mid L \text{ a hyperplane of } T_pM\},$$

if $r > m(m-1)$, for $a(r)$ set as

$$a(r) = \frac{(m-1)(r+m)(m^2 - m - r)}{mr},$$

where $r \in \mathbb{R}_+$ and $r \neq m(m-1)$.

Furthermore, denote by $\delta^*_\mathcal{C}(r; m-1)$ and $\hat{\delta}^*_\mathcal{C}(r; m-1)$ the *dual generalized normalized δ^*-Casorati curvatures* of the submanifold M, defined as follows:

$$\delta^*_\mathcal{C}(r; m-1)|_p = r\,\mathcal{C}^*|_p + a(r) \inf\{\mathcal{C}^*(L) \mid L \text{ a hyperplane of } T_pM\},$$

if $0 < r < m(m-1)$, and

$$\hat{\delta}^*_\mathcal{C}(r; m-1)|_p = r\,\mathcal{C}^*|_p + a(r) \sup\{\mathcal{C}^*(L) \mid L \text{ a hyperplane of } T_pM\},$$

if $r > m(m-1)$, for $a(r)$ set above.

A statistical submanifold (M, g, ∇) of $(\tilde{M}, \tilde{g}, \tilde{\nabla})$ is called *totally geodesic* with respect to the connection $\tilde{\nabla}$ if the second fundamental form h of M for $\tilde{\nabla}$ vanishes identically [40].

Let $\tilde{M}$ be an almost complex manifold with almost complex structure $J \in \Gamma(T\tilde{M}^{(1,1)})$. A quadruplet $(\tilde{M}, \tilde{\nabla}, \tilde{g}, J)$ is called a *holomorphic statistical manifold* if

1. $(\tilde{\nabla}, \tilde{g})$ is a statistical structure on $\tilde{M}$; and
2. ω is a $\tilde{\nabla}$-parallel 2-form on $\tilde{M}$,

where ω is defined by $\omega(X, Y) = \tilde{g}(X, JY)$, for any $X, Y \in \Gamma(T\tilde{M})$.

For a holomorphic statistical manifold, the following formula holds:

$$\tilde{g}(\tilde{S}(Z,W)JY, JX) = \tilde{g}(\tilde{S}(JZ, JW)Y, X) = \tilde{g}(\tilde{S}(Z,W)Y, X), \tag{8}$$

for any $X, Y, Z, W \in \Gamma(T\tilde{M})$.

A holomorphic statistical manifold $(\tilde{M}, \tilde{\nabla}, \tilde{g}, J)$ is said to be of *constant holomorphic sectional curvature* $c \in \mathbb{R}$ if the following formula holds [42]:

$$\tilde{S}(X,Y)Z = \frac{c}{4}\{\tilde{g}(Y,Z)X - \tilde{g}(X,Z)Y + \tilde{g}(JY,Z)JX - \tilde{g}(JX,Z)JY + 2\tilde{g}(X,JY)JZ\}, \tag{9}$$

for any $X, Y, Z \in \Gamma(T\tilde{M})$, where $\tilde{S}$ is the statistical curvature tensor field of $\tilde{M}$.

Remark 2 ([43]). *Let $(\tilde{M}, \tilde{g}, J)$ be a Kähler manifold. If we define a connection $\tilde{\nabla}$ as $\tilde{\nabla} = \nabla^{\tilde{g}} + K$, where $K \in \Gamma(T\tilde{M}^{(1,2)})$ satisfying the conditions*

$$K(X,Y) = K(Y,X), \tag{10}$$

$$\tilde{g}(K(X,Y), Z) = \tilde{g}(Y, K(X,Z)), \tag{11}$$

$$K(X, JY) = -JK(X,Y), \tag{12}$$

for any $X, Y, Z \in \Gamma(T\tilde{M})$, then $(\tilde{M}, \tilde{\nabla}, \tilde{g}, J)$ is a holomorphic statistical manifold.

Let M be an m-dimensional statistical submanifold of a holomorphic statistical manifold $(\tilde{M}, \tilde{\nabla}, \tilde{g}, J)$. For any vector field X tangent to M we can decompose

$$JX = PX + FX, \tag{13}$$

where PX and FX are the tangent component and the normal component, respectively, of JX. Given a local orthonormal frame $\{e_1, e_2, \cdots, e_m\}$ of M, then the squared norm of P is expressed by

$$\|P\|^2 = \sum_{i,j=1}^{m} g^2(Pe_i, e_j).$$

Next, we consider the constrained extremum problem

$$\min_{x \in M} f(x), \tag{14}$$

where M is a Riemannian submanifold of a Riemannian manifold $(\tilde{M}, \tilde{g})$, and $f : \tilde{M} \to \mathbb{R}$ is a function of differentiability class C^2.

Theorem 1 ([44]). *If M is complete and connected, $(\mathrm{grad} f)(p) \in T_p^\perp M$ for a point $p \in M$, and the bilinear form $\mathcal{A} : T_p M \times T_p M \to \mathbb{R}$ defined by*

$$\mathcal{A}(X, Y) = \mathrm{Hess}(f)(X, Y) + \tilde{g}(h^0(X, Y), \mathrm{grad} f), \tag{15}$$

is positive definite in p, then p is the optimal solution of the Problem (14).

Remark 3 ([44]). *If the bilinear form $\mathcal{A}$ defined by Equation (15) is positive semi-definite on the submanifold M, then the critical points of $f|M$ are global optimal solutions of the Problem (14).*

3. Main Inequalities

Theorem 2. *Let M be an m-dimensional statistical submanifold of a 2n-dimensional holomorphic statistical manifold $(\tilde{M}, \tilde{\nabla}, \tilde{g}, J)$ of constant holomorphic sectional curvature c. Then we have*

(i)

$$\begin{aligned} 2\tau &\leq \delta_{\mathcal{C}}^0(r; m-1) + m\mathcal{C}^0 - 2m^2 \|H^0\|^2 \\ &\quad + m^2 \tilde{g}(H, H^*) + \frac{3c}{4}\|P\|^2 + \frac{c}{4}m(m-1), \end{aligned} \tag{16}$$

for any real number r such that $0 < r < m(m-1)$, where $\delta_{\mathcal{C}}^0(r; m-1) = \frac{\delta_{\mathcal{C}}(r;m-1)+\delta_{\mathcal{C}}^*(r;m-1)}{2}$ and $\mathcal{C}^0 = \frac{\mathcal{C}+\mathcal{C}^*}{2}$; and

(ii)

$$\begin{aligned} 2\tau &\leq \hat{\delta}_{\mathcal{C}}^0(r; m-1) + m\mathcal{C}^0 - 2m^2 \|H^0\|^2 \\ &\quad + m^2 \tilde{g}(H, H^*) + \frac{3c}{4}\|P\|^2 + \frac{c}{4}m(m-1), \end{aligned} \tag{17}$$

for any real number r such that $r > m(m-1)$, where $\hat{\delta}_{\mathcal{C}}^0(r; m-1) = \frac{\hat{\delta}_{\mathcal{C}}(r;m-1)+\hat{\delta}_{\mathcal{C}}^*(r;m-1)}{2}$.

Moreover, the equality cases of Inequalities (16) and (17) hold identically at all points $p \in M$ if and only if the following condition is satisfied:

$$h + h^* = 0, \tag{18}$$

where h and h^* are the imbedding curvature tensors of the submanifold associated to the dual connections $\tilde{\nabla}$ and $\tilde{\nabla}^*$, respectively.

Proof. The relations (Equations (2)–(4)) imply

$$\begin{aligned} 2\tilde{g}(\tilde{S}(X,Y)Z, W) &= 2g(S(X,Y)Z, W) - \tilde{g}(h(Y,Z), h^*(X,W)) + \tilde{g}(h(X,Z), h^*(Y,W)) \\ &\quad - \tilde{g}(h^*(Y,Z), h(X,W)) + \tilde{g}(h^*(X,Z), h(Y,W)), \end{aligned} \tag{19}$$

where $X, Y, Z, W \in \Gamma(TM)$.

For $p \in M$, we choose $\{e_1, ..., e_m\}$ and $\{e_{m+1}, ..., e_{2n}\}$ orthonormal bases of $T_p M$ and $T_p^\perp M$, respectively. For $X = Z = e_i$ and $Y = W = e_j$ with $i, j \in \{1, ..., m\}$, from the Equation (19), it follows that

$$2\tau(p) = m^2 \tilde{g}(H, H^*) - \sum_{1 \leq i,j \leq m} \tilde{g}(h^*(e_i, e_j), h(e_i, e_j)) \qquad (20)$$
$$+ \frac{c}{4}(m^2 - m + 3\|P\|^2).$$

Denoting $2H^0 = H + H^*$ and $2\mathcal{C}^0 = \mathcal{C} + \mathcal{C}^*$, Equation (20) becomes

$$2\tau(p) = 2m^2 \|H^0\|^2 - \frac{m^2}{2}\|H\|^2 - \frac{m^2}{2}\|H^*\|^2$$
$$- 2m\mathcal{C}^0 + \frac{m}{2}(\mathcal{C} + \mathcal{C}^*) + \frac{c}{4}(m^2 - m + 3\|P\|^2). \qquad (21)$$

Let $\mathcal{P}$ be the quadratic polynomial defined by

$$\mathcal{P} = r\mathcal{C}^0 + a(r)\,\mathcal{C}^0(L) + \frac{m}{2}(\mathcal{C} + \mathcal{C}^*) - \frac{m^2}{2}(\|H\|^2 + \|H^*\|^2)$$
$$- 2\tau(p) + \frac{c}{4}(m^2 - m + 3\|P\|^2), \qquad (22)$$

where L is a hyperplane of $T_p M$.

We consider that the hyperplane L is spanned by the tangent vectors $e_1, ..., e_{m-1}$, without loss of generality. Therefore, we get

$$\mathcal{P} = \sum_{\alpha=m+1}^{2n} \left[\frac{2m+r}{m} \sum_{i,j=1}^{m} (h_{ij}^{0\alpha})^2 + a(r)\frac{1}{m-1} \sum_{i,j=1}^{m-1} (h_{ij}^{0\alpha})^2 - 2\left(\sum_{i=1}^{m} h_{ii}^{0\alpha}\right)^2 \right]. \qquad (23)$$

Then, Equation (23) yields

$$\mathcal{P} = \sum_{\alpha=m+1}^{2n} \Bigg\{ \left[\frac{2(2m+r)}{m} + \frac{2a(r)}{m-1}\right] \sum_{1 \leq i < j \leq m-1} (h_{ij}^{0\alpha})^2 + \left[\frac{2(2m+r)}{m} + \frac{2a(r)}{m-1}\right] \sum_{i=1}^{m-1} (h_{im}^{0\alpha})^2$$
$$+ \left(\frac{2m+r}{m} + \frac{a(r)}{m-1} - 2\right) \sum_{i=1}^{m-1} (h_{ii}^{0\alpha})^2$$
$$- 4 \sum_{1 \leq i < j \leq m} h_{ii}^{0\alpha} h_{jj}^{0\alpha} + \left(\frac{2m+r}{m} - 2\right) (h_{mm}^{0\alpha})^2 \Bigg\}$$
$$\geq \sum_{\alpha=m+1}^{2n} \left[\frac{r(m-1) + a(r)m}{m(m-1)} \sum_{i=1}^{m-1} (h_{ii}^{0\alpha})^2 + \left(\frac{r}{m}\right)(h_{mm}^{0\alpha})^2 - 4 \sum_{1 \leq i < j \leq m} h_{ii}^{0\alpha} h_{jj}^{0\alpha} \right].$$

Let f_α be a quadratic form defined by $f_\alpha : \mathbb{R}^m \to \mathbb{R}$ for any $\alpha \in \{m+1, ..., 2n\}$,

$$f_\alpha(h_{11}^{0\alpha}, h_{22}^{0\alpha}, ..., h_{mm}^{0\alpha}) = \sum_{i=1}^{m-1} \frac{r(m-1) + a(r)m}{m(m-1)} (h_{ii}^{0\alpha})^2$$
$$+ \frac{r}{m}(h_{mm}^{0\alpha})^2 - 4 \sum_{1 \leq i < j \leq m} h_{ii}^{0\alpha} h_{jj}^{0\alpha}.$$

We investigate the constrained extremum problem

$$\min f_\alpha$$

with the constraint

$$Q: h_{11}^{0\alpha} + h_{22}^{0\alpha} + \ldots + h_{mm}^{0\alpha} = k^\alpha,$$

where k^α is a real constant.

We obtain the system of first-order partial derivatives:

$$\begin{cases} \dfrac{\partial f_\alpha}{\partial h_{ii}^{0\alpha}} = 2\dfrac{r(m-1) + a(r)m}{m(m-1)} h_{ii}^{0\alpha} - 4\left(\sum_{k=1}^{m} h_{kk}^{0\alpha} - h_{ii}^{0\alpha}\right) = 0 \\ \dfrac{\partial f_\alpha}{\partial h_{mm}^{0\alpha}} = \dfrac{2r}{m} h_{mm}^{0\alpha} - 4\sum_{k=1}^{m-1} h_{kk}^{0\alpha} = 0, \end{cases}$$

for every $i \in \{1, \ldots, m-1\}$, $\alpha \in \{m+1, \ldots, 2n\}$.

It follows that the constrained critical point is

$$h_{ii}^{0\alpha} = \frac{2m(m-1)}{(m-1)(2m+r) + ma(r)} k^\alpha$$

$$h_{mm}^{0\alpha} = \frac{2m}{2m+r} k^\alpha,$$

for any $i \in \{1, \ldots, m-1\}$, $\alpha \in \{m+1, \ldots, 2n\}$.

For $p \in Q$, let $\mathcal{A}$ be a 2-form, $\mathcal{A}: T_pQ \times T_pQ \to \mathbb{R}$ defined by

$$\mathcal{A}(X, Y) = \text{Hess}(f_\alpha)(X, Y) + \langle h'(X, Y), (\text{grad} f_\alpha)(p)\rangle,$$

where h' is the second fundamental form of Q in $\mathbb{R}^{m+1}$ and $\langle \cdot, \cdot \rangle$ is the standard inner product on $\mathbb{R}^m$.

The Hessian matrix of f_α is given by

$$\text{Hess}(f_\alpha) = \begin{pmatrix} \lambda & -4 & \ldots & -4 & -4 \\ -4 & \lambda & \ldots & -4 & -4 \\ \vdots & \vdots & \ddots & \vdots & \vdots \\ -4 & -4 & \ldots & \lambda & -4 \\ -4 & -4 & \ldots & -4 & \frac{2r}{m} \end{pmatrix},$$

where $\lambda = 2\frac{(m-1)(r+2m) + ma(r)}{m(m-1)}$ is a real constant.

The condition $\sum_{i=1}^m X_i = 0$ is satisfied, for a vector field $X \in T_pQ$, as the hyperplane Q is totally geodesic in $\mathbb{R}^m$. Then, we achieve

$$\mathcal{A}(X, X) = \lambda \sum_{i=1}^{m-1} X_i^2 + \frac{2r}{m} X_m^2 - 8 \sum_{i,j=1(i\neq j)}^{m} X_i X_j$$

$$= \lambda \sum_{i=1}^{m-1} X_i^2 + \frac{2r}{m} X_m^2 + 4\left(\sum_{i=1}^{m} X_i\right)^2 - 8 \sum_{i,j=1(i\neq j)}^{m} X_i X_j$$

$$= \lambda \sum_{i=1}^{m-1} X_i^2 + \frac{2r}{m} X_m^2 + 4 \sum_{i=1}^{m} X_i^2$$

$$\geq 0.$$

Applying Remark 3, the critical point $(h_{11}^{0\alpha}, \ldots, h_{mm}^{0\alpha})$ of f_α is the global minimum point of the problem. Since $f_\alpha(h_{11}^{0\alpha}, \ldots, h_{mm}^{0\alpha}) = 0$, we get $\mathcal{P} \geq 0$.

We have then proved Inequalities (16) and (17), considering infimum and supremum, respectively, over all tangent hyperplanes L of T_pM.

In addition, we study the equality cases of Inequalities (16) and (17). First, we find out the critical points of $\mathcal{P}$

$$h^c = (h_{11}^{0\,m+1}, h_{12}^{0\,m+1}, \ldots, h_{m\,m}^{0\,m+1}, \ldots, h_{11}^{0\,2n}, \ldots, h_{m\,m}^{0\,2n})$$

as the solutions of following system of linear homogeneous equations:

$$\begin{cases} \dfrac{\partial \mathcal{P}}{\partial h_{ii}^{0\alpha}} = 2\left[\dfrac{2m+r}{m} + \dfrac{a(r)}{m-1} - 2\right] h_{ii}^{0\alpha} - 4 \sum_{k \neq i, k=1}^{m} h_{kk}^{0\alpha} = 0, \\[6pt] \dfrac{\partial \mathcal{P}}{\partial h_{mm}^{0\alpha}} = 2\dfrac{r}{m} h_{mm}^{0\alpha} - 4 \sum_{k=1}^{m-1} h_{kk}^{0\alpha} = 0, \\[6pt] \dfrac{\partial \mathcal{P}}{\partial h_{ij}^{0\alpha}} = 4\left[\dfrac{2m+r}{m} + \dfrac{a(r)}{m-1}\right] h_{ij}^{0\alpha} = 0, \quad i \neq j, \\[6pt] \dfrac{\partial \mathcal{P}}{\partial h_{im}^{0\alpha}} = 4\left[\dfrac{2m+r}{m} + \dfrac{a(r)}{m-1}\right] h_{im}^{0\alpha} = 0. \end{cases}$$

The critical points satisfy $h_{ij}^{0\alpha} = 0$, with $i,j \in \{1, \ldots, m\}$ and $\alpha \in \{m+1, \ldots, 2n\}$. On the other hand, we know that $\mathcal{P} \geq 0$ and $\mathcal{P}(h^c) = 0$, then the critical point h^c is a minimum point of $\mathcal{P}$. Consequently, the cases of equality hold in both Inequalities (16) and (17) if and only if $h_{ij}^{\alpha} = -h_{ij}^{*\alpha}$, for $i,j \in \{1, \ldots, m\}$, $\alpha \in \{m+1, \ldots, 2n\}$. □

Remark 4. *Under Equation (18), the submanifold M is totally geodesic with respect to the Levi–Civita connection $\tilde{\nabla}^0$. Then, the equality cases of Inequalities (16) and (17) hold for all unit tangent vectors at p if and only if p is a totally geodesic point with respect to the Levi–Civita connection.*

By virtue of Theorem 2, the generalized normalized δ-Casorati curvatures satisfy Inequalities (16) and (17). If the normalized δ-Casorati curvatures $\delta_C(m-1)$ and $\delta_C^*(m-1)$, respectively, $\hat{\delta}_C(m-1)$ and $\hat{\delta}_C^*(m-1)$ are involved, then we can state the following result.

Corollary 1. *Let M be an m-dimensional statistical submanifold of a 2n-dimensional holomorphic statistical manifold $(\tilde{M}, \tilde{\nabla}, \tilde{g}, J)$ of constant holomorphic sectional curvature c. Then, we have*

(i)

$$\rho \leq \delta_C^0(m-1) + \frac{1}{m-1} C^0 - \frac{2m}{m-1} \|H^0\|^2 \qquad (24)$$
$$+ \frac{m}{m-1} \tilde{g}(H, H^*) + \frac{3c}{4m(m-1)} \|P\|^2 + \frac{c}{4},$$

where $2\delta_C^0(m-1) = \delta_C(m-1) + \delta_C^*(m-1)$ and $2C^0 = C + C^*$, and

(ii)

$$\rho \leq \hat{\delta}_C^0(m-1) + \frac{1}{m-1} C^0 - \frac{2m}{m-1} \|H^0\|^2 \qquad (25)$$
$$+ \frac{m}{m-1} \tilde{g}(H, H^*) + \frac{3c}{4m(m-1)} \|P\|^2 + \frac{c}{4},$$

where $2\hat{\delta}_C^0(m-1) = \hat{\delta}_C(m-1) + \hat{\delta}_C^*(m-1)$.

Moreover, the equality cases of Inequalities (24) and (25) hold identically at all points if and only if h and h^* satisfy the condition in Equation (18), which implies that M is a totally geodesic submanifold with respect to the Levi–Civita connection.

4. An Example

Example 1. Let (x_1, x_2, y_1, y_2) be a standard system on $\mathbb{R}^4$, g the Euclidean metric. Define $t = (y_1^2 + y_2^2)/2$ ($t \geq 0$) and the functions u, v on $\mathbb{R}^4$ as

$$u(x_1, x_2, y_1, y_2) = a(t), \quad v(x_1, x_2, y_1, y_2) = b(t),$$

where a is a function $a : [0, \infty) \to (0, \infty)$, and $b(t) = -a(t)a'(t)(2ta'(t) - a(t))^{-1}$, assuming that $a(t) + 2tb(t) > 0$ for $t \geq 0$.

Let G be a g-natural metric on $\mathbb{R}^4$ and J a complex structure defined by Oproiu ([45]) such that $\mathbb{R}^4$ is Kählerian, as follows:

$$
\begin{aligned}
G &= (u + vy_1^2)dx_1 dx_1 + 2vy_1 y_2 dx_1 dx_2 + (u + vy_2^2) dx_2 dx_2 + \frac{u + vy_2^2}{u(u + 2tv)} dy_1 dy_1 \\
&\quad - 2\frac{vy_1 y_2}{u(u+2tv)} dy_1 dy_2 + \frac{u + vy_1^2}{u(u+2tv)} dy_2 dy_2,
\end{aligned}
\tag{26}
$$

$$
\begin{cases}
J\dfrac{\partial}{\partial x_1} = (u + vy_1^2)\dfrac{\partial}{\partial y_1} + vy_1 y_2 \dfrac{\partial}{\partial y_2}, \\[4pt]
J\dfrac{\partial}{\partial x_2} = vy_1 y_2 \dfrac{\partial}{\partial y_1} + (u + vy_2^2)\dfrac{\partial}{\partial y_2}, \\[4pt]
J\dfrac{\partial}{\partial y_1} = -\dfrac{u + vy_2^2}{u(u+2tv)}\dfrac{\partial}{\partial x_1} + \dfrac{vy_1 y_2}{u(u+2tv)}\dfrac{\partial}{\partial x_2}, \\[4pt]
J\dfrac{\partial}{\partial y_2} = \dfrac{vy_1 y_2}{u(u+2tv)}\dfrac{\partial}{\partial x_1} - \dfrac{u + vy_1^2}{u(u+2tv)}\dfrac{\partial}{\partial x_2}.
\end{cases}
\tag{27}
$$

Let the function u be defined as $u(x_1, x_2, y_1, y_2) = \frac{1+\sqrt{1+4t}}{2}$. Therefore, the function v becomes $v(x_1, x_2, y_1, y_2) = 1$. Then, for the metric G and the complex structure J, there exists a tensor field K such that $(\mathbb{R}^4, \tilde{\nabla} := \nabla^G + K, \tilde{g} := G, J)$ is a special Kähler manifold [46]. Notice that a holomorphic statistical structure of holomorphic curvature 0 is nothing but a special Kähler manifold [43].

In this respect, define a $(1, 2)$-tensor field K on $\mathbb{R}^4$:

$$K = \sum_{i,j,l=1}^{4} k_{ij}^{l} \frac{\partial}{\partial x^l} \otimes dx^i \otimes dx^j. \tag{28}$$

Let $\alpha_1, \ldots, \alpha_7$ be functions on $\mathbb{R}^4$ and denote $p := u + vy_1^2$, $q := u + vy_2^2$, $r := u + 2tv$, $s := vy_1 y_2$. Suppose that α_2 has the expression

$$\alpha_2 = \frac{1}{2}s(u_{y_1} + 2y_1) + \frac{1}{2}qu_{y_2}. \tag{29}$$

Moreover, α_1 and α_3 satisfy the equation

$$\left(\alpha_2 \frac{q}{sur} - \alpha_3 \frac{1}{ur} - \alpha_1 \frac{q}{s}\right)\frac{pur}{q} + \alpha_1 \frac{sur}{q} + \alpha_2 \frac{s}{q} = \frac{1}{2}p(u_{y_1} + 2y_1) + \frac{1}{2}su_{y_2}, \tag{30}$$

where $u_{y_1} := \frac{\partial u}{\partial y_1}$ and $u_{y_2} := \frac{\partial u}{\partial y_2}$.

If K performs the conditions in Equations (10)–(12) and also the conditions in Equations (29), (30), then we get $(\mathbb{R}^4, \tilde{\nabla} := \nabla^G + K, \tilde{g} := G, J)$ a special Kähler manifold [46] with K constructed as follows:

$$k_{14}^1 = k_{41}^1 = k_{13}^2 = k_{31}^2 = -k_{34}^3 = -k_{43}^3 = \alpha_1, \quad k_{11}^4 = k_{12}^3 = k_{21}^3 = \alpha_2, \quad k_{12}^4 = k_{21}^4 = k_{22}^3 = \alpha_3,$$

$$k_{24}^1 = k_{42}^1 = k_{23}^2 = k_{32}^2 = -k_{34}^4 = -k_{43}^4 = \alpha_4, \quad k_{22}^2 = -k_{24}^4 = -k_{42}^4 = \alpha_5,$$

$$k_{11}^1 = k_{12}^1 = k_{21}^1 = -k_{23}^3 = -k_{32}^3 = 0, \ k_{12}^2 = k_{21}^2 = -k_{24}^3 = -k_{42}^3 = \alpha_7 \frac{q}{s},$$

$$k_{33}^1 = \alpha_6, \ k_{11}^2 = k_{14}^3 = k_{41}^3 = 0, \ k_{14}^2 = k_{41}^2 = -\alpha_2 \frac{s}{urq} + \alpha_3 \frac{p}{urq} - \alpha_1 \frac{s}{q},$$

$$k_{23}^1 = k_{32}^1 = \alpha_2 \frac{q}{urp} - \alpha_4 \frac{s}{p} - \alpha_3 \frac{s}{urp}, \ k_{11}^3 = \alpha_1 \frac{2s^4 - u^2 r^2}{sq} + \alpha_2 \frac{ur + 2s^2}{sq} - \alpha_3 \frac{p}{q},$$

$$k_{22}^4 = -\alpha_2 \frac{q}{p} - \alpha_4 \frac{u^2 r^2}{sp} + \alpha_3 \frac{ur + 2s^2}{sp}, \ k_{13}^1 = k_{31}^1 = \alpha_2 \frac{q}{sur} - \alpha_3 \frac{1}{ur} - \alpha_1 \frac{q}{s},$$

$$k_{33}^1 = -\alpha_2 \frac{q}{urp} + \alpha_3 \frac{s}{urp} + \alpha_4 \frac{s}{p}, \ k_{44}^1 = -k_{24}^2 = -k_{42}^2 = \alpha_2 \frac{1}{ur} - \alpha_3 \frac{p}{sur} + \alpha_4 \frac{p}{s},$$

$$k_{44}^3 = \alpha_2 \frac{s}{qur} - \alpha_3 \frac{p}{qur} + \alpha_1 \frac{s}{q}, \ k_{33}^3 = -\alpha_2 \frac{q}{sur} + \alpha_3 \frac{1}{ur} + \alpha_1 \frac{q}{s},$$

$$k_{22}^1 = -k_{23}^4 = -k_{32}^4 = -\alpha_5 \frac{s}{p},$$

$$k_{34}^1 = k_{43}^1 = -\alpha_6 \frac{s}{q} - \alpha_5 \frac{s^2}{urpq},$$

$$k_{44}^1 = \alpha_6 \frac{s^2}{q^2} + \alpha_5 \frac{s(2s^2 + ur)}{urpq^2},$$

$$k_{13}^3 = k_{31}^3 = k_{14}^4 = k_{41}^4 = 0,$$

$$k_{13}^4 = k_{31}^4 = 0, \ k_{33}^2 = -\alpha_6 \frac{p}{s},$$

$$k_{34}^2 = k_{43}^2 = \alpha_6 \frac{p}{q} + \alpha_5 \frac{s}{urq},$$

$$k_{44}^2 = -\alpha_5 \frac{pq + s^2}{urq^2} - \alpha_6 \frac{sp}{q^2}.$$

Then, $\tilde{M} = (\mathbb{R}^4, \tilde{\nabla} := \nabla^G + K, G, J)$ is a holomorphic statistical manifold of holomorphic curvature 0.

Next, let M be any m-dimensional submanifold ($m < 4$) of $\tilde{M}$. Then, Inequalities (16) and (17) are satisfied. Moreover, the statistical submanifold M of $\tilde{M}$ attains equality in both these inequalities, provided that M is totally geodesic.

5. Conclusions

In this research study, we provided new solutions to the fundamental problem of finding simple relationships between various invariants (intrinsic and extrinsic) of the submanifolds. In this respect, we obtained inequalities involving the normalized δ-Casorati curvatures (extrinsic invariants) and the scalar curvature (intrinsic invariant) of statistical submanifolds in holomorphic statistical manifolds with constant holomorphic sectional curvature. In addition, we characterized the equality cases. These results may stimulate new research aimed at obtaining similar relationships in terms of various invariants, for statistical submanifolds in other ambient spaces.

Author Contributions: All authors have contributed equally to the study and preparation of the article. Conceptualization, all authors; Methodology, all authors; Validation, all authors; Investigation, all authors; Writing–original draft preparation, all authors; Writing–review and editing, all authors. All authors have read and agreed to the published version of the manuscript.

Funding: This research received no external funding.

Conflicts of Interest: The authors declare no conflict of interest.

References

1. Chen, B.Y. Riemannian submanifolds. In *Handbook of Differential Geometry*; Dillen, F., Verstraelen, L., Eds.; Elsevier North-Holland: Amsterdam, The Netherlands, 2000; Volume 1, pp. 187–418.
2. Chen, B.Y. Strings of Riemannian invariants, inequalities, ideal immersions and their applications. In Proceedings of the Third Pacific Rim Geometry Conference, Seoul, South Korea, 4–7 December 1996; International Press: Cambridge, UK, 1998; pp. 7–60.
3. Chen, B.Y. Some pinching and classification theorems for minimal submanifolds. *Arch. Math.* **1993**, *60*, 569–578. [CrossRef]
4. Chen, B.Y. A general inequality for submanifolds in complex-space-forms and its applications. *Arch. Math.* **1996**, *67*, 519–528. [CrossRef]
5. Oiagă, A.; Mihai, I. B.Y. Chen inequalities for slant submanifolds in complex space forms. *Demonstr. Math.* **1999**, *32*, 835–846. [CrossRef]
6. Oprea, T. Chen's inequality in Lagrangian case. *Colloq. Math.* **2007**, *108*, 163–169. [CrossRef]
7. Chen, B.Y. *Pseudo-Riemannian Geometry, δ-Invariants and Applications*; World Scientific: Singapore, 2011.
8. Chen, B.Y.; Dillen, F.; der Veken, J.V.; Vrancken, L. Classification of $\delta(2, n-2)$-ideal Lagrangian submanifolds in n-dimensional complex space forms. *J. Math. Anal. Appl.* **2018**, *458*, 1456–1485. [CrossRef]
9. Decu, S.; Haesen, S.; Verstraelen, L. Optimal inequalities involving Casorati curvatures. *Bull. Transilv. Univ. Braşov Ser. B Suppl.* **2007**, *14*, 85–93.
10. Decu, S.; Haesen, S.; Verstraelen, L. Optimal inequalities characterising quasi-umbilical submanifolds. *J. Inequal. Pure Appl. Math.* **2008**, *9*.
11. Casorati, F. Mesure de la courbure des surfaces suivant l'idée commune. *Acta Math.* **1890**, *14*, 95–110. [CrossRef]
12. Koenderink, J.J. *Shadows of Shapes*; De Clootcrans Press: Utrecht, The Netherlands, 2012.
13. Verstraelen, P. A geometrical description of visual perception-The Leuven Café Erasmus model and Bristol Café Wall illusion. *Kragujevac J. Math.* **2005**, *28*, 7–17.
14. Haesen, S.; Kowalczyk, D.; Verstraelen, L. On the extrinsic principal directions of Riemannian submanifolds. *Note Mat.* **2009**, *29*, 41–53.
15. Decu, S.; Verstraelen, L. A note of the isotropical geometry of production surfaces. *Kragujevac J. Math.* **2013**, *37*, 217–220.
16. He, G.; Liu, H.; Zhang, L. Optimal inequalities for the Casorati curvatures of submanifolds in generalized space forms endowed with semi-symmetric non-metric connections. *Symmetry* **2016**, *8*, 113. [CrossRef]
17. Lee, C.W.; Yoon, D.W.; Lee, J.W. Optimal inequalities for the Casorati curvatures of submanifolds of real space forms endowed with semi-symmetric metric connections. *J. Inequal. Appl.* **2014**, *2014*, 327. [CrossRef]
18. Park, K. Inequalities for the Casorati curvatures of real hypersurfaces in some Grassmannians. *Taiwan. J. Math.* **2018**, *22*, 63–77. [CrossRef]
19. Zhang, P.; Zhang, L. Casorati Inequalities for Submanifolds in a Riemannian Manifold of Quasi-Constant Curvature with a Semi-Symmetric Metric Connection. *J. Inequal. Appl.* **2016**, *8*, 19. [CrossRef]
20. Zhang, P.; Zhang, L. Inequalities for Casorati curvatures of submanifolds in real space forms. *Adv. Geom.* **2016**, *16*, 329–335. [CrossRef]
21. Lee, C.; Lee, J.; Vîlcu, G.E. Optimal inequalities for the normalized δ-Casorati curvatures of submanifolds in Kenmotsu space forms. *Adv. Geom.* **2017**, *17*, 355–362. [CrossRef]
22. Vîlcu, G. An optimal inequality for Lagrangian submanifolds in complex space forms involving Casorati curvature. *J. Math. Anal. Appl.* **2018**, *465*, 1209–1222. [CrossRef]
23. Aquib, M.; Lee, J.; Vîlcu, G.E.; Yoon, D. Classification of Casorati ideal Lagrangian submanifolds in complex space forms. *Differ. Geom. Appl.* **2019**, *63*, 30–49. [CrossRef]
24. Suceavă, B.; Vajiac, M. Estimates of B.-Y. Chen's $\mathring{\delta}$-Invariant in Terms of Casorati Curvature and Mean Curvature for Strictly Convex Euclidean Hypersurfaces. *Int. Electron. J. Geom.* **2019**, *12*, 26–31.
25. Brubaker, N.; Suceavă, B. A Geometric Interpretation of Cauchy-Schwarz inequality in terms of Casorati Curvature. *Int. Electron. J. Geom.* **2018**, *11*, 48–51.
26. Amari, S. *Differential-Geometrical Methods in Statistics*; Lecture Notes in Statistics; Berger, J., Fienberg, S., Gani, J., Krickeberg, K., Olkin, I., Singer, B., Eds.; Springer: Berlin, Germany, 1985; Volume 28.

27. Boyom, N. Foliations-Webs-Hessian Geometry-Information Geometry-Entropy and Cohomology. *Entropy* **2016**, *18*, 433. [CrossRef]
28. Cheng, Y.; Wang, X.; Moran, B. Optimal Nonlinear Estimation in Statistical Manifolds with Application to Sensor Network Localization. *Entropy* **2017**, *19*, 308. [CrossRef]
29. Zhang, J. Nonparametric Information Geometry: From Divergence Function to Referential-Representational Biduality on Statistical Manifolds. *Entropy* **2013**, *15*, 5384–5418. [CrossRef]
30. Cuingnet, R.; Glaunès, J.; Chupin, M.; Benali, H.; Colliot, O. Anatomical regularization on Statistical Manifolds for the Classification of Patients with Alzheimer's Disease. In *Machine Learning in Medical Imaging. MLMI 2011*; Suzuki, K., Wang, F., Shen, D., Yan, P., Eds.; Lecture Notes in Comput. Sci.; Springer: Berlin/Heidelber, Germany, 2011; Volume 7009.
31. Aydin, M.; Mihai, A.; Mihai, I. Some inequalities on submanifolds in statistical manifolds of constant curvature. *Filomat* **2015**, *29*, 465–477. [CrossRef]
32. Aydin, M.; Mihai, A.; Mihai, I. Generalized Wintgen inequality for statistical submanifolds in statistical manifolds of constant curvature. *Bull. Math. Sci.* **2017**, *7*, 155–166. [CrossRef]
33. Lee, C.; Lee, J. Inequalities on Sasakian Statistical Manifolds in Terms of Casorati Curvatures. *Mathematics* **2018**, *6*, 259. [CrossRef]
34. Aquib, M.; Shahid, M. Generalized normalized δ-Casorati curvature for statistical submanifolds in quaternion Kaehler-like statistical space forms. *J. Geom.* **2018**, *109*, 13. [CrossRef]
35. Vîlcu, A.; Vîlcu, G. Statistical manifolds with almost quaternionic structures and quaternionic Kaehler-like statistical submersions. *Entropy* **2015**, *17*, 6213–6228. [CrossRef]
36. Aytimur, H.; Kon, M.; Mihai, A.; Ozgur, C.; Takano, K. Chen Inequalities for Statistical Submanifolds of Kaehler-Like Statistical Manifolds. *Mathematics* **2019**, *7*, 1202. [CrossRef]
37. Aquib, M.; Boyom, M.; Shahid, M.; Vîlcu, G. The First Fundamental Equation and Generalized Wintgen-Type Inequalities for Submanifolds in Generalized Space Forms. *Mathematics* **2019**, *7*, 1151. [CrossRef]
38. Chen, B.Y.; Mihai, A.; Mihai, I. A Chen First Inequality for Statistical Submanifolds in Hessian Manifolds of Constant Hessian Curvatures. *Results Math.* **2019**, *74*, 165. [CrossRef]
39. Siddiqui, A.; Chen, B.Y.; Bahadir, O. Statistical Solitons and Inequalities for Statistical Warped Product Submanifolds. *Mathematics* **2019**, *7*, 797. [CrossRef]
40. Furuhata, H.; Hasegawa, I. Submanifold theory in holomorphic statistical manifolds. In *Geometry of Cauchy-Riemann Submanifolds*; Dragomir, S., Shahid, M.H., Al-Solamy, F.R., Eds.; Springer Science+Business Media: Singapore, 2016; pp. 179–214.
41. Vos, P. Fundamental equations for statistical submanifolds with applications to the Barlett correction. *Ann. Inst. Statist. Math.* **1989**, *41*, 429–450. [CrossRef]
42. Furuhata, H.; Hasegawa, I.; Okuyama, Y.; Sato, K. Kenmotsu statistical manifolds and warped product. *J. Geom.* **2017**, *108*, 1175–1191. [CrossRef]
43. Furuhata, H. Hypersurfaces in statistical manifolds. *Differ. Geom. Appl.* **2009**, *27*, 420–429. [CrossRef]
44. Oprea, T. *Constrained Extremum Problems in Riemannian Geometry*; University of Bucharest Publishing House: Bucharest, Romania, 2006.
45. Oproiu, V. Some new geometric structures on the tangent bundle. *Publ. Math. Debrecen* **1999**, *55*, 261–281.
46. Milijevic, M. Totally real statistical submanifolds. *Interdiscip. Inform. Sci.* **2015**, *21*, 87–96. [CrossRef]

© 2020 by the authors. Licensee MDPI, Basel, Switzerland. This article is an open access article distributed under the terms and conditions of the Creative Commons Attribution (CC BY) license (http://creativecommons.org/licenses/by/4.0/).

Article

Quantum Integral Inequalities of Simpson-Type for Strongly Preinvex Functions

Yongping Deng [1], Muhammad Uzair Awan [2] and Shanhe Wu [1,*]

1. Department of Mathematics, Longyan University, Longyan 364012, China
2. Department of Mathematics, Government College University, Faisalabad 38000, Pakistan
* Correspondence: shanhewu@gmail.com

Received: 29 June 2019; Accepted: 13 August 2019; Published: 16 August 2019

Abstract: In this paper, we establish a new q-integral identity, the result is then used to derive two q-integral inequalities of Simpson-type involving strongly preinvex functions. Some special cases of the obtained results are also considered, it is shown that several new and previously known results can be derived via generalized strongly preinvex functions and quantum integrals.

Keywords: q-integral inequality; strongly preinvex function; Simpson inequality; quantum integral

MSC: 26D10; 26D15; 26A51; 34A08

1. Introduction

Quantum calculus or q-calculus is often known as "calculus without limits" and was first developed by Jackson in the early twentieth century, but the history of quantum calculus can be traced back to some much earlier work done by Euler and Jacobi et al. (see [1]). Over the recent decade, the investigation of q-calculus has attracted the interest of many researchers, because it has been found to have a lot of applications in mathematics and physics. As is known to us, q-calculus can be treated as a bridge between mathematics and physics, it is a significant tool for researchers working in analytic number theory, noncommutative geometry, or theoretical physics. In quantum calculus, we obtain the q-analogues of mathematical objects which can be recaptured as $q \to 1^-$. It has been noticed that quantum calculus is a subfield of timescale calculus. Timescale calculus provides a unified framework for studying dynamic equations on both the discrete and continuous domains. In quantum calculus, we are concerned with a specific timescale, called the q-timescale (see [1–4]).

The concept of convexity has been extended in several directions, since these generalized versions have significant applications in different fields of pure and applied sciences. We only point out that convexity was recently used in differential geometry to completely classify ideal Casorati submanifolds in complex space forms (see [5–8]). One of the convincing examples on extensions of convexity is the introduction of invex function, which was introduced by Hanson [9]. This concept is particularly interesting from an optimization viewpoint, since it provides a broader setting to study the optimization and mathematical programming problems. Such optimization problems have recently been considered in Riemannian geometry by an original choice of a set of quadratic programming problems. Since then, some classes of generalized convex functions, such as the preinvex function, strongly α-invex function, and strongly α-preinvex function, were put forward successively, see [10–16].

In this paper, the quantum calculus and the strongly preinvex function are subtly linked together via integral inequalities. It is well known that the theory of inequality plays a fundamental role in pure and applied mathematics and has extensive applications. Apart from the larger number of research results of inequalities in classical analysis, there are considerable works on the study of inequalities for q-calculus, particularly the study of inequalities related to quantum integral (q-integral), for example, q-Hermite–Hadamard integral inequality, q-Cauchy–Schwarz integral inequality, q-Hölder integral inequality, q-Ostrowski integral inequality, etc. For more details, we refer the interested reader to [17–23] and the references cited therein.

The purpose of this paper is to establish several q-integral inequalities of Simpson-type via strongly preinvex functions. The classical Simpson inequality is described as follows:

$$\left|\frac{1}{6}\left[\phi(\alpha)+4\phi\left(\frac{\alpha+\beta}{2}\right)+\phi(\beta)\right]-\frac{1}{\beta-\alpha}\int_{\alpha}^{\beta}\phi(v)dv\right|\leq\frac{1}{1280}\left\|\phi^{(4)}\right\|_{\infty}(\beta-\alpha)^{4}, \quad (1)$$

where the mapping $\phi : [\alpha, \beta] \to \mathbb{R}$ is four times continuously differentiable, and $\|\phi^{(4)}\|_\infty = \sup_{v\in(\alpha,\beta)} |\phi^{(4)}(v)| < \infty$ (see [24]).

The paper is organized as follows: In Sections 2 and 3, we shall introduce some notions and properties on strongly preinvex functions and q-calculus. As an auxiliary result, we present an identity associated with q-integral. In Section 4, with the help of the auxiliary result, we will establish our main results. At the end of the paper, some examples are provided to illustrate the applications of our main results.

2. Preliminaries

Let us recall some preliminary concepts and results.

Definition 1 ([15]). *A set $K_\eta \subset \mathbb{R}^n$ is said to be invex with respect to bifunction $\eta(.,.): \mathbb{R}^n \times \mathbb{R}^n \to \mathbb{R}^n$, if*

$$u + \lambda \eta(v,u) \in K_\eta, \quad \forall\ u,v \in K_\eta,\ \lambda \in [0,1].$$

Definition 2 ([15]). *A function ϕ on the invex set $K_\eta \subset \mathbb{R}^n$ is said to be preinvex with respect to bifunction $\eta(.,.): \mathbb{R}^n \times \mathbb{R}^n \to \mathbb{R}^n$, if*

$$\phi(u + \lambda \eta(v,u)) \leq (1-\lambda)\phi(u) + \lambda\phi(v), \quad \forall u,v \in K_\eta,\ \lambda \in [0,1].$$

Definition 3 ([16]). *A function ϕ on the invex set $K_\eta \subset \mathbb{R}^n$ is said to be strongly preinvex with respect to bifunction $\eta(.,.): \mathbb{R}^n \times \mathbb{R}^n \to \mathbb{R}^n$, and modulus $\mu > 0$, if*

$$\phi(u + \lambda \eta(v,u)) \leq (1-\lambda)\phi(u) + \lambda\phi(v) - \mu\lambda(1-\lambda)\eta^2(v,u), \quad \forall u,v \in K_\eta,\ \lambda \in [0,1].$$

Here, we introduce a new definition which combines the preinvex functions and the strongly preinvex functions given above.

Definition 4. *A function ϕ on the invex set $K_\eta \subset \mathbb{R}^n$ is said to be generalized strongly preinvex with respect to bifunction $\eta(.,.): \mathbb{R}^n \times \mathbb{R}^n \to \mathbb{R}^n$ and modulus $\mu \geq 0$, if*

$$\phi(u + \lambda \eta(v,u)) \leq (1-\lambda)\phi(u) + \lambda\phi(v) - \mu\lambda(1-\lambda)\eta^2(v,u), \quad \forall u,v \in K_\eta,\ \lambda \in [0,1].$$

Clearly, if $\mu = 0$, then the class of generalized strongly preinvex functions reduces to the class of preinvex functions as defined in Definition 2.

In the following, we recall some basic properties of q-calculus.

Let $J = [a,b] \subseteq \mathbb{R}$ be an interval and $0 < q < 1$ be a constant. The q-derivative of a function $\phi : J \to \mathbb{R}$ at a point $u \in J$ on $[a,b]$ is defined as follows:

Definition 5 ([25]). *Let $\phi : J = [a,b] \to \mathbb{R}$ be a continuous function and let $u \in J$. Then, the q-derivative of ϕ on J at u is defined as*

$$_aD_q\phi(u) = \frac{\phi(u) - \phi(qu + (1-q)a)}{(1-q)(u-a)}, \quad u \neq a. \quad (2)$$

Definition 6 ([25]). *Let $\phi : J = [a,b] \to \mathbb{R}$ is a continuous function. A second-order q-derivative on J, which is denoted as ${}_aD_q^2\phi$, provided that ${}_aD_q\phi$ is q-differentiable on J with ${}_aD_q^2\phi = {}_aD_q({}_aD_q\phi) : J \to \mathbb{R}$. Similarly higher order q-derivative on J is defined by ${}_aD_q^n\phi : J \to \mathbb{R}$.*

In [25], Tariboon and Ntouyas defined the q-integral as follows:

Definition 7 ([25]). *Let $\phi : J = [a,b] \to \mathbb{R}$ be a continuous function. Then, the q-integral on J is defined as:*

$$\int_a^u \phi(v) \, {}_ad_qv = (1-q)(u-a)\sum_{n=0}^{\infty} q^n \phi(q^n u + (1-q^n)a), \tag{3}$$

for $u \in J$.

The following results are useful in the computation of q-integral in subsequent section.

Proposition 1 ([25]). *Let $f, g : J = [a,b] \to \mathbb{R}$ be continuous functions, $c \in \mathbb{R}$. Then, for $x \in J$,*

$$\int_a^x (f(v) + g(v)) \, {}_ad_qv = \int_a^x f(v) \, {}_ad_qv + \int_a^x f(v) \, {}_ad_qv,$$

$$\int_a^x cf(v) \, {}_ad_qv = c\int_a^x f(v) \, {}_ad_qv,$$

$$\int_\xi^x f(v) \, {}_ad_qv = \int_a^x f(v) \, {}_ad_qv - \int_a^\xi f(v) \, {}_ad_qv, \ \xi \in (a,x).$$

Proposition 2 ([25]). *For q-integral, we have the following identities*

$$\int_a^u 1 \, {}_ad_qv = u - a,$$

$$\int_a^u v \, {}_ad_qv = \frac{(u-a)(u+qa)}{1+q},$$

$$\int_a^u (v-a)^\tau \, {}_ad_qv = \left(\frac{1-q}{1-q^{\tau+1}}\right)(u-a)^{\tau+1}, \ \tau \neq -1,$$

$$\int_\xi^u (v-\xi) \, {}_ad_qv = \frac{u^2 - (1+q)u\xi + q\xi^2}{1+q} - \frac{a(1-q)(u-\xi)}{1+q}, \ \xi \in (a,u).$$

3. A Key Lemma

In this section, we present an identity associated with q-integral, which plays an important role in establishing our main results.

Lemma 1. *Let $f : I = [a, a + \eta(b,a)] \to \mathbb{R}$ be a q-differentiable function on I with $\eta(b,a) > 0$. If ${}_aD_q f$ is integrable on I and $0 < q < 1$, then*

$$\frac{1}{6}\left[f(a) + 4f\left(\frac{2a + \eta(b,a)}{2}\right) + f(a + \eta(b,a))\right] - \frac{1}{\eta(b,a)}\int_a^{a+\eta(b,a)} f(t) \, {}_ad_qt$$

$$= \eta(b,a)\int_0^1 \Psi(t,q) \, {}_aD_q f(a + t\eta(b,a)) \, {}_0d_qt, \tag{4}$$

where

$$\Psi(t,q) = \begin{cases} qt - \dfrac{1}{6}, & \text{if } 0 \le t < \dfrac{1}{2}, \\ qt - \dfrac{5}{6}, & \text{if } \dfrac{1}{2} \le t \le 1. \end{cases}$$

Proof. Let

$$Q_1 = \int_0^{\frac{1}{2}} \left(qt - \frac{1}{6}\right) {}_aD_q f(a + t\eta(b,a)) \, {}_0d_q t,$$

$$Q_2 = \int_{\frac{1}{2}}^1 \left(qt - \frac{5}{6}\right) {}_aD_q f(a + t\eta(b,a)) \, {}_0d_q t,$$

then

$$\int_0^1 \Psi(t,q) \, {}_aD_q f(a + t\eta(b,a)) \, {}_0d_q t = Q_1 + Q_2.$$

Utilizing the Definitions 5 and 7, and the properties of q-derivative and q-integral described in Propositions 1, a direct computation gives

$$Q_1 = \int_0^{\frac{1}{2}} qt \, {}_aD_q f(a + t\eta(b,a)) \, {}_0d_q t - \int_0^{\frac{1}{2}} \frac{1}{6} \, {}_aD_q f(a + t\eta(b,a)) \, {}_0d_q t$$

$$= \int_0^{\frac{1}{2}} q \frac{f(a + t\eta(b,a)) - f(a + qt\eta(b,a))}{(1-q)\eta(b,a)} \, {}_0d_q t$$

$$- \frac{1}{6} \int_0^{\frac{1}{2}} \frac{f(a + t\eta(b,a)) - f(a + qt\eta(b,a))}{(1-q)t\eta(b,a)} \, {}_0d_q t$$

$$= \frac{1}{2} \sum_{n=0}^{\infty} q^{n+1} \frac{f\left(\frac{2a + q^n \eta(b,a)}{2}\right) - f\left(\frac{2a + q^{n+1} \eta(b,a)}{2}\right)}{\eta(b,a)}$$

$$- \frac{1}{6} \sum_{n=0}^{\infty} \frac{f\left(\frac{2a + q^n \eta(b,a)}{2}\right) - f\left(\frac{2a + q^{n+1} \eta(b,a)}{2}\right)}{\eta(b,a)}$$

$$= \frac{1}{2}\left[\frac{f\left(\frac{2a + \eta(b,a)}{2}\right)}{\eta(b,a)} - \sum_{n=0}^{\infty}(1-q)q^n \frac{f\left(\frac{2a + q^n \eta(b,a)}{2}\right)}{\eta(b,a)}\right] - \frac{1}{6} \cdot \frac{f\left(\frac{2a + \eta(b,a)}{2}\right) - f(a)}{\eta(b,a)}$$

$$= \frac{1}{3} \cdot \frac{f\left(\frac{2a + \eta(b,a)}{2}\right)}{\eta(b,a)} + \frac{1}{6} \cdot \frac{f(a)}{\eta(b,a)} - \frac{1}{2}\sum_{n=0}^{\infty}(1-q)q^n \frac{f\left(\frac{2a + q^n \eta(b,a)}{2}\right)}{\eta(b,a)}$$

$$= \frac{1}{3} \cdot \frac{f\left(\frac{2a + \eta(b,a)}{2}\right)}{\eta(b,a)} + \frac{1}{6} \cdot \frac{f(a)}{\eta(b,a)} - \frac{1}{\eta(b,a)} \int_0^{\frac{1}{2}} f(a + t\eta(b,a)) \, {}_0d_q t.$$

On the other hand, one has

$$Q_2 = \int_{\frac{1}{2}}^1 qt \, {}_aD_q f(a + t\eta(b,a)) \, {}_0d_q t - \int_{\frac{1}{2}}^1 \frac{5}{6} \, {}_aD_q f(a + t\eta(b,a)) \, {}_0d_q t$$

$$= \int_0^1 qt \, {}_aD_q f(a + t\eta(b,a)) \, {}_0d_q t - \int_0^1 \frac{5}{6} \, {}_aD_q f(a + t\eta(b,a)) \, {}_0d_q t$$

$$- \left(\int_0^{\frac{1}{2}} qt \, {}_aD_q f(a + t\eta(b,a)) \, {}_0d_q t - \int_0^{\frac{1}{2}} \frac{5}{6} \, {}_aD_q f(a + t\eta(b,a)) \, {}_0d_q t\right).$$

Since

$$\int_0^1 qt \, {}_aD_q f(a + t\eta(b,a)) \, {}_0d_q t - \int_0^1 \frac{5}{6} \, {}_aD_q f(a + t\eta(b,a)) \, {}_0d_q t$$

$$= \int_0^1 q \frac{f(a + t\eta(b,a)) - f(a + qt\eta(b,a))}{(1-q)\eta(b,a)} \, {}_0d_q t$$

$$-\frac{5}{6}\int_0^1 \frac{f(a+t\eta(b,a))-f(a+qt\eta(b,a))}{(1-q)t\eta(b,a)}\,_0d_q t$$

$$=\sum_{n=0}^{\infty} q^{n+1}\frac{f(a+q^n\eta(b,a))-f(a+q^{n+1}\eta(b,a))}{\eta(b,a)}$$

$$-\frac{5}{6}\sum_{n=0}^{\infty}\frac{f(a+q^n\eta(b,a))-f(a+q^{n+1}\eta(b,a))}{\eta(b,a)}$$

$$=\frac{f(a+\eta(b,a))}{\eta(b,a)}-\sum_{n=0}^{\infty}(1-q)q^n\frac{f(a+q^n\eta(b,a))}{\eta(b,a)}-\frac{5}{6}\cdot\frac{f(a+\eta(b,a))-f(a)}{\eta(b,a)}$$

$$=\frac{1}{6}\cdot\frac{f(a+\eta(b,a))}{\eta(b,a)}+\frac{5}{6}\cdot\frac{f(a)}{\eta(b,a)}-\sum_{n=0}^{\infty}(1-q)q^n\frac{f(a+q^n\eta(b,a))}{\eta(b,a)}$$

$$=\frac{1}{6}\cdot\frac{f(a+\eta(b,a))}{\eta(b,a)}+\frac{5}{6}\cdot\frac{f(a)}{\eta(b,a)}-\frac{1}{\eta(b,a)}\int_0^1 f(a+t\eta(b,a))\,_0d_q t$$

and

$$\int_0^{\frac{1}{2}} qt\,_aD_q f(a+t\eta(b,a))\,_0d_q t - \int_0^{\frac{1}{2}}\frac{5}{6}\,_aD_q f(a+t\eta(b,a))\,_0d_q t$$

$$=\frac{1}{2}\left[\frac{f\left(\frac{2a+\eta(b,a)}{2}\right)}{\eta(b,a)}-\sum_{n=0}^{\infty}(1-q)q^n\frac{f\left(\frac{2a+q^n\eta(b,a)}{2}\right)}{\eta(b,a)}\right]-\frac{5}{6}\cdot\frac{f\left(\frac{2a+\eta(b,a)}{2}\right)-f(a)}{\eta(b,a)}$$

$$=-\frac{1}{3}\cdot\frac{f\left(\frac{2a+\eta(b,a)}{2}\right)}{\eta(b,a)}+\frac{5}{6}\cdot\frac{f(a)}{\eta(b,a)}-\frac{1}{\eta(b,a)}\int_0^{\frac{1}{2}} f(a+t\eta(b,a))\,_0d_q t,$$

we obtain

$$Q_2 = \frac{1}{6}\cdot\frac{f(a+\eta(b,a))}{\eta(b,a)}+\frac{1}{3}\cdot\frac{f\left(\frac{2a+\eta(b,a)}{2}\right)}{\eta(b,a)}-\frac{1}{\eta(b,a)}\int_0^1 f(a+t\eta(b,a))\,_0d_q t$$

$$+\frac{1}{\eta(b,a)}\int_0^{\frac{1}{2}} f(a+t\eta(b,a))\,_0d_q t.$$

Thus,

$$\int_0^1 \Psi(t,q)\,_aD_q f(a+t\eta(b,a))\,_0d_q t = Q_1 + Q_2$$

$$=\frac{1}{6}\cdot\frac{f(a+\eta(b,a))+f(a)}{\eta(b,a)}+\frac{2}{3}\cdot\frac{f\left(\frac{2a+\eta(b,a)}{2}\right)}{\eta(b,a)}-\frac{1}{\eta(b,a)}\int_0^1 f(a+t\eta(b,a))\,_0d_q t$$

$$=\frac{1}{6}\cdot\frac{f(a+\eta(b,a))+f(a)}{\eta(b,a)}+\frac{2}{3}\cdot\frac{f\left(\frac{2a+\eta(b,a)}{2}\right)}{\eta(b,a)}-\frac{1}{\eta^2(b,a)}\int_a^{a+\eta(b,a)} f(t)\,_ad_q t,$$

which leads to the desired identity (4). The proof of Lemma 1 is complete. □

4. Main Results

We are in a position to establish the q-integral inequalities of Simpson-type for strongly preinvex functions.

Theorem 1. Let $f: I = [a, a + \eta(b,a)] \to \mathbb{R}$ be a q-differentiable function on I with $\eta(b,a) > 0$. If $|{}_aD_qf|$ is an integrable and a generalized strongly preinvex function with modulus $\mu \geq 0$ and $0 < q < 1$, then

$$\left| \frac{1}{6} \left[f(a) + 4f\left(\frac{2a + \eta(b,a)}{2}\right) + f(a + \eta(b,a)) \right] - \frac{1}{\eta(b,a)} \int_a^{a+\eta(b,a)} f(t) \, {}_ad_qt \right| \tag{5}$$
$$\leq \eta(b,a) \left[(A_1(q) + A_4(q))|{}_aD_qf(a)| + (A_2(q) + A_5(q))|{}_aD_qf(b)| - \mu(A_3(q) + A_6(q))\eta^2(b,a) \right],$$

where $A_1(q), A_2(q), A_3(q), A_4(q), A_5(q)$, and $A_6(q)$ are given by

$$A_1(q) = \begin{cases} \frac{1 - 4q^3}{24(1+q)(1+q+q^2)}, & 0 < q < \frac{1}{3}, \\ \frac{1 + 12q + 12q^2 + 36q^3}{216(1+q)(1+q+q^2)}, & \frac{1}{3} \leq q < 1, \end{cases}$$

$$A_2(q) = \begin{cases} \frac{1 - 2q - 2q^2}{24(1+q)(1+q+q^2)}, & 0 < q < \frac{1}{3}, \\ \frac{18q^2 + 18q - 7}{216(1+q)(1+q+q^2)}, & \frac{1}{3} \leq q < 1, \end{cases}$$

$$A_3(q) = \begin{cases} \frac{1 - 2q - 2q^3 - 4q^4}{48(1+q)(1+q^2)(1+q+q^2)}, & 0 < q < \frac{1}{3}, \\ \frac{108q^4 + 54q^3 + 12q^2 + 54q - 17}{1296(1+q)(1+q^2)(1+q+q^2)}, & \frac{1}{3} \leq q < 1, \end{cases}$$

$$A_4(q) = \begin{cases} \frac{-5 + 8q + 8q^2 - 8q^3}{24(1+q)(1+q+q^2)}, & 0 < q < \frac{5}{6}, \\ \frac{12q^2 + 12q + 5}{216(1+q)(1+q+q^2)}, & \frac{5}{6} \leq q < 1, \end{cases}$$

$$A_5(q) = \begin{cases} \frac{5 - 2q - 2q^2}{8(1+q)(1+q+q^2)}, & 0 < q < \frac{5}{6}, \\ \frac{18q^2 + 18q + 25}{216(1+q)(1+q+q^2)}, & \frac{5}{6} \leq q < 1, \end{cases}$$

$$A_6(q) = \begin{cases} \frac{5 - 2q + 28q^2 - 2q^3 - 12q^4}{48(1+q)(1+q^2)(1+q+q^2)}, & 0 < q < \frac{5}{6}, \\ \frac{108q^4 - 54q^3 + 96q^2 - 54q + 115}{1296(1+q)(1+q^2)(1+q+q^2)}, & \frac{5}{6} \leq q < 1. \end{cases}$$

Proof. Using Lemma 1 and the assumption condition that $|{}_aD_qf|$ is a generalized strongly preinvex function, we have

$$\left| \frac{1}{6} \left[f(a) + 4f\left(\frac{2a + \eta(b,a)}{2}\right) + f(a + \eta(b,a)) \right] - \frac{1}{\eta(b,a)} \int_a^{a+\eta(b,a)} f(t) \, {}_ad_qt \right|$$
$$= \left| \eta(b,a) \int_0^1 \Psi(t,q) \, {}_aD_qf(a + t\eta(b,a)) \, {}_0d_qt \right|$$

$$
\begin{aligned}
&= \eta(b,a) \left| \int_0^{\frac{1}{2}} \left(qt - \frac{1}{6}\right) {}_aD_qf(a+t\eta(b,a))\, {}_0d_qt \right.\\
&\quad + \left. \int_{\frac{1}{2}}^1 \left(qt - \frac{5}{6}\right) {}_aD_qf(a+t\eta(b,a))\, {}_0d_qt \right| \\
&\leq \eta(b,a) \left[\int_0^{\frac{1}{2}} \left|qt - \frac{1}{6}\right| |{}_aD_qf(a+t\eta(b,a))|\, {}_0d_qt \right.\\
&\quad + \left. \int_{\frac{1}{2}}^1 \left|qt - \frac{5}{6}\right| |{}_aD_qf(a+t\eta(b,a))|\, {}_0d_qt \right] \\
&\leq \eta(b,a) \left[\int_0^{\frac{1}{2}} \left|qt - \frac{1}{6}\right| \Big((1-t)|{}_aD_qf(a)| + t|{}_aD_qf(b)| - \mu t(1-t)\eta^2(b,a)\Big)\, {}_0d_qt \right.\\
&\quad + \left. \int_{\frac{1}{2}}^1 \left|qt - \frac{5}{6}\right| \Big((1-t)|{}_aD_qf(a)| + t|{}_aD_qf(b)| - \mu t(1-t)\eta^2(b,a)\Big)\, {}_0d_qt \right] \\
&= \eta(b,a) \left[|{}_aD_qf(a)| \left(\int_0^{\frac{1}{2}} (1-t)\left|qt - \frac{1}{6}\right| {}_0d_qt + \int_{\frac{1}{2}}^1 (1-t)\left|qt - \frac{5}{6}\right| {}_0d_qt \right) \right.\\
&\quad + |{}_aD_qf(b)| \left(\int_0^{\frac{1}{2}} t\left|qt - \frac{1}{6}\right| {}_0d_qt + \int_{\frac{1}{2}}^1 t\left|qt - \frac{5}{6}\right| {}_0d_qt \right) \\
&\quad \left. - \mu\eta^2(b,a) \left(\int_0^{\frac{1}{2}} t(1-t)\left|qt - \frac{1}{6}\right| {}_0d_qt + \int_{\frac{1}{2}}^1 t(1-t)\left|qt - \frac{5}{6}\right| {}_0d_qt \right) \right].
\end{aligned}
$$

In view of the Definitions 5 and 7, and Propositions 1 and 2, a direct calculation gives

$$
A_1(q) = \int_0^{\frac{1}{2}} (1-t)\left|qt - \frac{1}{6}\right| {}_0d_qt = \begin{cases} \frac{1-4q^3}{24(1+q)(1+q+q^2)}, & 0 < q < \frac{1}{3}, \\[4pt] \frac{1+12q+12q^2+36q^3}{216(1+q)(1+q+q^2)}, & \frac{1}{3} \leq q < 1, \end{cases}
$$

$$
A_2(q) = \int_0^{\frac{1}{2}} t\left|qt - \frac{1}{6}\right| {}_0d_qt = \begin{cases} \frac{1-2q-2q^2}{24(1+q)(1+q+q^2)}, & 0 < q < \frac{1}{3}, \\[4pt] \frac{18q^2+18q-7}{216(1+q)(1+q+q^2)}, & \frac{1}{3} \leq q < 1, \end{cases}
$$

$$
A_3(q) = \int_0^{\frac{1}{2}} t(1-t)\left|qt - \frac{1}{6}\right| {}_0d_qt = \begin{cases} \frac{1-2q-2q^3-4q^4}{48(1+q)(1+q^2)(1+q+q^2)}, & 0 < q < \frac{1}{3}, \\[4pt] \frac{108q^4+54q^3+12q^2+54q-17}{1296(1+q)(1+q^2)(1+q+q^2)}, & \frac{1}{3} \leq q < 1, \end{cases}
$$

$$
A_4(q) = \int_{\frac{1}{2}}^1 (1-t)\left|qt - \frac{5}{6}\right| {}_0d_qt = \begin{cases} \frac{-5+8q+8q^2-8q^3}{24(1+q)(1+q+q^2)}, & 0 < q < \frac{5}{6}, \\[4pt] \frac{12q^2+12q+5}{216(1+q)(1+q+q^2)}, & \frac{5}{6} \leq q < 1, \end{cases}
$$

$$A_5(q) = \int_{\frac{1}{2}}^1 t \left| qt - \frac{5}{6} \right| {}_0 d_q t = \begin{cases} \frac{5-2q-2q^2}{8(1+q)(1+q+q^2)}, & 0 < q < \frac{5}{6}, \\[6pt] \frac{18q^2+18q+25}{216(1+q)(1+q+q^2)}, & \frac{5}{6} \le q < 1, \end{cases}$$

$$A_6(q) = \int_{\frac{1}{2}}^1 t(1-t) \left| qt - \frac{5}{6} \right| {}_0 d_q t = \begin{cases} \frac{5-2q+28q^2-2q^3-12q^4}{48(1+q)(1+q^2)(1+q+q^2)}, & 0 < q < \frac{5}{6}, \\[6pt] \frac{108q^4-54q^3+96q^2-54q+115}{1296(1+q)(1+q^2)(1+q+q^2)}, & \frac{5}{6} \le q < 1. \end{cases}$$

Hence, we deduce the required inequality (5). This completes the proof of Theorem 1. □

Theorem 2. *Let $f : I = [a, a + \eta(b, a)] \to \mathbb{R}$ be a q-differentiable function on I with $\eta(b, a) > 0$. If $|{}_aD_q|^r$ is an integrable and a generalized strongly preinvex function with modulus $\mu \ge 0$, $r > 1$ and $0 < q < 1$, then*

$$\left| \frac{1}{6} \left[f(a) + 4f\left(\frac{2a + \eta(b,a)}{2} \right) + f(a + \eta(b,a)) \right] - \frac{1}{\eta(b,a)} \int_a^{a+\eta(b,a)} f(t) \, {}_a d_q t \right| \qquad (6)$$

$$\le \eta(b,a) \left[(B_1(q))^{1-\frac{1}{r}} \left(A_1(q) |{}_a D_q f(a)|^r + A_2(q) |{}_a D_q f(b)|^r - \mu A_3(q) \eta^2(b,a) \right)^{\frac{1}{r}} \right.$$

$$\left. + (B_2(q))^{1-\frac{1}{r}} \left(A_4(q) |{}_a D_q f(a)|^r + A_5(q) |{}_a D_q f(b)|^r - \mu A_6(q) \eta^2(b,a) \right)^{\frac{1}{r}} \right],$$

where

$$B_1(q) = \begin{cases} \frac{1-2q}{12(1+q)}, & 0 < q < \frac{1}{3}, \\[6pt] \frac{6q-1}{36(1+q)}, & \frac{1}{3} \le q < 1, \end{cases}$$

$$B_2(q) = \begin{cases} \frac{5-4q}{12(1+q)}, & 0 < q < \frac{5}{6}, \\[6pt] \frac{4q-5}{12(1+q)}, & \frac{5}{6} \le q < 1, \end{cases}$$

$A_1(q), A_2(q), A_3(q), A_4(q), A_5(q),$ and $A_6(q)$ are given by the same expressions as described in Theorem 1.

Proof. Using Lemma 1 and the Hölder inequality, one has

$$\left| \frac{1}{6} \left[f(a) + 4f\left(\frac{2a + \eta(b,a)}{2} \right) + f(a + \eta(b,a)) \right] - \frac{1}{\eta(b,a)} \int_a^{a+\eta(b,a)} f(x) \, {}_a d_q x \right|$$

$$= \left| \eta(b,a) \int_0^1 \Psi(t,q) \, {}_a D_q f(a + t\eta(b,a)) \, {}_0 d_q t \right|$$

$$= \eta(b,a) \left| \int_0^{\frac{1}{2}} \left(qt - \frac{1}{6} \right) {}_a D_q f(a + t\eta(b,a)) \, {}_0 d_q t \right.$$

$$\left. + \int_{\frac{1}{2}}^1 \left(qt - \frac{5}{6} \right) {}_a D_q f(a + t\eta(b,a)) \, {}_0 d_q t \right|$$

$$\leq \eta(b,a)\left[\int_0^{\frac{1}{2}}\left|qt-\frac{1}{6}\right|\,|_aD_qf(a+t\eta(b,a))|\,_0d_qt\right.$$

$$\left.+\int_{\frac{1}{2}}^1\left|qt-\frac{5}{6}\right|\,|_aD_qf(a+t\eta(b,a))|\,_0d_qt\right]$$

$$\leq \eta(b,a)\left[\left(\int_0^{\frac{1}{2}}\left|qt-\frac{1}{6}\right|\,_0d_qt\right)^{1-\frac{1}{r}}\left(\int_0^{\frac{1}{2}}\left|qt-\frac{1}{6}\right|\,|_aD_qf(a+t\eta(b,a))|^r\,_0d_qt\right)^{\frac{1}{r}}\right.$$

$$\left.+\left(\int_{\frac{1}{2}}^1\left|qt-\frac{5}{6}\right|\,_0d_qt\right)^{1-\frac{1}{r}}\left(\int_{\frac{1}{2}}^1\left|qt-\frac{5}{6}\right|\,|_aD_qf(a+t\eta(b,a))|^r\,_0d_qt\right)^{\frac{1}{r}}\right]$$

$$\leq \eta(b,a)\left[\left(\int_0^{\frac{1}{2}}\left|qt-\frac{1}{6}\right|\,_0d_qt\right)^{1-\frac{1}{r}}\right.$$

$$\times \left(\int_0^{\frac{1}{2}}\left|qt-\frac{1}{6}\right|\left[(1-t)|_aD_qf(a)|^r+t|_aD_qf(b)|^r-\mu t(1-t)\eta^2(b,a)\right]\,_0d_qt\right)^{\frac{1}{r}}$$

$$+\left(\int_{\frac{1}{2}}^1\left|qt-\frac{5}{6}\right|\,_0d_qt\right)^{1-\frac{1}{r}}$$

$$\left.\times \left(\int_{\frac{1}{2}}^1\left|qt-\frac{5}{6}\right|\left[(1-t)|_aD_qf(a)|^r+t|_aD_qf(b)|^r-\mu t(1-t)\eta^2(b,a)\right]\,_0d_qt\right)^{\frac{1}{r}}\right]$$

By direct computation, we find

$$B_1(q)=\int_0^{\frac{1}{2}}\left|qt-\frac{1}{6}\right|\,_0d_qt=\begin{cases}\frac{1-2q}{12(1+q)}, & 0<q<\frac{1}{3},\\ \frac{6q-1}{36(1+q)}, & \frac{1}{3}\leq q<1,\end{cases}$$

$$B_2(q)=\int_{\frac{1}{2}}^1\left|qt-\frac{5}{6}\right|\,_0d_qt=\begin{cases}\frac{5-4q}{12(1+q)}, & 0<q<\frac{5}{6},\\ \frac{4q-5}{12(1+q)}, & \frac{5}{6}\leq q<1,\end{cases}$$

and obtain the integral expressions of $A_1(q), A_2(q), A_3(q), A_4(q), A_5(q)$, and $A_6(q)$, which have the same formulas as those given in Theorem 1. This completes the proof of Theorem 2. □

5. Applications

It is worth noting that in Definition 4 for $\mu=0$, the generalized strongly preinvex functions reduce to the preinvex functions. Moreover, if we put $\eta=v-u$ in Definition 2, then the preinvex functions reduce to the classical convex functions. Besides, the quantum integral inequalities would lead to the corresponding Riemann integral inequalities by taking the limit $q\to 1^-$. Thus, several new and previously known results can be derived from Theorems 1 and 2 as special cases. Here, we illustrate the applications of our main results by three examples.

Example 1. *Recently, Zhang and Du et al. [26] investigated the quantum integral inequalities for convex functions, they established the following inequality:*

$$\left| \frac{1}{3} \left[\frac{qf(a)+f(b)}{1+q} + 2f\left(\frac{qa+b}{1+q}\right) \right] - \frac{1}{b-a} \int_a^b f(t)\,_ad_qt \right| \tag{7}$$
$$\leq \min\left\{ \mathcal{H}_1\left(\frac{1}{3}, \frac{1}{1+q}, 1, 1\right), \mathcal{H}_2\left(\frac{1}{3}, \frac{1}{1+q}, 1, 1\right) \right\},$$

where $f : [a,b] \to \mathbb{R}$ is a q-differentiable function and $|\,_aD_qf|$ is an integrable and convex function with $0 < q < 1$, the expressions of $\mathcal{H}_1$ and $\mathcal{H}_2$ are given by [26] (Theorem 3.2).

Further, in [26], the authors derived a remarkable inequality from (7), as follows:

$$\left| \frac{1}{3}\left[\frac{f(a)+f(b)}{2} + 2f\left(\frac{a+b}{2}\right) \right] - \frac{1}{b-a}\int_a^b f(t)\,dt \right| \leq \frac{5(b-a)}{72}\left[|f'(a)| + |f'(b)|\right], \tag{8}$$

where $f : [a,b] \to \mathbb{R}$ is a differentiable function, and $|f'|$ is an integrable and convex function on $[a,b]$.

In the following, we show a new result analogous to the inequality (7), which can be obtained directly by taking $\mu = 0$ in Theorem 1.

Corollary 1. *Let $f : I = [a, a+\eta(b,a)] \to \mathbb{R}$ be a q-differentiable function on I with $\eta(b,a) > 0$. If $|\,_aD_qf|$ is an integrable and preinvex function, $0 < q < 1$, then*

$$\left| \frac{1}{3}\left[\frac{f(a)+f(a+\eta(b,a))}{2} + 2f\left(\frac{2a+\eta(b,a)}{2}\right) \right] - \frac{1}{\eta(b,a)}\int_a^{a+\eta(b,a)} f(t)\,_ad_qt \right| \tag{9}$$
$$\leq \eta(b,a)\left[(A_1(q) + A_4(q))|\,_aD_qf(a)| + (A_2(q) + A_5(q))|\,_aD_qf(b)|\right],$$

where $A_1(q), A_2(q), A_4(q),$ and $A_5(q)$ are the coefficients as described in Theorem 1.

Putting $\eta(b,a) = b - a$ in Corollary 1, it follows that

Corollary 2. *Let $f : [a,b] \to \mathbb{R}$ be a q-differentiable function. If $|\,_aD_qf|$ is an integrable and convex function, $0 < q < 1$, then*

$$\left| \frac{1}{3}\left[\frac{f(a)+f(b)}{2} + 2f\left(\frac{a+b}{2}\right) \right] - \frac{1}{b-a}\int_a^b f(t)\,_ad_qt \right| \tag{10}$$
$$\leq (b-a)\left[(A_1(q) + A_4(q))|\,_aD_qf(a)| + (A_2(q) + A_5(q))|\,_aD_qf(b)|\right],$$

where $A_1(q), A_2(q), A_4(q),$ and $A_5(q)$ are the coefficients as described in Theorem 1.

Remark 1. *In Corollary 2, if we take the limit $q \to 1^-$ in (10) and use the basic properties of q-derivative and q-integral ([25], see also [1])*

$$\lim_{q\to 1^-}\,_aD_qf(t) = f'(t), \quad \lim_{q\to 1^-}\int_a^b f(t)\,_ad_qt = \int_a^b f(t)\,dt,$$

along with the equalities

$$\lim_{q\to 1^-}(A_1(q) + A_4(q)) = \lim_{q\to 1^-}\left(\frac{1 + 12q + 12q^2 + 36q^3}{216(1+q)(1+q+q^2)} + \frac{12q^2 + 12q + 5}{216(1+q)(1+q+q^2)}\right) = \frac{5}{72},$$

$$\lim_{q\to 1^-}(A_2(q)+A_5(q)) = \lim_{q\to 1^-}\left(\frac{18q^2+18q-7}{216(1+q)(1+q+q^2)}+\frac{18q^2+18q+25}{216(1+q)(1+q+q^2)}\right) = \frac{5}{72},$$

then we obtain the inequality

$$\left|\frac{1}{3}\left[\frac{f(a)+f(b)}{2}+2f\left(\frac{a+b}{2}\right)\right]-\frac{1}{b-a}\int_a^b f(t)\,dt\right| \le \frac{5(b-a)}{72}[|f'(a)|+|f'(b)|]. \tag{11}$$

This is exactly the above-mentioned inequality (8) due to Zhang and Du et al. [26].

Example 2. *In a recent paper [27], Tunç, Göv, and Balgeçti established a Simpson-type quantum integral inequality for convex functions ([27] Theorem 1), as follows:*

$$\left|\frac{1}{6}\left[f(a)+4f\left(\frac{a+b}{2}\right)+f(b)\right]-\frac{1}{b-a}\int_a^b f(t)\,_a d_q t\right|$$
$$\le \frac{(b-a)}{12}\left[\frac{2q^2+2q+1}{q^3+2q^2+2q+1}|_a D_q f(b)|+\frac{1}{3}\cdot\frac{6q^3+4q^2+4q+1}{q^3+2q^2+2q+1}|_a D_q f(a)|\right], \tag{12}$$

where $f:[a,b]\to\mathbb{R}$ is a continuous function, $|_a D_q f|$ is a convex and integrable function with $0 < q < 1$.

Remark 2. *Before we describe the related result of inequality (12), we should point out that in (12) there is an error occurring in the coefficients of $|_a D_q f(b)|$ and $|_a D_q f(a)|$. The mistakes arise from the calculations of quantum integrals in [27] (Lemmas 4 and 5), the details are as follows:*

As an auxiliary for establishing the inequality (12), in [27] (Lemmas 4 and 5), the authors gave the following results involving q-integrals ($0 < q < 1$):

$$\int_0^{\frac{1}{2}}(1-t)\left|qt-\frac{1}{6}\right|\,_0 d_q t = \frac{36q^3+12q^2+12q+1}{216(q^3+2q^2+2q+1)}, \tag{13}$$

$$\int_{\frac{1}{2}}^1 (1-t)\left|qt-\frac{5}{6}\right|\,_0 d_q t = \frac{12q^2+12q+5}{216(q^3+2q^2+2q+1)}. \tag{14}$$

However, the equality (13) is incorrect for the case of $0 < q < \frac{1}{3}$; and the equality (14) is incorrect for the case of $0 < q < \frac{5}{6}$, which can be observed by direct computation of q-integrals. In fact, by the formulas and algorithms for q-integrals stated in Propositions 1 and 2, when $0 < q < \frac{1}{3}$, we have

$$\int_0^{\frac{1}{2}}(1-t)\left|qt-\frac{1}{6}\right|\,_0 d_q t$$
$$= \int_0^{\frac{1}{2}}(1-t)\left(\frac{1}{6}-qt\right)\,_0 d_q t$$
$$= \int_0^{\frac{1}{2}}\left(qt^2-\frac{1}{6}t-qt+\frac{1}{6}\right)\,_0 d_q t$$
$$= q\int_0^{\frac{1}{2}}t^2\,_0 d_q t - \left(\frac{1}{6}+q\right)\int_0^{\frac{1}{2}}t\,_0 d_q t + \frac{1}{6}\int_0^{\frac{1}{2}}1\,_0 d_q t$$
$$= q\frac{1}{8(q+q^2+1)}-\left(\frac{1}{6}+q\right)\frac{1}{4(1+q)}+\frac{1}{12}$$
$$= \frac{1-4q^3}{24(q^3+2q^2+2q+1)}.$$

When $0 < q < \frac{5}{6}$, we have

$$\int_{\frac{1}{2}}^{1} (1-t) \left| qt - \frac{5}{6} \right| {}_0 d_q t$$

$$= \int_{\frac{1}{2}}^{1} (1-t) \left(\frac{5}{6} - qt \right) {}_0 d_q t$$

$$= \int_{\frac{1}{2}}^{1} \left(qt^2 - \frac{1}{6}t - qt + \frac{1}{6} \right) {}_0 d_q t$$

$$= q \int_{\frac{1}{2}}^{1} t^2 \, {}_0 d_q t - \left(\frac{5}{6} + q \right) \int_{\frac{1}{2}}^{1} t \, {}_0 d_q t + \frac{5}{6} \int_{\frac{1}{2}}^{1} 1 \, {}_0 d_q t$$

$$= q \left(\int_{0}^{1} t^2 \, {}_0 d_q t - \int_{0}^{\frac{1}{2}} t^2 \, {}_0 d_q t \right) - \left(\frac{5}{6} + q \right) \left(\int_{0}^{1} t \, {}_0 d_q t - \int_{0}^{\frac{1}{2}} t \, {}_0 d_q t \right) + \frac{5}{12}$$

$$= q \left(\frac{1}{q+q^2+1} - \frac{1}{8(q+q^2+1)} \right) - \left(\frac{5}{6} + q \right) \left(\frac{1}{1+q} - \frac{1}{4(1+q)} \right) + \frac{5}{12}$$

$$= \frac{-5 + 8q + 8q^2 - 8q^3}{24(q^3 + 2q^2 + 2q + 1)}.$$

In the same way, one can verify that the equality (13) is valid for $\frac{1}{3} \le q < 1$, the equality (14) is valid for $\frac{5}{6} \le q < 1$.

In the following we provide a modified version of inequality (12).

Corollary 3. *Let $f : [a, b] \to \mathbb{R}$ be a q-differentiable function. If $|{}_a D_q f|$ is an integrable and convex function, $0 < q < 1$, then*

$$\left| \frac{1}{3} \left[\frac{f(a) + f(b)}{2} + 2f\left(\frac{a+b}{2} \right) \right] - \frac{1}{b-a} \int_a^b f(t) \, {}_a d_q t \right| \quad (15)$$

$$\le (b-a) \left[C_1(q) |{}_a D_q f(a)| + C_2(q) |{}_a D_q f(b)| \right],$$

where $C_1(q)$ and $C_2(q)$ are given by

$$C_1(q) = \begin{cases} \frac{-3q^3 + 2q^2 + 2q - 1}{6(q^3 + 2q^2 + 2q + 1)}, & 0 < q < \frac{1}{3}, \\ \frac{-9q^3 + 21q^2 + 21q - 11}{54(q^3 + 2q^2 + 2q + 1)}, & \frac{1}{3} \le q < \frac{5}{6}, \\ \frac{6q^3 + 4q^2 + 4q + 1}{36(q^3 + 2q^2 + 2q + 1)}, & \frac{5}{6} \le q < 1. \end{cases}$$

$$C_2(q) = \begin{cases} \frac{-q^2 - q + 2}{3(q^3 + 2q^2 + 2q + 1)}, & 0 < q < \frac{1}{3}, \\ \frac{-9q^2 - 9q + 32}{54(q^3 + 2q^2 + 2q + 1)}, & \frac{1}{3} \le q < \frac{5}{6}, \\ \frac{2q^2 + 2q + 1}{12(q^3 + 2q^2 + 2q + 1)}, & \frac{5}{6} \le q < 1. \end{cases}$$

Proof. Using Corollary 2 and performing a simple calculation in the expressions $C_1(q) = A_1(q) + A_4(q)$ and $C_2(q) = A_2(q) + A_5(q)$, where $A_1(q), A_2(q), A_4(q)$, and $A_5(q)$ are the coefficients from Theorem 1, we obtain the inequality (15). □

Example 3. *We provide an estimation of upper bound for the q-integral $\int_a^{a+\eta(b,a)} f(t) \, {}_a d_q t$.*

Corollary 4. *Let $f : I = [a, a + \eta(b, a)] \to \mathbb{R}$ be a q-differentiable function on I with $\eta(b, a) > 0$. If $|{}_a D_q f|$ is an integrable and generalized strongly preinvex function with modulus $\mu \ge 0$ and $0 < q < 1$, then*

$$\left| \int_a^{a+\eta(b,a)} f(t)\,_ad_qt \right| \leq \eta(b,a) \left| \frac{1}{6}\left[f(a) + 4f\left(\frac{2a+\eta(b,a)}{2}\right) + f(a+\eta(b,a)) \right] \right| \tag{16}$$
$$+ \eta^2(b,a)\left[(A_1(q)+A_4(q))|\,_aD_qf(a)| + (A_2(q)+A_5(q))|\,_aD_qf(b)| - \mu(A_3(q)+A_6(q))\eta^2(b,a) \right],$$

where $A_1(q), A_2(q), A_3(q), A_4(q), A_5(q)$ and $A_6(q)$ are the coefficients as described in Theorem 1.

Proof. Note that

$$\left| \frac{1}{\eta(b,a)} \int_a^{a+\eta(b,a)} f(t)\,_ad_qt \right| \leq \left| \frac{1}{6}\left[f(a) + 4f\left(\frac{2a+\eta(b,a)}{2}\right) + f(a+\eta(b,a)) \right] \right|$$
$$+ \left| \frac{1}{\eta(b,a)} \int_a^{a+\eta(b,a)} f(t)\,_ad_qt - \frac{1}{6}\left[f(a) + 4f\left(\frac{2a+\eta(b,a)}{2}\right) + f(a+\eta(b,a)) \right] \right|.$$

Utilizing Theorem 1, one has

$$\left| \frac{1}{\eta(b,a)} \int_a^{a+\eta(b,a)} f(t)\,_ad_qt \right| \leq \left| \frac{1}{6}\left[f(a) + 4f\left(\frac{2a+\eta(b,a)}{2}\right) + f(a+\eta(b,a)) \right] \right|$$
$$+ \eta(b,a)\left[(A_1(q)+A_4(q))|\,_aD_qf(a)| + (A_2(q)+A_5(q))|\,_aD_qf(b)| - \mu(A_3(q)+A_6(q))\eta^2(b,a) \right].$$

Multiplying both sides of the above inequality by $\eta(b,a)$ leads to the desired inequality (16). □

Author Contributions: Y.D., M.U.A., and S.W. finished the proofs of the main results and the writing work. Both authors contributed equally to the writing of this paper. All authors read and approved the final manuscript.

Funding: This work was supported by the Teaching Reform Project of Longyan University (Grant No. 2017JZ02) and the Teaching Reform Project of Fujian Provincial Education Department (Grant No. FBJG20180120).

Acknowledgments: The authors would like to express sincere appreciation to the editors and the anonymous reviewers for their valuable comments and suggestions which helped to improve the manuscript.

Conflicts of Interest: The authors declare no conflict of interest.

References

1. Kac, V.; Cheung, P. *Quantum Calculus*; Springer: New York, NY, USA, 2002.
2. Ernst, T. *A Comprehensive Treatment of q-Calculus*; Birkhäuser: Basel, Switzerland; New York, NY, USA, 2012.
3. Ernst, T. A method for q-calculus. *J. Nonlinear Math. Phys.* **2003**, *10*, 487–525. [CrossRef]
4. Alp, N.; Sarikaya, M.Z.; Kunt, M.; Iscan, I. q-Hermite Hadamard inequalities and quantum estimates for midpoint type inequalities via convex and quasi-convex functions. *J. King Saud Univ. Sci.* **2018**, *30*, 193–203. [CrossRef]
5. Lee, J.; Vîlcu, G.E. Inequalities for generalized normalized δ-Casorati curvatures of slant submanifolds in quaternionic space forms. *Taiwan J. Math.* **2015**, *19*, 691–702. [CrossRef]
6. Vîlcu, G.E. An optimal inequality for Lagrangian submanifolds in complex space forms involving Casorati curvature. *J. Math. Anal. Appl.* **2018**, *465*, 1209–1222. [CrossRef]
7. Aquib, M.; Lee, J.E.; Vîlcu, G.E.; Yoon, D.W. Classification of Casorati ideal Lagrangian submanifolds in complex space forms. *Differ. Geom. Appl.* **2019**, *63*, 30–49. [CrossRef]
8. Vîlcu, A.D.; Vîlcu, G.E. On quasi-homogeneous production functions. *Symmetry* **2019**, *11*, 976. [CrossRef]
9. Hanson, M.A. On sufficiency of the Kuhn-Tucker conditions. *J. Math. Anal. Appl.* **1981**, *80*, 545–550. [CrossRef]
10. Jeyakumar, V. Strong and weak invexity in mathematical programming. *Methods Oper. Res.* **1985**, *55*, 109–125.
11. Ben-Israel, A.; Mond, B. What is invexity? *J. Austral. Math. Soc. Ser. B* **1986**, *28*, 1–9. [CrossRef]
12. Weir, T.; Mond, B. Preinvex functions in multiple objective optimization. *J. Math. Anal. Appl.* **1988**, *136*, 29–38. [CrossRef]

13. Jeyakumar, V.; Mond, B. On generalized convex mathematical programming. *J. Aust. Math. Soc. Ser. B* **1992**, *34*, 43–53. [CrossRef]
14. Noor, M.A. Generalized convex functions. *Pan-Am. Math. J.* **1994**, *4*, 73–89.
15. Mohan, S.R.; Neogy, S.K. On invex sets and preinvex functions. *J. Math. Anal. Appl.* **1995**, *189*, 901–908. [CrossRef]
16. Noor, M.A.; Noor, K.I. Some characterizations of strongly preinvex functions. *J. Math. Anal. Appl.* **2006**, *316*, 697–706. [CrossRef]
17. Awan, M.U.; Cristescu, G.; Noor, M.A.; Riahi, L. Upper and lower bounds for Riemann type quantum integrals of preinvex and preinvex dominated functions. *UPB Sci. Bull. Ser. A* **2017**, *79*, 33–44.
18. Cristescu, G.; Noor, M.A.; Awan, M.U. Bounds of the second degree cumulative frontier gaps of functions with generalized convexity. *Carpath. J. Math.* **2015**, *31*, 173–180.
19. Niculescu, C.P.; Persson, L.E. *Convex Functions and Their Applications, A Contemporary Approach*, 2nd ed.; CMS Books in Mathematics; Springer: New York, NY, USA, 2018; Volume 23.
20. Cristescu, G.; Lupsa, L. *Non-Connected Convexities and Applications*; Kluwer Academic Publishers: Dordrecht, The Netherlands, 2002.
21. Noor, M.A.; Noor, K.I.; Awan, M.U. Some quantum integral inequalities via preinvex functions. *Appl. Math.Comput.* **2015**, *269*, 242–251. [CrossRef]
22. Noor, M.A.; Noor, K.I.; Awan, M.U. Quantum Ostrowski inequalities for q-differentiable convex functions. *J. Math. Inequal.* **2016**, *10*, 1013–1018. [CrossRef]
23. Riahi, L.; Awan, M.U.; Noor, M.A. Some complementary q-bounds via different classes of convex functions. *UPB Sci. Bull. Ser. A* **2017**, *79*, 171–182.
24. Sarikaya, M.Z.; Set, E.; Ozdemir, M.E. On new inequalities of Simpson's type for s-convex functions. *Comput. Math. Appl.* **2010**, *60*, 2191–2199. [CrossRef]
25. Tariboon, J.; Ntouyas, S.K. Quantum integral inequalities on finite intervals. *J. Inequal. Appl.* **2014**, *2014*, 121. [CrossRef]
26. Zhang, Y.; Du, T.S.; Wang, H.; Shen, Y.J. Different types of quantum integral inequalities via α, m-convexity. *J. Inequal. Appl.* **2018**, *2018*, 264. [CrossRef] [PubMed]
27. Tunç, M.; Göv, E.; Balgeçti, S. Simpson type quantum integral inequalities for convex functions. *Miskolc Math. Notes* **2018**, *19*, 649–664. [CrossRef]

© 2019 by the authors. Licensee MDPI, Basel, Switzerland. This article is an open access article distributed under the terms and conditions of the Creative Commons Attribution (CC BY) license (http://creativecommons.org/licenses/by/4.0/).

Article

A Note on Minimal Hypersurfaces of an Odd Dimensional Sphere

Sharief Deshmukh [1,*] and Ibrahim Al-Dayel [2]

1. Department of Mathematics, College of Science, King Saud University, P.O.Box-2455, Riyadh 11451, Saudi Arabia
2. Department of Mathematics and Statistics, College of Science, Imam Muhammad Ibn Saud Islamic University, P.O. Box-65892, Riyadh 11566, Saudi Arabia; iaaldayel@imamu.edu.sa
* Correspondence: shariefd@ksu.edu.sa

Received: 17 January 2020; Accepted: 11 February 2020; Published: 21 February 2020

Abstract: We obtain the Wang-type integral inequalities for compact minimal hypersurfaces in the unit sphere S^{2n+1} with Sasakian structure and use these inequalities to find two characterizations of minimal Clifford hypersurfaces in the unit sphere S^{2n+1}.

Keywords: clifford minimal hypersurfaces; sasakian structure; integral inequalities; reeb function; contact vector field

MSC: 53C40; 53C42; 53C25

---

1. Introduction

Let M be a compact minimal hypersurface of the unit sphere S^{n+1} with shape operator A. In his pioneering work, Simons [1] has shown that on a compact minimal hypersurface M of the unit sphere S^{n+1} either $A = 0$ (totally geodesic), or $\|A\|^2 = n$, or $\|A\|^2(p) > n$ for some point $p \in M$, where $\|A\|$ is the length of the shape operator. This work was further extended in [2] and for compact constant mean curvature hypersurfaces in [3]. If for every point p in M, the square of the length of the second fundamental form of M is n, then it is known that M must be a subset of a Clifford minimal hypersurface

$$S^l\left(\sqrt{\frac{l}{n}}\right) \times S^m\left(\sqrt{\frac{m}{n}}\right),$$

where l, m are positive integers, $l + m = n$ (cf. Theorem 3 in [4]). Note that this result was independently proven by Lawson [2] and Chern, do Carmo, and Kobayashi [5]. One of the interesting questions in differential geometry of minimal hypersurfaces of the unit sphere S^{n+1} is to characterize minimal Clifford hypersurfaces. Minimal hypersurfaces have also been studied in (cf. [6–8]). In [2], bounds on Ricci curvature are used to find a characterization of the minimal Clifford hypersurfaces in the unit sphere S^4. Similarly in [3,9–11], the authors have characterized minimal Clifford hypersurfaces in the odd-dimensional unit spheres S^3 and S^5 using constant contact angle. Wang [12] studied compact minimal hypersurfaces in the unit sphere S^{n+1} with two distinct principal curvatures, one of them being simple and obtained the following integral inequality,

$$\int_M \|A\|^2 \leq n Vol(M),$$

where $Vol(M)$ is the volume of M. Moreover, he proved that equality in the above inequality holds if and only if M is the Clifford hypersurface,

$$S^1\left(\sqrt{\frac{1}{n}}\right) \times S^m\left(\sqrt{\frac{n-1}{n}}\right).$$

In this paper, we are interested in studying compact minimal hypersurfaces of the unit sphere S^{2n+1} using the Sasakian structure (φ, ξ, η, g) (cf. [13]) and finding characterizations of minimal Clifford hypersurface of S^{2n+1}. On a compact minimal hypersurface M of the unit sphere S^{2n+1}, we denote by N the unit normal vector field and define a smooth function $f = g(\xi, N)$, which we call the *Reeb function* of the minimal hypersurface M. Also, on the hypersurface M, we have a smooth vector field $v = \varphi(N)$, which we call the *contact vector field* of the hypersurface (v being orthogonal to ξ belongs to contact distribution). Instead of demanding two distinct principal curvatures one being simple, we ask the contact vector field v of the minimal hypersurface in S^{2n+1} to be conformal vector field and obtain an inequality similar to Wang's inequality and show that the equality holds if and only if M is isometric to a Clifford hypersurface. Indeed we prove

Theorem 1. *Let M be a compact minimal hypersurface of the unit sphere S^{2n+1} with Reeb function f and contact vector field v a conformal vector field on M. Then,*

$$\int_M (1-f^2)\|A\|^2 \leq 2n \int_M \left(1-f^2\right)$$

and the equality holds if and only if M is isometric to the Clifford hypersurface $S^l\left(\sqrt{\frac{l}{2n}}\right) \times S^m\left(\sqrt{\frac{m}{2n}}\right)$, where $l+m=2n$.

Also in [12], Wang studied embedded compact minimal non-totally geodesic hypersurfaces in S^{n+1} those are symmetric with respect to $n+2$ pair-wise orthogonal hyperplanes of R^{n+2}. If M is such a hypersurface, then it is proved that

$$\int_M \|A\|^2 \geq nVol(M),$$

and the equality holds precisely if M is a Clifford hypersurface. Note that compact embedded hypersurface has huge advantage over the compact immersed hypersurface, as it divides the ambient unit sphere S^n into two connected components.

In our next result, we consider compact immersed minimal hypersurface M of the unit sphere S^{2n+1} such that the Reeb function f is a constant along the integral curves of the contact vector field v and show that above inequality of Wang holds, and we get another characterization of minimal Clifford hypersurface in the unit sphere S^{2n+1}. Precisely, we prove the following.

Theorem 2. *Let M be a compact minimal hypersurface of the unit sphere S^{2n+1} with Reeb function f a constant along the integral curves of the contact vector field v. Then,*

$$\int_M \|A\|^2 \geq 2nVol(M)$$

and the equality holds if and only if M is isometric to the Clifford hypersurface $S^l\left(\sqrt{\frac{l}{2n}}\right) \times S^m\left(\sqrt{\frac{m}{2n}}\right)$, where $l+m=2n$.

2. Preliminaries

Recall that conformal vector fields play an important role in the geometry of a Riemannian manifolds. A conformal vector field v on a Riemannian manifold (M, g) has local flow consisting of conformal transformations, which is equivalent to

$$\pounds_v g = 2\rho g. \tag{1}$$

The smooth function ρ appearing in Equation (1) defined on M is called the potential function of the conformal vector field v. We denote by (φ, ξ, η, g) the Sasakian structure on the unit sphere S^{2n+1} as a totally umbilical real hypersurface of the complex space form $(C^{n+1}, \bar{J}, \langle,\rangle)$, where $\bar{J}$ is the complex structure and $\langle,\rangle$ is the Euclidean Hermitian metric. The Sasakian structure (φ, ξ, η, g) on S^{2n+1} consists of a $(1,1)$ skew symmetric tensor field φ, a smooth unit vector field ξ, a smooth 1-form η dual to ξ, and the induced metric g on S^{2n+1} as real hypersurface of C^{n+1} and they satisfy (cf. [13])

$$\varphi^2 = -I + \eta \otimes \xi, \ \eta \circ \varphi = 0, \ \eta(\xi) = 1, \ g(\varphi X, \varphi Y) = g(X,Y) - \eta(X)\eta(Y), \tag{2}$$

and

$$\left(\overline{\nabla}\varphi\right)(X,Y) = g(X,Y)\xi - \eta(Y)X, \quad \overline{\nabla}_X \xi = -\varphi X, \tag{3}$$

where X, Y are smooth vector fields, $\overline{\nabla}$ is Riemannian connection on S^{2n+1} and the covariant derivative

$$\left(\overline{\nabla}\varphi\right)(X,Y) = \overline{\nabla}_X \varphi Y - \varphi\left(\overline{\nabla}_X Y\right).$$

We dente by N and A the unit normal and the shape operator of the hypersurface M of the unit sphere S^{2n+1}. We denote the induced metric on the hypersurface M by the same letter g and denote by ∇ the Riemannian connection on the hypersurface M with respect to the induced metric g. Then, the fundamental equations of hypersurface are given by (cf. [14])

$$\overline{\nabla}_X Y = \nabla_X Y + g(AX, Y), \quad \overline{\nabla}_X N = -AX, \quad X, Y \in \mathfrak{X}(M), \tag{4}$$

$$R(X,Y)Z = g(Y,Z)X - g(X,Z)Y + g(AY,Z)AX - g(AX,Z)AY, \tag{5}$$

$$(\nabla A)(X,Y) = (\nabla A)(Y,X), \quad X, Y \in \mathfrak{X}(M), \tag{6}$$

where $\mathfrak{X}(M)$ is the Lie algebra of smooth vector fields and $R(X,Y)Z$ is the curvature tensor field of the hypersurface M. The Ricci tensor of the minimal hypersurface M of the unit sphere S^{2n+1} is given by

$$Ric(X,Y) = (2n-1)g(X,Y) - g(AX, AY), \quad X, Y \in \mathfrak{X}(M) \tag{7}$$

and

$$\sum_{i=1}^{2n}(\nabla A)(e_i, e_i) = 0 \tag{8}$$

holds for a local orthonormal frame $\{e_1, \ldots, e_{2n}\}$ on the minimal hypersurface M.

Using the Sasakian structure (φ, ξ, η, g) on the unit sphere S^{2n+1}, we analyze the induced structure on a hypersurface M of S^{2n+1}. First, we have a smooth function f on the hypersurface M defined by $f = g(\xi, N)$, which we call the *Reeb function* of the hypersurface M, where N is the unit normal vector field. As the operator φ is skew symmetric, we get a vector field $v = \varphi N$ defined on M, which we call the *contact vector field* of the hypersurface M. Note that the vector field v is orthogonal to ξ, and therefore lies in the contact distribution of the Sasakian manifold S^{2n+1}. We denote by $u = \xi^T$ the tangential component of ξ to the hypersurface M and, consequently, we have $\xi = u + fN$. Let α and β be smooth 1-forms on M dual to the vector fields u and v, respectively, that is, $\alpha(X) = g(X, u)$ and $\beta(X) = g(X, v)$, $X \in \mathfrak{X}(M)$. For $X \in \mathfrak{X}(M)$, we set $JX = (\varphi X)^T$ the tangential component of φX to the hypersurface, which gives a skew symmetric $(1,1)$ tensor field J on the hypersurface M. It follows that $\varphi X = JX - \beta(X)N$. Thus, we get a structure $(J, u, v, \alpha, \beta, f, g)$ on the hypersurface M and using properties in Equations (2) and (3) of the Sasakian structure (φ, ξ, η, g) on the unit sphere S^{2n+1} and Equation (4), it is straightforward to see that the structure $(J, u, v, \alpha, \beta, f, g)$ on the hypersurface M has the properties described in the following Lemma.

Lemma 1. *Let M be a hypersurface of the unit sphere S^{2n+1}. Then, M admits the structure $(J, u, v, \alpha, \beta, f, g)$ satisfying*

(i) $J^2 = -I + \alpha \otimes u + \beta \otimes v$,
(ii) $Ju = -fv$, $Jv = fu$,
(iii) $g(JX, JY) = g(X, Y) - \alpha(X)\alpha(Y) - \beta(X)\beta(Y)$,
(iv) $\nabla_X u = -JX + fAX$, $\nabla_X v = -fX - JAX$,
(v) $(\nabla J)(X, Y) = g(X, Y)u - \alpha(Y)X + g(AX, Y)v - \beta(Y)AX$,
(vi) $\nabla f = -Au + v$,
(vii) $\|u\|^2 = \|v\|^2 = (1 - f^2)$, $g(u, v) = 0$,

where ∇f is the gradient of the Reeb function f.

Let Δf be the Laplacian of the Reeb function f of the minimal hypersurface M of the unit sphere S^{2n+1} defined by $\Delta f = \text{div} \nabla f$. Then using Lemma 1 and $\frac{1}{2}\Delta f^2 = f\Delta f + \|\nabla f\|^2$ and Equations (6) and (8), we get the following:

Lemma 2. *Let M be a minimal hypersurface of the unit sphere S^{2n+1}. Then, the Reeb function f satisfies*

(i) $\Delta f = -\left(2n + \|A\|^2\right) f$,
(ii) $\frac{1}{2}\Delta f^2 = -\left(2n + \|A\|^2\right) f^2 + \|\nabla f\|^2$.

On the hypersurface M of the unit sphere S^{2n+1}, we define a $(1,1)$ tensor field $\Psi = JA - AJ$, then it follows that $g(\Psi X, Y) = g(X, \Psi Y)$, $X, Y \in \mathfrak{X}(M)$, that is, Ψ is symmetric and that $tr\Psi = 0$. Next, we prove the following:

Lemma 3. *Let M be a compact minimal hypersurface of the unit sphere S^{2n+1}. Then,*

$$\int_M \left(1 - f^2\right) \|A\|^2 = \int_M \left(2n - 2n(2n+1)f^2 + \frac{1}{2}\|\Psi\|^2\right).$$

Proof. Using Equation (7), we have $Ric(v, v) = (2n - 1)\|v\|^2 - \|Av\|^2$. Now, using Lemma 1, we get

$$(\pounds_v g)(X, Y) = -2fg(X, Y) - g(\Psi X, Y),$$

which on using the fact that $tr\Psi = 0$, gives

$$|\pounds_v g|^2 = 8nf^2 + \|\Psi\|^2.$$

Also, using (iii) of Lemma 1, we have

$$\|JA\|^2 = \|A\|^2 - \|Au\|^2 - \|Av\|^2,$$

which together with second equation in (iv) of Lemma 1 and the fact that $trJA = 0$, implies

$$\|\nabla v\|^2 = 2nf^2 + \|A\|^2 - \|Au\|^2 - \|Av\|^2.$$

Note that second equation in (iv) of Lemma 1 also gives

$$\mathrm{div}\, v = -2nf.$$

Now, inserting above values in the following Yano's integral formula (cf. [15])

$$\int_M \left(Ric(v,v) + \frac{1}{2}|\pounds_v g|^2 - \|\nabla v\|^2 - (\mathrm{div}\, v)^2 \right) = 0,$$

we get

$$\int_M \left((2n-1)\|v\|^2 + 2nf^2 + \frac{1}{2}\|\Psi\|^2 - \|A\|^2 + \|Au\|^2 - 4n^2 f^2 \right) = 0. \tag{9}$$

Also, (vi) of Lemma 1, gives $Au = v - \nabla f$, that is, $\|Au\|^2 = \|v\|^2 + \|\nabla f\|^2 - 2v(f)$, which on using $\mathrm{div}(fv) = v(f) + f\mathrm{div}\, v = v(f) - 2nf^2$, gives

$$\|Au\|^2 = \|v\|^2 + \|\nabla f\|^2 - 2\mathrm{div}(fv) - 4nf^2.$$

Inserting above value of $\|Au\|^2$ in Equation (9), yields

$$\int_M \left(2n\|v\|^2 - 2nf^2 + \frac{1}{2}\|\Psi\|^2 - \|A\|^2 + \|\nabla f\|^2 - 4n^2 f^2 \right) = 0. \tag{10}$$

Integrating (ii) of Lemma 2, we get

$$\int_M \|\nabla f\|^2 = \int_M \left(2n + \|A\|^2 \right) f^2,$$

which together with $\|v\|^2 = 1 - f^2$ and Equation (10) proves the integral formula. □

Lemma 4. *Let M be a minimal hypersurface of the unit sphere S^{2n+1}. Then, the contact vector field v is a conformal vector field if and only if $JA = AJ$.*

Proof. Suppose that $AJ = JA$. Then, using Lemma 1 and symmetry of shape operator A and skew symmetry of the operator J, we have

$$(\pounds_v g)(X,Y) = g(\nabla_X v, Y) + g(\nabla_Y v, X) = -2fg(X,Y), \quad X \in \mathfrak{X}(M),$$

which proves that v is a conformal vector field with potential function $-f$. Conversely, suppose v is conformal vector field with potential function ρ. Then, using Equation (1), we have

$$(\pounds_v g)(X,Y) = g(\nabla_X v, Y) + g(\nabla_Y v, X) = 2\rho g(X,Y),$$

which on using Lemma 1, gives

$$g(-JAX - fX, Y) + g(-JAY - fY, X) = 2\rho g(X,Y),$$

that is,

$$g(AJX - JAX, Y) = 2(\rho + f)g(X,Y).$$

Choosing a local orthonormal frame $\{e_1, \ldots, e_{2n}\}$ on the minimal hypersurface M and taking $X = Y = e_i$ in above equation and summing, we get $\rho = -f$. This gives $g(AJX - JAX, Y) = 0$, $X, Y \in \mathfrak{X}(M)$, that is, $AJ = JA$. □

Lemma 5. *Let M be a minimal hypersurface of the unit sphere S^{2n+1}. If the contact vector field v is a conformal vector field on M, then*

$$Au = \frac{\|A\|^2}{2n} v.$$

Proof. Suppose v is a conformal vector field. Then, by Lemma 4, we have $JA = AJ$. Note that for the Hessian operator A_f of the Reeb function f using Lemma 1, we have

$$A_f(X) = \nabla_X \nabla f = \nabla_X(v - Au) = -JAX - fX - \nabla_X Au, \quad X \in \mathfrak{X}(M),$$

which on using (vi) of Lemma 1, gives

$$A_f(X) = -f(X + A^2 X) - (\nabla A)(X, u).$$

Taking covariant derivative in above equation gives

$$\begin{aligned}(\nabla A_f)(X, Y) &= -X(f)((Y + A^2 Y) - f(\nabla A^2)(X, Y) - (\nabla^2 A)(X, Y, u) \\ &\quad + (\nabla A)(Y, JX) - f(\nabla A)(Y, AX),\end{aligned}$$

where we used (iv) of Lemma 1. Now, on taking a local orthonormal frame $\{e_1, \ldots, e_{2n}\}$ on the minimal hypersurface M and taking $X = Y = e_i$ in above equation and summing, we get

$$\begin{aligned}\sum_{i=1}^{2n}(\nabla A_f)(e_i, e_i) &= -\nabla f - A^2 \nabla f - f \sum_{i=1}^{2n}(\nabla A^2)(e_i, e_i) - \sum_{i=1}^{2n}(\nabla^2 A)(e_i, e_i, u) \\ &\quad + \sum_{i=1}^{2n}(\nabla A)(e_i, Je_i) - f \sum_{i=1}^{2n}(\nabla A)(e_i, Ae_i).\end{aligned}$$

Note that for the minimal hypersurface, we have

$$\begin{aligned}\sum_{i=1}^{2n}(\nabla A)(e_i, Ae_i) &= \sum_{i=1}^{2n}\left(\nabla_{e_i} A^2 e_i - A((\nabla A))(e_i, e_i) + A(\nabla_{e_i} e_i)\right) \\ &= \sum_{i=1}^{2n}(\nabla A^2)(e_i, e_i).\end{aligned}$$

Thus, the previous equation takes the form

$$\sum_{i=1}^{2n}(\nabla A_f)(e_i, e_i) = -\nabla f - A^2 \nabla f - 2f \sum_{i=1}^{2n}(\nabla A^2)(e_i, e_i) - \sum_{i=1}^{2n}(\nabla^2 A)(e_i, e_i, u) + \sum_{i=1}^{2n}(\nabla A)(e_i, Je_i). \quad (11)$$

Now, using the definition of Hessian operator, we have

$$R(X, Y)\nabla f = (\nabla A_f)(X, Y) - (\nabla A_f)(Y, X),$$

which gives

$$Ric(Y, \nabla f) = g\left(Y, \sum_{i=1}^{2n}(\nabla A_f)(e_i, e_i)\right) - Y(\Delta f)$$

and we conclude

$$Q(\nabla f) = -\nabla(\Delta f) + \sum_{i=1}^{2n}(\nabla A_f)(e_i, e_i), \quad (12)$$

where Q is the Ricci operator defined by $Ric(X,Y) = g(QX,Y)$, $X,Y \in \mathfrak{X}(M)$. Using (i) of Lemma 2, we have

$$\nabla(\Delta f) = -2n\nabla f - \|A\|^2 \nabla f - f\nabla \|A\|^2$$

and, consequently, using $Q(X) = (2n-1)X - A^2 X$ (outcome of Equation (7)), the Equation (12) takes the form

$$\sum_{i=1}^{2n} \left(\nabla A_f\right)(e_i, e_i) = (2n-1)\nabla f - A^2(\nabla f) - 2n\nabla f - \|A\|^2 \nabla f - f\nabla \|A\|^2,$$

that is,

$$\sum_{i=1}^{2n} \left(\nabla A_f\right)(e_i, e_i) = -\nabla f - A^2(\nabla f) - \|A\|^2 \nabla f - f\nabla \|A\|^2. \tag{13}$$

Also, note that

$$\begin{aligned}
X\left(\|A\|^2\right) &= X\left(\sum_{i=1}^{2n} g(Ae_i, Ae_i)\right) = 2\sum_{i=1}^{2n} g\left((\nabla A)(X, e_i), Ae_i\right) \\
&= 2\sum_{i=1}^{2n} g\left(X, (\nabla A)(e_i, Ae_i)\right),
\end{aligned}$$

where we have used Equation (6) and symmetry of the shape operator A. Therefore, the gradient of the function $\|A\|^2$ is

$$\nabla \|A\|^2 = 2\sum_{i=1}^{2n} (\nabla A)(e_i, Ae_i),$$

and, consequently, Equation (13), takes the form

$$\sum_{i=1}^{2n} \left(\nabla A_f\right)(e_i, e_i) = -\nabla f - A^2(\nabla f) - \|A\|^2 \nabla f - 2f\sum_{i=1}^{2n} (\nabla A)(e_i, Ae_i). \tag{14}$$

Using Equations (11) and (14), we conclude

$$-\|A\|^2 \nabla f = -\sum_{i=1}^{2n} \left(\nabla^2 A\right)(e_i, e_i, u) + \sum_{i=1}^{2n} (\nabla A)(e_i, Je_i). \tag{15}$$

Now, using Equations (6) and (8) and the Ricci identity, we have

$$\sum_{i=1}^{2n} \left(\nabla^2 A\right)(e_i, e_i, u) = \sum_{i=1}^{2n} \left(\nabla^2 A\right)(e_i, u, e_i) = \sum_{i=1}^{2n} \left(R(e_i, u)Ae_i - AR(e_i, u)e_i\right),$$

which on using Equation (5) and $tr A = 0$ gives

$$\sum_{i=1}^{2n} \left(\nabla^2 A\right)(e_i, e_i, u) = -\|A\|^2 Au + 2nAu. \tag{16}$$

Also, using $JA = AJ$, we have

$$\begin{aligned}
\sum_{i=1}^{2n} (\nabla A)(e_i, Je_i) &= \sum_{i=1}^{2n} \left(\nabla_{e_i} JAe_i - A\left((\nabla J)(e_i, e_i) + J(\nabla_{e_o} e_i)\right)\right) \\
&= \sum_{i=1}^{2n} \left((\nabla J)(e_i, Ae_i) - A((\nabla J)(e_i, e_i))\right),
\end{aligned}$$

which on using (v) of Lemma 1, yields

$$\sum_{i=1}^{2n} (\nabla A)(e_i, Je_i) = \|A\|^2 v - 2nAu. \tag{17}$$

Finally, using (vi) of Lemma 1 and Equations (16) and (17) in Equation (15), we get

$$-\|A\|^2(-Au + v) = \|A\|^2 Au - 2nAu + \|A\|^2 v - 2nAu$$

and this proves the Lemma. □

3. Proof of Theorem 1

As the contact vector field v is a conformal vector field by Lemma 4, we have $JA = AJ$, that is, $\Psi = 0$. Then Lemma 3 implies

$$\int_M \left(1 - f^2\right) \|A\|^2 = \int_M \left(2n - 2n(2n+1)f^2\right),$$

that is,

$$\int_M \left(1 - f^2\right) \|A\|^2 = \int_M \left(2n(1 - f^2) - 4nf^2\right). \tag{18}$$

Therefore, we get the inequality

$$\int_M \left(1 - f^2\right) \|A\|^2 \leq \int_M 2n(1 - f^2).$$

Moreover, if the equality holds, then by Equation (18), we get $f = 0$, which in view of (vi), (vii) of Lemma 1, we conclude that $Au = v$ and that the contact vector field v is a unit vector field. As v is a conformal vector field, combining $Au = v$ with Lemma 5, we get $\|A\|^2 v = 2nv$, that is, $\|A\|^2 = 2n$. Therefore, M is a Clifford hypersurface (cf. [5]).

The converse is trivial.

4. Proof of Theorem 2

As the Reeb function f is a constant along the integral curves of the contact vector field v, we have $v(f) = 0$. Note that $\text{div}(fv) = v(f) + f\text{div}v = -2nf^2$, which on integration gives $f = 0$, and consequently, the contact vector field v is a unit vector field. Then Lemma 3, implies

$$\int_M \|A\|^2 = \int_M \left(2n + \frac{1}{2}\|\Psi\|^2\right), \tag{19}$$

which proves the inequality

$$\int_M \|A\|^2 \geq 2n\text{Vol}(M).$$

If the equality holds, then by Equation (4.1), we get that $\Psi = 0$, that is, $JA = AJ$. Thus, by Lemma 4, the contact vector field v is a conformal vector field. Using Lemma 5, we get $\|A\|^2 = 2n$. Therefore, M is a Clifford hypersurface (cf. [5]).

The converse is trivial.

Author Contributions: Conceptualization, S.D. and I.A.-D.; methodology, S.D.; software, I.A.-D.; validation, S.D. and I.A.-D.; formal analysis, S.D.; investigation, I.A.-D.; resources, S.D.; data curation, I.A.-D.; writing—original draft preparation, S.D. and I.A.-D.; writing—review and editing, S.D. and I.A.-D.; visualization, I.A.-D.; supervision, S.D. All authors have read and agreed to the published version of the manuscript.

Funding: This research received no external funding.

Acknowledgments: This work is supported by King Saud University, Deanship of Scientific Research, College of Science Research Center.

Conflicts of Interest: The authors declare no conflict of interest.

References

1. Simons, J. Minimal varieties in riemannian manifolds. *Ann. Math.* **1968**, *88*, 62–105. [CrossRef]
2. Lawson, H.B., Jr. Local rigidity theorems for minimal hypersurfaces. *Ann. Math.* **1969**, *89*, 187–197. [CrossRef]
3. Min, S.H.; Seo, K. A characterization of Clifford hypersurfaces among embedded constant mean curvature hypersurfaces in a unit sphere. *Math. Res. Lett.* **2017**, *24*, 503–534. [CrossRef]
4. Perdomo, O. Rigidity of minimal hypersurface of spheres with constant Ricci curvature. *Rev. Colomb. Mat.* **2004**, *38*, 73–85.
5. Chern, S.S.; Carmo, M.D.; Kobayashi, S. Minimal submanifolds of a sphere with second fundamental form of constant length. In *Functional Analysis and Related Fields*; Springer: New York, NY, USA, 1970; pp. 59–75.
6. Lei, L.; Xu, H.; Xu, Z. On Chern's conjecture for minimal hypersurface in spheres. *arXiv* **2017**, arXiv:1712.01175.
7. Perdomo, O. Another proof for the rigidity of Clifford minimal hypersurfaces of S^n. *Mat. Ins. Univ.* **2005**, *13*, 1–6.
8. Sun, H.; Ogiue, K. Minimal hypersurfaces of unit sphere. *Tohoku Math. J.* **1997**, *149*, 423–429. [CrossRef]
9. Haizhong, L. A characterization of Clifford minimal hypersurfaces in S^3. *PAMS* **1995**, *123*, 3183–3187. [CrossRef]
10. Montes, R.R.; Verderesi, J.A. Minimal surfaces in S^3 with constant contact angle. *Monatsh. Math.* **2009**, *157*, 379–386. [CrossRef]
11. Montes, R.R.; Verderesi, J.A. Contact angle for immersed surfaces in S^{2n+1}. *Diff. Geom. Appl.* **2007**, *25*, 2–100. [CrossRef]
12. Wang, Q. Rigidity of Clifford minimal hypersurfaces. *Monatsh. Math.* **2003**, *140*, 163–167. [CrossRef]
13. Blair, D.E. *Contact Manifolds in Riemannian Geometry*; Lecture Notes in Mathematics; Springer: Berlin, Germany, 1976; Volume 509.
14. Chen, B.-Y. *Total Mean Curvature and Submanifolds of Finite Type, Volume 1 of Series in Pure Mathematics*; World Scientific Publishing Co.: Singapore, 1984.
15. Yano, K. *Integral Formulas in Riemannian Geometry*; Marcel Dekker: New York, NY, USA, 1970.

© 2020 by the authors. Licensee MDPI, Basel, Switzerland. This article is an open access article distributed under the terms and conditions of the Creative Commons Attribution (CC BY) license (http://creativecommons.org/licenses/by/4.0/).

Article

The Minimal Perimeter of a Log-Concave Function

Niufa Fang [1] and Zengle Zhang [2,*]

[1] Chern Institute of Mathematics, Nankai University, Tianjin 300071, China; fangniufa@nankai.edu.cn
[2] Key Laboratory of Group and Graph Theories and Applications, Chongqing University of Arts and Sciences, Chongqing 402160, China
* Correspondence: zhangzengle128@163.com

Received: 23 July 2020; Accepted: 11 August 2020; Published: 14 August 2020

Abstract: Inspired by the equivalence between isoperimetric inequality and Sobolev inequality, we provide a new connection between geometry and analysis. We define the minimal perimeter of a log-concave function and establish a characteristic theorem of this extremal problem for log-concave functions analogous to convex bodies.

Keywords: isoperimetric problem; minimal perimeter; log-concave functions; isotropic measure

1. Introduction

The isoperimetric inequality is an important inequality in geometry which originated from the well-known isoperimetric problem. The isoperimetric inequality has a profound influence on each branches of mathematics. The breakthrough works of Federer and Fleming [1] and Mazya [2] discovered independently the connection between the isoperimetric problem and the Sobolev embedding problem. They established the sharp Sobolev inequality by using the isoperimetric inequality. This exciting connection has motivated a number of studies in recent years about interactions of geometric and analytic inequalities. In this paper, we further study the connection between geometry and analysis.

Let us recall some facts about convex bodies. Let K be a convex body (i.e., compact, convex subset with non-empty interior) in the n-dimensional Euclidean space $\mathbb{R}^n$, the family $\{TK : T \in \mathrm{SL}(n)\}$ of its positions are studied by many mathematicians. Introducing the right position of the unit ball K_X of a finite dimensional normed space X is one of the main problems in the asymptotic theory. There exist many celebrated positions for different purposes, for example isotropic position, M-position, John's position, the ℓ-position and so on, see [3,4].

Our purpose is to study the isotropic position of log-concave functions. Hence, we first recall some geometric backgrounds and these are our motivations. Let K be a convex body in $\mathbb{R}^n$ with centroid at the origin and volume equal to one. A convex body K is in isotropic position if

$$\int_K \langle x, \theta \rangle^2 dx = L_K^2, \quad \forall \theta \in S^{n-1},$$

where $\langle \cdot, \cdot \rangle$ is the usual inner product in $\mathbb{R}^n$ and S^{n-1} is the unit sphere in $\mathbb{R}^n$. It's worth noting that every convex body K with volume one has an isotropic position, and this position is uniqueness (up to an orthogonal transformation), see, e.g., [3]. Isotropic positions have been used to study the classical convexity problems, for example, the minimal surface area of a convex body and its extension [5,6], the minimal mean width of a convex body and its extension [7,8]. Other contributions include e.g., [9–11] among others.

We recall two specific examples on isotropic positions. Let K be a convex body and denote by $S(K)$ its surface area. If $S(K) \leq S(TK)$ for every $T \in \mathrm{SL}(n)$, then K has minimal surface area (see, e.g., [5]).

Petty [5] obtained the following characterization of the minimal surface area position: a convex body K has minimal surface area if and only if its surface area measure σ_K is isotropic, i.e.,

$$\int_{S^{n-1}} \langle u, \theta \rangle^2 d\sigma_K(u) = \frac{S(K)}{n}, \quad \forall \theta \in S^{n-1}.$$

As a second example, the minimal mean width will be recalled which was defined by Giannopoulos and Milman [7]. Let K be a convex body in $\mathbb{R}^n$, the mean width $w(K)$ of K is define as

$$w(K) = 2 \int_{S^{n-1}} h_K(u) d\sigma(u),$$

where $h_K(u) := \sup_{y \in K} \langle u, y \rangle$ is the support function of K and σ is the rotationally invariant probability measure on S^{n-1}. For every $T \in \mathrm{SL}(n)$, if $w(K) \leq w(TK)$ then K has minimal mean width (see, e.g., [7]). Giannopoulos and Milman [7] showed that if the support function of K is twice continuously differentiable, then K has minimal mean width if and only if the measure $d\nu_K = h_K d\sigma$ is isotropic, i.e.,

$$\int_{S^{n-1}} h_K(u) \langle u, \theta \rangle^2 d\sigma(u) = \frac{w(K)}{2n}, \quad \forall \theta \in S^{n-1}.$$

Within the last few years, many geometric results have been generalized to their corresponding functional versions, including but not limited to the functional version Blaschke-Santaló inequality and its reverse [12–16], the functional affine surface areas [17–19], Minkowski problem for functions [20–22], and analytic inequalities with geometric background [23–28].

In this paper, we consider the log-concave functions in $\mathbb{R}^n$. A function $f : \mathbb{R}^n \to \mathbb{R}$ is log-concave if for any $x, y \in \mathbb{R}^n$ and $\lambda \in [0, 1]$, it holds

$$f(\lambda x + (1 - \lambda) y) \geq f(x)^\lambda f(y)^{1-\lambda}. \tag{1}$$

A typical example of log-concave functions is the characteristic function of convex bodies, $\mathbf{1}_K$ (which is defined as $\mathbf{1}_K(x) = 1$ when $x \in K$ and $\mathbf{1}_K(x) = 0$ when $x \notin K$). Let $J(f)$ denote the total mass functional of $f : \mathbb{R}^n \to \mathbb{R}$, namely

$$J(f) = \int_{\mathbb{R}^n} f(x) dx.$$

For any $t > 0$ and log-concave functions $f, g : \mathbb{R}^n \to \mathbb{R}$, Colesanti and Fragalà [21] defined the first variation of J at f along g as

$$\delta J(f, g) = \lim_{t \to 0^+} \frac{J(f \oplus t \cdot g) - J(f)}{t}, \tag{2}$$

where $t \cdot g(x) = g^t(x/t)$ for $t > 0$ and $x \in \mathbb{R}^n$, and $f \oplus g$ the Asplund sum of functions f and g, i.e.,

$$[f \oplus g](x) = \sup_{x = x_1 + x_2} f(x_1) g(x_2), \quad x \in \mathbb{R}^n. \tag{3}$$

It was proved that if f and g are restricted to a subclass of log-concave functions, then the first variation $\delta J(f, g)$ precisely turns out to be L_p mixed volume of convex bodies (see Proposition 3.13 in [21]). In particular, the perimeter of f is defined as (see [21])

$$P(f) = \delta J(f, \gamma_n),$$

where $\gamma_n(x) = e^{-\frac{\|x\|^2}{2}}$ is the Gaussian function and $\|x\|$ is the Euclidean norm of $x \in \mathbb{R}^n$.

Motivated by the work of Giannopoulos and Milman [7], we consider the extremal problems of log-concave functions instead of convex bodies, and our purpose is to discuss the possibility of an isometric approach to these questions. We introduce the notion of minimal perimeters of log-concave functions. Assume that f is a log-concave function, we call f has minimal perimeter if $P(f) \leq P(f \circ T)$ for every $T \in \mathrm{SL}(n)$. Furthermore, we derive the following characteristic theorem of the minimal perimeter.

Theorem 1. *If $f : \mathbb{R}^n \to [0, \infty)$ is a log-concave function, then f has minimal perimeter if and only if*

$$\frac{tr(T)}{n} P(f) = \frac{1}{2} \int_{\mathbb{R}^n} \langle x, Tx \rangle d\mu_f(x) \tag{4}$$

for every $T \in \mathrm{GL}(n)$. Here $tr(T)$ denotes the trace of T, and $\mu_f = (\nabla u)_\sharp (f \mathcal{H}^n)$ is a Borel measure on $\mathbb{R}^n$ (where $\mathcal{H}^n$ is the n-dimensional Hausdorff measure and $u = -\log f$).

Theorem 1 implies that the log-concave function f has minimal perimeter if and only if $\mu_f(\cdot)$ is isotropic, and provides a further example of the connections between the theory of convex bodies and that of functions.

We remark that our works belong to the asymptotic theory of log-concave functions which parallel to that of convex bodies. From a geometric and analytic view of point to study convex bodies is the asymptotic theory of convex bodies which emphasize the dependence of various parameters on the dimension. Isotropic positions for convex bodies play important roles in the asymptotic theory of convex bodies. We are not aware of the related results for log-concave functions. Hence, our work in this paper presents a new connection between convex bodies and log-concave functions and it also leads to a new topic in the study of geometry of log-concave functions. We hope that our work provides some useful tools or ideas in the development of geometry of log-concave functions.

2. Preliminaries

In this section, we provide some preliminaries and notations required for functions. More details can be found in [3,4].

A function $u : \mathbb{R}^n \to \mathbb{R} \cup \{+\infty\}$ is convex if

$$u((1-\lambda)x + \lambda y) \leq (1-\lambda)u(x) + \lambda u(y)$$

for any $x, y \in \mathbb{R}^n$ and $\lambda \in [0,1]$. Let

$$\mathrm{dom}(u) = \{x \in \mathbb{R}^n : u(x) \in \mathbb{R}\}.$$

Since the convexity of u, $\mathrm{dom}(u)$ is a convex set. If $\mathrm{dom}(u) \neq \emptyset$, then u is said proper. The function u is called of class $\mathcal{C}^2$ if it is twice differentiable on $\mathrm{int}(\mathrm{dom}(u))$, with a positive definite Hessian matrix. The Fenchel conjugate of u is the convex function defined by

$$u^*(y) = \sup_{x \in \mathbb{R}^n} \{\langle x, y \rangle - u(x)\}, \quad \forall y \in \mathbb{R}^n.$$

Clearly, $u(x) + u^*(y) \geq \langle x, y \rangle$ for all $x, y \in \mathbb{R}^n$. The equality holds if and only if $x \in \mathrm{dom}(u)$ and y is in the subdifferential of u at x. Hence, one can checked that

$$u^*(\nabla u(x)) + u(x) = \langle x, \nabla u(x) \rangle.$$

From the definition of log-concave functions (1), we known that a log-concave function $f : \mathbb{R}^n \to \mathbb{R}$ has the form $f(x) = e^{-u(x)}$ where $u : \mathbb{R}^n \to \mathbb{R} \cup \{+\infty\}$ is convex. Writing

$$\mathcal{L} = \left\{ u : \mathbb{R}^n \to \mathbb{R} \cup \{+\infty\} \,\Big|\, u \text{ is proper and convex}, \lim_{\|x\| \to +\infty} u(x) = +\infty \right\},$$

$$\mathcal{A} = \left\{ f : \mathbb{R}^n \to \mathbb{R} \,\Big|\, f = e^{-u}, u \in \mathcal{L} \right\}.$$

For $u, v \in \mathcal{L}$, the inf-convolution of u, v is defined by

$$(u \square v)(x) = \inf_{y \in \mathbb{R}^n} [u(y) + v(y - x)], \quad \forall x \in \mathbb{R}^n, \tag{5}$$

and the right scalar multiplication ut is defined by

$$(ut)(x) := \begin{cases} tu\left(\frac{x}{t}\right), & \text{if } t > 0 \\ I_{\{0\}}, & \text{if } t = 0. \end{cases}$$

Note that these operations are convexity preserving, and $I_{\{0\}}$ acts as the identity element in (5). It is proved that $(us \square vt)(x) \in \mathcal{L}$ for $u, v \in \mathcal{L}$ and $s, t \geq 0$ (see [21]). Let $f = e^{-u}, g = e^{-v} \in \mathcal{A}$ and $t > 0$. Form (5), the Asplund sum (defined in (3)) can be rewritten as

$$f \oplus g = e^{-[u \square v]}, \tag{6}$$

and $t \cdot g = e^{-vt}$. Let $f = e^{-u} \in \mathcal{A}$. The support function, h_f, of f is defined as (see, e.g., [28])

$$h_f(x) = u^*(x). \tag{7}$$

We recall that a probability measure μ is called isotropic if it satisfies $\int_{\mathbb{R}^n} x \, d\mu(x) = 0$ and

$$\int_{\mathbb{R}^n} \langle x, \theta \rangle^2 \, d\mu(x) = \frac{1}{n}, \quad \forall \theta \in S^{n-1}. \tag{8}$$

For a measure μ with $\int_{\mathbb{R}^n} x \, d\mu(x) = 0$, the following claims are equivalent (see, e.g., [3]):

(a) μ is isotropic;
(b) For any $T \in \mathrm{GL}(n)$, one has

$$\int_{\mathbb{R}^n} \langle x, Tx \rangle \, d\mu(x) = \frac{\mathrm{tr}(T)}{n};$$

(c)

$$\int_{\mathbb{R}^n} x_i x_j \, d\mu(x) = \frac{1}{n} \delta_{ij} \quad \text{for all} \quad i, j = 1, \cdots, n.$$

3. Minimal Perimeter of Log-Concave Functions

In this section, we consider the properties of the minimal perimeter of log-concave functions. Let $f = e^{-u} \in \mathcal{A}$, the perimeter $P(f)$ has an integral expression (see [21,29]):

$$P(f) = \frac{1}{2} \int_{\mathbb{R}^n} \frac{\|\nabla f\|^2}{f} \, dx. \tag{9}$$

We define that f has minimal perimeter if $P(f) \leq P(f \circ T)$ for every $T \in SL(n)$. This is, if f has minimal perimeter, then

$$\int_{\mathbb{R}^n} \frac{\|\nabla f\|^2}{f} dx \leq \int_{\mathbb{R}^n} \frac{\|\nabla (f \circ T)\|^2}{f \circ T} dx,$$

for any $T \in SL(n)$.

The Borel measure μ_f on $\mathbb{R}^n$ of a log-concave function $f = e^{-u}$ is defined by (see [21])

$$\mu_f = (\nabla u)_{\sharp}(f\mathcal{H}^n).$$

Here $\mathcal{H}^n$ denotes the n-dimensional Hausdorff measure. For any continuous function $g : \mathbb{R}^n \to \mathbb{R}$, one has

$$\int_{\mathbb{R}^n} g(y) d\mu_f(y) = \int_{\mathbb{R}^n} g(\nabla u(x)) f(x) dx. \tag{10}$$

The Borel measure μ_f plays the same role for f as the surface area measure for the convex body.

Proposition 1. *If $f \in \mathcal{A}$, then*

$$P(f) = \int_{\mathbb{R}^n} h_{\gamma_n}(x) d\mu_f(x).$$

Proof. From Equations (9), (10) and (7), we have

$$\begin{aligned} P(f) &= \frac{1}{2} \int_{\mathbb{R}^n} \frac{\|\nabla f\|^2}{f} dx \\ &= \int_{\mathbb{R}^n} f(x) \frac{\|\nabla u(x)\|^2}{2} dx \\ &= \int_{\mathbb{R}^n} f(x) h_{\gamma_n}(\nabla u(x)) dx \\ &= \int_{\mathbb{R}^n} h_{\gamma_n}(x) d\mu_f(x). \end{aligned}$$

□

We recall that the gauge function of a convex body K is defined by

$$\|x\|_K = \min\{\alpha \geq 0 : x \in \alpha K\}. \tag{11}$$

It is clear that

$$\|x\|_K = 1 \quad \text{whenever} \quad x \in \partial K, \tag{12}$$

where ∂K is the boundary of K.

We note that the minimal perimeter of a log-concave function f is equivalent to considering the minimization problem:

$$\min_{T \in SL(n)} P(f \circ T). \tag{13}$$

For $T \in SL(n)$, we write γ_T for $\gamma_n \circ T$. From (9) and the fact that $\nabla_x(f \circ T) = T^t \nabla_{Tx} f$ for $T \in SL(n)$ and $x \in \mathbb{R}^n$, we have

$$\begin{aligned} P(f \circ T) = \delta J(f \circ T, \gamma_n) &= \frac{1}{2} \int_{\mathbb{R}^n} \frac{\|\nabla_x f \circ T\|^2}{f(Tx)} dx \\ &= \frac{1}{2} \int_{\mathbb{R}^n} \frac{\|T^t \nabla_{Tx} f(Tx)\|^2}{f(Tx)} dx \\ &= \frac{1}{2} \int_{\mathbb{R}^n} \frac{\|T^t \nabla f(x)\|^2}{f(x)} dx \\ &= \delta J(f, \gamma_T). \end{aligned}$$

Therefore, we can reformulate problem (13) as follows:

$$\min\{\delta J(f, \gamma_T) : T \in SL(n)\}. \tag{14}$$

Proposition 2. *There exists a unique (un to an orthogonal matrix) $T_0 \in SL(n)$ such that it solves the minimization problem (14).*

Proof. We can limit our attention to $T \in SL(n)$ when T is a positive definite symmetric matrix, since any $T \in SL(n)$ can be represented in the form $T = PQ$ where $P \in SL(n)$ is a positive definite symmetric matrix and Q is an orthogonal matrix. In this case, we can write the function $\gamma_T(x) = \gamma_n(Tx) = e^{-\frac{\|Tx\|^2}{2}}$ as $\gamma_T(x) = e^{-\frac{\|x\|_{E^\circ}^2}{2}}$, where $E = T^t B$ is an origin-centered ellipsoid and E° is the polar body of E defined as $E^\circ = \{x \in \mathbb{R}^n : \langle x, y \rangle \leq 1 \text{ for all } y \in E\}$. There exists a $z_E \in S^{n-1}$ such that the diameter of E satisfies $\mathrm{diam}\,(E) \frac{|\langle z_E, x \rangle|}{2} \leq \|x\|_{E^\circ}$. Let $\{T_k\}_k \in SL(n)$ be a minimizing sequence for the problem (14), namely,

$$\lim_{k \to \infty} \delta J(f, \gamma_{T_k}) = \min\{\delta J(f, \gamma_T) : T \in SL(n), T \text{ is a positive definite symmetric matrix}\}. \tag{15}$$

From (15) and the fact that $\min\{\delta J(f, \gamma_T) : T \in SL(n)\} \leq \delta J(f, \gamma_n)$, we have

$$\begin{aligned} \frac{\mathrm{diam}\,(E_k^\circ)^2}{8} \min_{z \in S^{n-1}} \int_{\mathbb{R}^n} |\langle z, x \rangle|^2 d\mu_f(x) &\leq \int_{\mathbb{R}^n} \frac{\|x\|_{E_k}^2}{2} d\mu_f(x) \\ &= \delta J(f, \gamma_{T_k}) \\ &\leq \delta J(f, \gamma_n). \end{aligned}$$

Since $\|x\|_{E_k} < \mathrm{diam}\,(E_k^\circ) \|x\|$, therefore the upper bound of the convex function $\left(\frac{\|T_k x\|^2}{2}\right)^*$ is depended only on f. According to Theorem 10.9 in [30], there exist a function γ_{T_0} such that the Legendre conjugate of a minimizing sequence of functions for problem (14) converge to γ_{T_0}. Due to Theorem 11.34 in [31], we known that a minimizing sequence of functions for problem (14) converge to $\gamma_{T_0}^*$. According to the dominated convergence theorem, there exists a solution to problem (14).

Next, we prove the uniqueness of T_0. Assume there are two different solutions $T_1, T_2 \in SL(n)$ to the considered problem which satisfy $T_1 \neq aT_2$ for all $a > 0$. If there exists a $a_0 > 0$ such that $T_1 = a_0 T_2$, then

$$\delta J(f, \gamma_{T_1}) = \delta J(f, \gamma_{a_0 T_2}) = \delta J(f, \gamma_{T_2}) \implies a_0 = 1.$$

This contradicts to $T_1 \neq T_2$. The Minkowski inequality for symmetric positive definite matrices shows that

$$\det\left(\frac{T_1 + T_2}{2}\right)^{\frac{1}{n}} > \frac{1}{2} \det(T_1)^{\frac{1}{n}} + \frac{1}{2} \det(T_2)^{\frac{1}{n}} = 1.$$

Let
$$T_3 = \det(\tfrac{T_1^{-1}+T_2^{-1}}{2})^{\frac{1}{n}} \left(\tfrac{T_1^{-1}+T_2^{-1}}{2}\right)^{-1}.$$

Then $T_3 \in SL(n)$ and $\gamma_{T_3} < \gamma_{(\frac{T_1^{-1}+T_2^{-1}}{2})^{-1}}$, i.e., $h_{\gamma_{T_3}} < h_{\gamma_{(\frac{T_1^{-1}+T_2^{-1}}{2})^{-1}}}$. This deduces that

$$\begin{aligned}
\delta J(f, \gamma_{T_3}) &= \int_{\mathbb{R}^n} h_{\gamma_{T_3}}(x) d\mu_f(x) \\
&< \int_{\mathbb{R}^n} h_{\gamma_{(\frac{T_1^{-1}+T_2^{-1}}{2})^{-1}}}(x) d\mu_f(x) \\
&= \int_{\mathbb{R}^n} \frac{\|\frac{T_1^{-1}+T_2^{-1}}{2}x\|^2}{2} d\mu_f(x).
\end{aligned}$$

By the convexity of the square of the Euclidean normal, we have

$$\begin{aligned}
\delta J(f, \gamma_{T_3}) &< \int_{\mathbb{R}^n} \frac{\tfrac{1}{2}\|T_1^{-1}x\|^2 + \tfrac{1}{2}\|T_2^{-1}x\|^2}{2} d\mu_f(x) \\
&= \tfrac{1}{2}\int_{\mathbb{R}^n} h_{\gamma_{T_1}}(x) d\mu_f(x) + \tfrac{1}{2}\int_{\mathbb{R}^n} h_{\gamma_{T_2}}(x) d\mu_f(x) \\
&= \tfrac{1}{2}\delta J(f, \gamma_{T_1}) + \tfrac{1}{2}\delta J(f, \gamma_{T_2}) \\
&= \delta J(f, \gamma_{T_1}) \\
&= \delta J(f, \gamma_{T_2}).
\end{aligned}$$

However, from the assumption on T_1 and T_2, we have

$$\delta J(f, \gamma_{T_3}) \geq \delta J(f, \gamma_{T_1}) = \delta J(f, \gamma_{T_2}),$$

which is a contradiction. □

Proposition 2 implies that the minimal perimeter of log-concave functions is well-defined. Namely,

Corollary 1. *For a log-concave function $f : \mathbb{R}^n \to \mathbb{R}$, there exists a unique (up to an orthogonal matrix) $T_0 \in SL(n)$ such that $f \circ T_0$ has minimal perimeter.*

Next we are in the position to consider the proof of Theorem 1.

Proof of Theorem 1. Let $T \in GL(n)$, and $\varepsilon > 0$ be a suitably small real number. Then

$$T_\varepsilon = [\det(I + \varepsilon T)]^{-\frac{1}{n}} (I + \varepsilon T)^t \in SL(n),$$

and this implies that $P(f) \leq P(f \circ T_\varepsilon)$, i.e.,

$$\int_{\mathbb{R}^n} \frac{\|\nabla f\|^2}{f} dx \leq \int_{\mathbb{R}^n} \frac{\|\nabla (f \circ T_\varepsilon)\|^2}{f \circ T_\varepsilon} dx.$$

By the fact that $\nabla_x(u \circ T) = T^t \nabla_{Tx} u$, then

$$\int_{\mathbb{R}^n} \frac{\|\nabla f\|^2}{f} dx \leq \int_{\mathbb{R}^n} \frac{\|\nabla(f \circ T_\varepsilon)\|^2}{f \circ T_\varepsilon} dx$$

$$= \int_{\mathbb{R}^n} \frac{\|T_\varepsilon^t \nabla_{T_\varepsilon x}(f(T_\varepsilon x))\|^2}{f(T_\varepsilon x)} dx$$

$$= \int_{\mathbb{R}^n} \frac{\|T_\varepsilon^t \nabla f\|^2}{f} dx,$$

i.e.,

$$[\det(I + \varepsilon T)]^{\frac{2}{n}} \int_{\mathbb{R}^n} \frac{\|\nabla f\|^2}{f} dx \leq \int_{\mathbb{R}^n} \frac{\|(I + \varepsilon T)\nabla f\|^2}{f} dx.$$

Because

$$\|(I + \varepsilon T)\nabla f\|^2 = \|\nabla f\|^2 + 2\varepsilon \langle \nabla f, T \nabla f \rangle + o(\varepsilon^2)$$

and

$$[\det(I + \varepsilon T)]^{\frac{2}{n}} = 1 + 2\varepsilon \frac{\text{tr}(T)}{n} + o(\varepsilon^2),$$

when letting $\varepsilon \to 0^+$, we obtain

$$\frac{\text{tr}(T)}{n} \int_{\mathbb{R}^n} \frac{\|\nabla f\|^2}{f} dx \leq \int_{\mathbb{R}^n} \frac{\langle \nabla f, T \nabla f \rangle}{f} dx. \tag{16}$$

Replacing T by $-T$ in (16), we conclude that there must be equality in (4) for every $T \in \text{GL}(n)$.

On the other hand, assume that (4) is satisfied and let $T \in \text{SL}(n)$. Since $\frac{\text{tr}T}{n} \geq 1$ for symmetric positive-definite metric, (9) and $\nabla_x(f \circ T) = T^t \nabla_{Tx} f$, we have

$$P(f \circ T) = \frac{1}{2} \int_{\mathbb{R}^n} \frac{\|\nabla(f \circ T)\|^2}{f \circ T} dx$$

$$= \frac{1}{2} \int_{\mathbb{R}^n} \frac{\|T^t \nabla_{Tx} f\|^2}{f \circ T} dx$$

$$= \frac{1}{2} \int_{\mathbb{R}^n} \frac{\langle T \nabla f, T \nabla f \rangle}{f} dx$$

$$= \frac{1}{2} \int_{\mathbb{R}^n} \frac{\langle \nabla f, T^t T \nabla f \rangle}{f} dx \tag{17}$$

$$= \frac{1}{2} \frac{\text{tr}(T^t T)}{n} \int_{\mathbb{R}^n} \frac{\|\nabla f\|^2}{f} dx$$

$$\geq \frac{1}{2} \int_{\mathbb{R}^n} \frac{\|\nabla f\|^2}{f} dx$$

$$= P(f).$$

This shows that f has minimal perimeter. Moreover, the equality in (17) holds only if T is the identity matrix. This prove that the uniqueness of the minimal perimeter position (up to $U \in O(n)$). □

Corollary 2. *From Theorem 1 and the definition of isotropic measure, the log-concave function $f \in \mathcal{A}$ has minimal perimeter if and only if $\frac{1}{J(f)} \mu_f$ is an isotropic measure.*

Next, we prove that Theorem 1 recovers the L_2 surface area measure of K, $dS_2(K, \cdot) = h_K(\cdot)^{-1} d\sigma_K(\cdot)$, is an isotropic measure on S^{n-1}.

Corollary 3. *Let K be a convex body in $\mathbb{R}^n$ containing the origin in its interior. If $f(x) = e^{-\|x\|_K}$ for $x \in \mathbb{R}^n$, then*

$$P(f) = \frac{1}{2}\Gamma(n)S_2(K),$$

and Theorem 1 includes the fact that L_2 surface area measure of K, $S_2(K, \cdot)$, is an isotropic measure.

Proof. For a convex body K in $\mathbb{R}$, let $\overline{V}_K$ denote the normalized cone volume measure of K, which is given by

$$d\overline{V}_K(z) = \frac{\langle z, \nu_K(z)\rangle}{nV(K)} d\mathcal{H}^{n-1}(z) \quad \text{for} \quad z \in \partial K.$$

Here $\nu_K(z)$ is the outer unit normal of K at the boundary point z, and $\mathcal{H}^{n-1}$ is the $(n-1)$ dimensional Hausdorff measure. For any $x \in \mathbb{R}^n$, we write $x = rz$, with $z \in \partial K$ and $dx = nV(K)r^{n-1}drd\overline{V}_K(z)$. Since the map $x \mapsto \nabla\|x\|_K$ is 0-homogeneous, and (12), we have

$$\begin{aligned}
P(f) &= \frac{1}{2}\int_{\mathbb{R}^n}\|\nabla\|x\|_K\|^2 e^{-\|x\|_K}dx \\
&= \frac{1}{2}nV(K)\int_0^\infty r^{n-1}\int_{\partial K}\|\nabla\|z\|_K\|^2 e^{-r} d\overline{V}_K(z)dr \\
&= \frac{1}{2}\Gamma(n)nV(K)\int_{\partial K}\|\nabla\|z\|_K\|^2 d\overline{V}_K(z),
\end{aligned}$$

where $\Gamma(\cdot)$ is the Gamma function. We need the fact that $\nabla\|z\|_K = \frac{\nu_K(z)}{\langle z, \nu_K(z)\rangle}$ when $z \in \partial K$ (see, e.g., [4]). Therefore,

$$\begin{aligned}
P(f) &= \frac{1}{2}\Gamma(n)nV(K)\int_{\partial K}\|\nabla\|z\|_K\|^2 d\overline{V}_K(z) \\
&= \frac{1}{2}\Gamma(n)\int_{\partial K} h_K(\nu_K(z))^{-1} d\mathcal{H}^{n-1}(z) \\
&= \frac{1}{2}\Gamma(n)S_2(K).
\end{aligned}$$

From the fact that the map $x \mapsto \nabla\|x\|_K$ is 0-homogeneous, (12) and (10), we have

$$\begin{aligned}
\frac{1}{2}\int_{\mathbb{R}^n}\langle x, Tx\rangle d\mu_f(x) &= \frac{1}{2}\int_{\mathbb{R}^n}\langle \nabla\|x\|_K, T\nabla\|x\|_K\rangle e^{-\|x\|_K}dx \\
&= \frac{1}{2}nV(K)\int_0^\infty r^{n-1}\int_{\partial K}\langle \nabla\|z\|_K, T\nabla\|z\|_K\rangle e^{-r}d\overline{V}_K(z)dr \\
&= \frac{1}{2}\Gamma(n)\int_{\partial K}\langle \nu_K(z), T\nu_K(z)\rangle h_K(\nu_K(z))^{-1} d\mathcal{H}^{n-1}(z) \\
&= \frac{1}{2}\Gamma(n)\int_{S^{n-1}}\langle u, Tu\rangle dS_2(K, u).
\end{aligned}$$

Hence, (4) implies that

$$\frac{\text{tr}(T)}{n}S_2(K) = \int_{S^{n-1}}\langle u, Tu\rangle dS_2(K, u)$$

for every $T \in \text{GL}(n)$. This means that the L_2 surface area measure of K, $S_2(K, \cdot)$, is an isotropic measure on S^{n-1}. □

4. Conclusions

Many outstanding works showed that the log-concave function is closely linked to the convex body. This paper presents a new connection between the theory of convex bodies and that of log-concave functions. We study the minimal perimeter of a log-concave function which can be viewed as a functional version of the minimal L_2 surface area measure of a convex body. A characteristic theorem (Theorem 1) shows that a log-concave function f has minimal perimeter if and only if the Borel measure $\frac{1}{J(f)}\mu_f(\cdot)$ is isotropic. The work done in this paper is mainly to propose a special position for log-concave functions and provides a new idea for the study of optimal problems for log-concave functions.

Author Contributions: Conceptualization, N.F. and Z.Z.; methodology, N.F. and Z.Z.; formal analysis, N.F. and Z.Z.; investigation, N.F. and Z.Z.; writing—original draft preparation, N.F. and Z.Z.; writing—review and editing, N.F. and Z.Z.; visualization, N.F. and Z.Z.; supervision, N.F. and Z.Z.; project administration, N.F. and Z.Z.; funding acquisition, N.F. and Z.Z. All authors have read and agreed to the published version of the manuscript.

Funding: This research was funded by China Postdoctoral Science Foundation (No.2019M651001) and the Science and Technology Research Program of Chongqing Municipal Education Commission (No. KJQN201901312).

Acknowledgments: The authors would like to thank the reviewers for valuable comments that helped improve the manuscript considerably.

Conflicts of Interest: The authors declare no conflict of interest.

References

1. Federer, H.; Fleming, W. Normal and integral currents. *Ann. Math.* **1960**, *72*, 458–520. [CrossRef]
2. Mazya, V. Classes of domains and imbedding theorems for function spaces. *Soviet Math. Dokl.* **1960**, *1*, 882–885.
3. Brazitikos, S.; Giannopoulos, A.; Valettas, P.; Vritsiou, B. *Geometry of Isotropic Convex Bodies*; American Mathematical Soc.: Providence, RI, USA, 2014.
4. Schneider, R. *Convex Bodies: The Brunn-Minkowski Theory, Encyclopedia of Mathematics and its Applications*, 2nd ed.; Cambridge University Press: Cambridge, UK, 2014.
5. Petty, C.M. Surface area of a convex body under affine transformations. *Proc. Am. Math. Soc.* **1961**, *12*, 824–828. [CrossRef]
6. Zou, D.; Xiong, G. The minimal Orlicz surface area. *Adv. Appl. Math.* **2014**, *61*, 25–45. [CrossRef]
7. Giannopoulos, A.A.; Milman, V.D. Extremal problems and isotropic positions of convex bodies. *Israel J. Math.* **2000**, *117*, 29–60. [CrossRef]
8. Yuan, J.; Leng, G.; Cheung, W. Convex bodies with minimal p-mean width. *Houston J. Math.* **2016**, *36*, 499–511.
9. Bastero, J.; Romance, M. Positions of convex bodies associated to extremal problems and isotropic measure. *Adv. Math.* **2004**, *184*, 64–88. [CrossRef]
10. Giannopoulos, A.A.; Papadimitrakis, M. Isotropic surface area measure. *Mathematika* **1999**, *46*, 1–13. [CrossRef]
11. Milman, E. On the mean-width of isotropic convex bodies. *Int. Math. Res. Not.* **2015**, *11*, 3408–3423.
12. Artstein-Avidan, S.; Klartag, B.; Milman, V.D. The Santaló point of a function, and a functional form of Santaló inequality. *Mathematika* **2004**, *51*, 33–48. [CrossRef]
13. Artstein-Avidan, S.; Slomka, B.A. A note on Santaló inequality for the polarity transform and its reverse. *Proc. Am. Math. Soc.* **2015**, *143*, 1693–1704. [CrossRef]
14. Fradelizi, M.; Meyer, M. Some functional forms of Blaschke-Santaló inequality. *Math. Z.* **2007**, *256*, 379–395. [CrossRef]
15. Ball, K. Isometric Problems in l_p and Sections of Convex Sets. Ph.D. Dissertation, University of Cambridge, Cambridge, UK, 1986.
16. Lehec, J. A simple proof of the functional Santaló inequality. *C. R. Acad. Sci. Paris. Sér. I* **2009**, *347*, 55–58. [CrossRef]
17. Artstein-Avidan, S.; Klartag, B.; Schütt, C.; Werner, E.M. Functional affine-isoperimetry and an inverse logarithmic Sobolev inequality. *J. Funct. Anal.* **2012**, *262*, 4181–4204. [CrossRef]
18. Caglar, U.; Werner, E.M. Divergence for s-concave and log concave functions. *Adv. Math.* **2014**, *257*, 219–247. [CrossRef]

19. Caglar, U.; Werner, E.M. Mixed f-divergence and inequalities for log concave functions. *Proc. Lond. Math. Soc.* **2015**, *110*, 271–290. [CrossRef]
20. Santambrogio, F. Dealing with moment measures via entropy and optimal transport. *J. Funct. Anal.* **2016**, *271*, 418–436. [CrossRef]
21. Colesanti, A.; Fragalà, I. The first variation of the total mass of log-concave functions and related inequalities. *Adv. Math.* **2013**, *244*, 708–749. [CrossRef]
22. Cordero-Erausquin, D.; Klartag, B. Moment measures. *J. Funct. Anal.* **2015**, *268*, 3834–3866. [CrossRef]
23. Fang, N.; Zhou, J. LYZ ellipsoid and Petty projection body for log-concave functions. *Adv. Math.* **2018**, *340*, 914–959. [CrossRef]
24. Klartag, B.; Milman, V.D. Geometry of log-concave functions and measures. *Geom. Dedicata* **2005**, *112*, 169–182. [CrossRef]
25. Lin, Y. Affine Orlicz Pólya-Szegö principle for log-concave functions. *J. Funct. Anal.* **2017**, *273*, 3295–3326. [CrossRef]
26. Milman, V.D.; Rotem, L. Mixed integral and related inequalities. *J. Funct. Anal.* **2013**, *264*, 570–604. [CrossRef]
27. Rotem, L. On the mean width of log-concave functions. In *Geometric Aspects of Functional Analysis*; Springer: Berlin/Heidelberg, Germany, 2012; pp. 355–372.
28. Rotem, L. Support functions and mean width for α-concave functions. *Adv. Math.* **2013**, *243*, 168–186. [CrossRef]
29. Rotem, L. Surface area measures of log-concave functions. *arXiv* **2020**, arXiv:2006.16933.
30. Rockafellar, R.T. *Convex Analysis*; Princeton University Press: Princeton, NJ, USA, 1970.
31. Rockafellar, R.T.; Wets, R.J.-B. *Variational Analysis, Grundlehren der Mathematischen Wissenschaften*; Springer: Berlin, Germany, 1998; Volume 317.

© 2020 by the authors. Licensee MDPI, Basel, Switzerland. This article is an open access article distributed under the terms and conditions of the Creative Commons Attribution (CC BY) license (http://creativecommons.org/licenses/by/4.0/).

Article
Ricci Curvature Inequalities for Skew CR-Warped Product Submanifolds in Complex Space Forms

Meraj Ali Khan [1],* and Ibrahim Aldayel [2]

[1] Department of Mathematics, University of Tauk, Tabuk 71491, Saudi Arabia
[2] Department of Mathematics, College of Science, AL Imam Mohammad Ibn Saud Islamic University, Riyadh 11566, Saudi Arabia; iaaldayel@imamu.edu.sa
* Correspondence: m_khan@ut.edu.sa; Tel.: +966-582903874

Received: 19 July 2020; Accepted: 3 August 2020; Published: 7 August 2020

Abstract: The fundamental goal of this study was to achieve the Ricci curvature inequalities for a skew CR-warped product (SCR W-P) submanifold isometrically immersed in a complex space form (CSF) in the expressions of the squared norm of mean curvature vector and warping functions (W-F). The equality cases were likewise examined. In particular, we also derived Ricci curvature inequalities for CR-warped product (CR W-P) submanifolds. To sustain this study, an example of these submanifolds is provided.

Keywords: Ricci curvature; skew CR-warped product submanifolds; complex space form; CR-warped product submanifolds; semi slant warped product submanifolds

1. Introduction

There have been several studies in the past to demonstrate the geometries of submanifolds in the settings of almost Hermitian (A-H) and almost contact metric (A-C M) manifolds. By the operation of the almost complex structure J, the tangent space of a submanifold of an almost Hermitian manifold can be classified into holomorphic and totally real submanifolds. The notion of CR-submanifolds was introduced and studied by A. Bejancu [1] in 1981 as a generalization of holomorphic and totally real submanifolds. Thus, as to have a more profound knowledge of the geometry of CR-submanifolds of almost Hermitian "AH" manifolds, Chen [2] further explored these submanifolds and provided many fundamental results. In 1990 Chen [3] instigated a generalized class of submanifolds, namely, slant submanifolds. Moreover, advances in the geometry of CR-submanifolds and slant submanifolds stimulated various authors to search for the class of submanifolds which unifies the properties of all previously discussed submanifolds. In this context, N. Papaghuic [4] introduced the notion of semi-slant submanifolds in the framework of almost-Hermitian manifolds and showed that submanifolds belonging to this class enjoy many of the desired properties. Later, the contact variant of semi-slant submanifolds was studied by Cabrerizo et al. [5]. Recently, B. Sahin [6] investigated another class of submanifolds in the setting of almost Hermitian manifolds and he called these submanifolds Hemi-slant submanifolds. This class includes the CR-submanifolds and slant submanifolds.

In 1990, Ronsse [7] started the study of skew CR-submanifolds in the setting of almost Hermitian manifolds. Skew CR-submanifolds contain the classes of CR-submanifolds, semi-slant submanifolds and Hemi-slant submanifolds.

The acknowledgment of warped product manifolds appeared after the methodology of Bishop and O'Neill [8] on the manifolds of non positive curvature. By analyzing the way that a Riemannian product of manifolds cannot have non positive curvature, they represented warped product (W-P) manifolds for the class of manifolds of non-positive curvature which is characterized as follows:

Let $(S_1, \langle,\rangle_1)$ and $(S_2, \langle,\rangle_2)$ be two Riemannian manifolds with Riemannian metrics $\langle,\rangle_1$ and $\langle,\rangle_2$ respectively and g be a smooth positive function on S_1. If $\pi : S_1 \times S_2 \to S_1$ and $\eta : S_1 \times S_2 \to S_2$ are the projection maps given by $\pi(x,y) = x$ and $\eta(x,y) = y$ for every $(x,y) \in S_1 \times S_2$, then the W-P manifold is the product manifold $S_1 \times S_2$ holding the Riemannian structure such that

$$\langle U_1, U_2 \rangle = \langle \pi_* U_1, \pi_* U_2 \rangle_1 + (g \circ \pi)^2 \langle \eta_* U_1, \eta_* U_2 \rangle_2,$$

for all $U_1, U_2 \in TS$. The function g is called the *warping function* (W-F) of the warped product (W-P) manifold. If the W-F is constant, then the W-P is a trivial, i.e., simply Riemannian product. Further, if $U_1 \in TS_1$ and $U_2 \in TS_2$, then from Lemma 7.3 of [8], we have the following well-known result

$$D_{U_1} U_2 = D_{U_2} U_1 = \left(\frac{U_1 g}{g}\right) U_2, \tag{1}$$

where D is the Levi-Civita connection on S. In the light of the fact that W-P manifolds have various uses in physics and the theory of relativity [9], this has been a subject of broad interest. The idea of displaying the space-time close to black holes admits the W-P manifolds [10]. Schwartzschild space-time $T \times_k S^2$, is a model of W-P, wherein the base $T = R \times R^+$ is a half plane $k > 0$ and the fiber S^2 is the unit sphere. A cosmological model to show the universe as space-time, known as the Robertson–Walker model, is a W-P manifold [11].

Some common properties of W-P manifolds were concentrated on in [8]. B.-Y. Chen [12] played out an outward investigation of W-P submanifolds in a Kaehler manifold. From that point forward, numerous geometers have investigated W-P manifolds in various settings such as almost complex and almost contact manifolds, and different existence results have been researched (see the survey article [13–16]). Recently, B. Sahin [17] contemplated SCR W-P submanifolds in Kaehler manifolds and got some essential outcomes. Further, these submanifolds were explored by Haidar and Thakur in the context of cosymplectic manifolds [18].

In 1999, Chen [19] discovered a relationship between Ricci curvature and a squared mean curvature vector for a discretionary Riemannian manifold. More precisely, Chen proved the following theorem

Theorem 1. *Let $\phi : S^t \to \tilde{S}^m(c)$ be an isometric immersion of a $t-$ dimensional Riemannian manifold into a Riemannian space form $\tilde{S}^m(c)$.*

1. *For each unit tangent vector $\chi \in T_p S^t$, we have*

$$\|\Pi\|^2(p) \geq \frac{4}{t^2} \{R^S(\chi) - (t-1)c\}$$

   where $\|\Pi\|^2(p)$ is the squared mean curvature and $R^S(\chi)$ the Ricci curvature of S^t at χ.
2. *If $\Pi(p) = 0$, then the unit tangent vector χ at p satisfies the equality case of (1) if and only if χ lies in the relative null space $\mathcal{N}_p$ at p.*
3. *The equality case holds identically for all unit tangent vectors at x if and only if either p is a totally geodesic point or $t = 2$ and p is a totally umbilical point.*

Theorem 1 was generalized for semi-slant submanifolds in Sasakian space form by Cioroboiu and Chen [20]. Further, D. W. Yoon [21] studied Chen Ricci inequality for slant submanifols in the framework of cosymplectic space forms. Motivated by Chen [19], Mihai and Ožgur [22] studied Chen Ricci inequality for real space forms with semi-symmetric connections. In [23] M. M. Tripathi formulated an improved relationship between Ricci curvature and squared mean curvature. More recently, Ali et al. [24] generalized Chen Ricci inequality for warped product submanifolds in spheres and provided some applications in mechanics and mathematical physics.

The class of SCR W-P submanifolds is rich in its geometric behavior; it contains classes of CR-warped product submanifolds, semi-slant warped product submanifolds and hemi-slant warped product submanifolds. In the literature it was found that Ricci curvature for these warped product submanifolds in complex space forms has not been studied. In other words, we can say that Theorem 1 is an open problem for skew CR-warped product submanifolds in the setting of complex space forms.

In this study our point is to establish a connection between Ricci curvature and squared mean curvature for SCR W-P submanifolds in the setting of complex space forms.

2. Preliminaries

Let $\bar{S}$ be an A-H manifold with an almost complex structure J and a Hermitian metric $\langle,\rangle$, i.e., $J^2 = -I$ and $\langle JU_1, JU_2\rangle = \langle U_1, U_2\rangle$, for all vector fields U_1, U_2 on $\bar{S}$. If J is parallel with respect to the Levi-Civita connection $\bar{D}$ on $\bar{S}$, that is

$$(\bar{D}_{U_1} J)U_2 = 0, \qquad (2)$$

for all $U_1, U_2 \in T\bar{S}$, then $(\bar{S}, J, \langle,\rangle, \bar{D})$ is called a *Kaehler manifold* (K-M).

A K-M $\bar{S}$ is called a *CSF* if it has constant holomorphic sectional curvature c denoted by $\bar{S}(c)$. The curvature tensor of the CSF $\bar{S}(c)$ is given by

$$\bar{R}(U_1, U_2, U_3, U_4) = \frac{c}{4}[\langle U_2, U_3\rangle\langle U_1, U_4\rangle - \langle U_1, U_3\rangle\langle U_2, U_4\rangle + \langle U_1, JU_3\rangle\langle JU_2, U_4\rangle \\ - \langle U_2, JU_3\rangle\langle JU_1, U_4\rangle + 2\langle U_1, JU_2\rangle\langle JU_3, U_4\rangle], \qquad (3)$$

for any $U_1, U_2, U_3, U_4 \in T\bar{S}$.

Let S be a $n-$dimensional Riemannian manifold isometrically immersed in a $m-$ dimensional Riemannian manifold $\bar{S}$. Then, the Gauss and Weingarten formulas are $\bar{D}_{U_1} U_2 = D_{U_1} U_2 + \Gamma(U_1, U_2)$ and $\bar{D}_{U_1}\xi = -A_\xi U_1 + D^\perp_{U_1}\xi$ respectively, for all $U_1, U_2 \in TS$ and $\xi \in T^\perp S$, where D is the induced Levi-Civita connection on S, ξ is a vector field normal to S, Γ is the second fundamental form of S, $D^\perp$ is the normal connection in the normal bundle $T^\perp S$ and A_ξ is the shape operator of the second fundamental form. The second fundamental form Γ and the shape operator are related by the following formula

$$\langle \Gamma(U_1, U_2), \xi\rangle = \langle A_\xi U_1, U_2\rangle. \qquad (4)$$

The Gauss equation is given by

$$R(U_1, U_2, U_3, U_4) = \bar{R}(U_1, U_2, U_3, U_4) + \langle \Gamma(U_1, U_4), \Gamma(U_2, U_3)\rangle - \langle \Gamma(U_1, U_3), \Gamma(U_2, U_4)\rangle, \qquad (5)$$

for all $U_1, U_2, U_3, U_4 \in TS$, where $\bar{R}$ and R are the curvature tensors of $\bar{S}$ and S respectively.

For any $U_1 \in TS$ and $\xi \in T^\perp S$, JU_1 and $J\xi$ can be decomposed as follows.

$$JU_1 = PU_1 + FU_1 \qquad (6)$$

and

$$J\xi = t\xi + f\xi, \qquad (7)$$

where PU_1 (resp. $t\xi$) is the tangential and FU_1 (resp. $f\xi$) is the normal component of JU_1 (resp. $J\xi$).

It is evident that $\langle JU_1, U_2\rangle = \langle PU_1, U_2\rangle$ for any $U_1, U_2 \in T_xS$; this implies that $\langle PU_1, Y_2\rangle + \langle U_1, PU_2\rangle = 0$. Thus, P^2 is a symmetric operator on the tangent space T_xS, for any $x \in S$. The eigenvalues of P^2 are real and diagonalizable. Moreover, for each $x \in S$, one can observe

$$L_x^\lambda = Ker\{P^2 + \lambda^2(x)I\}_x,$$

where I denotes the identity transformation on T_xS, and $\lambda(x) \in [0, 1]$ such that $-\lambda^2(x)$ is an eigenvalue of $P^2(x)$. Further, it is easy to observe that $KerF = L_x^1$ and $KerP = L_x^0$, where L_x^1 is the maximal holomorphic sub space of T_xS and L_x^0 is the maximal totally real subspace of T_xS; these distributions

are denoted by L and $L^\perp$ respectively. If $-\lambda_1^2(x), \ldots, -\lambda_k^2(x)$ are the eigenvalues of P^2 at x, then $T_x S$ can be decomposed as
$$T_x S = L_x^{\lambda_1} \oplus L_x^{\lambda_2} \oplus \ldots L_x^{\lambda_k}.$$

Every $L_x^{\lambda_i}$, $1 \leq i \leq k$ is a $P-$invariant subspace of $T_x S$. Moreover, if $\lambda_i \neq 0$, then $L_x^{\lambda_i}$ is even dimensional the submanifold S of a Kaehler manifold $\tilde{S}$ is a generic submanifold if there exists an integer k and functions λ_i $1 \leq i \leq k$ defined on S with $\lambda_i \in (0,1)$ such that

(i) Each $-\lambda_i^2(x), 1 \leq i \leq k$, is a distinct eigenvalue of P^2 with
$$T_x S = L_x^T \oplus L_x^\perp \oplus L_x^{\lambda_1} \oplus \ldots, \oplus L_x^{\lambda_k}$$
for any $x \in S$.

(ii) The distributions of $L_x^T, L_x^\perp$ and $L_x^{\lambda_i}$, $1 \leq i \leq k$ are independent of $x \in S$.

If in addition, each λ_i is constant on S, then S is called a skew CR-submanifold [7]. It is significant to recount that CR-submanifolds are a particular class of skew CR-submanifold for which $k = 1, L^T = \{0\}, L^\perp = \{0\}$ and λ_1 is constant. If $L^\perp = \{0\}, L^1 \neq \{0\}$ and $k = 1$, then S is a semi-slant submanifold, whereas if $L = \{0\}, L^\perp \neq \{0\}$ and $k = 1$, then S is a hemi-slant submanifold.

Definition 1. *A submanifold S of an A-H manifold $\tilde{S}$ is said to be a "skew CR-submanifold of order 1" if S is a skew CR-submanifold with $k = 1$ and λ_1 is constant.*

We have the following characterization

Theorem 2. *Reference [3] let S be a submanifold of an A-H manifold $\tilde{S}$. Then S is a slant if and only if there exists a constant $\lambda \in [0, 1]$ such that*
$$P^2 = -\lambda I.$$
Furthermore, if θ is a slant angle, then $\lambda = \cos^2 \theta$.

For any orthonormal basis $\{e_1, e_2, \ldots, e_t\}$ of the tangent space $T_x S$, the mean curvature vector $\Pi(x)$ and its squared norm are defined as follows.

$$\Pi(x) = \frac{1}{t} \sum_{i=1}^{t} \Gamma(e_i, e_i), \quad \|\Pi\|^2 = \frac{1}{t^2} \sum_{i,j=1}^{t} \langle \Gamma(e_i, e_i), \Gamma(e_j, e_j) \rangle, \tag{8}$$

where t is the dimension of S. If $\Gamma = 0$ then the submanifold is said to be totally geodesic and minimal if $\Pi = 0$. If $\Gamma(U_1, U_2) = \langle U_1, U_2 \rangle \Pi$ for all $U_1, U_2 \in TS$, then S is called totally umbilical (T-U).

The scalar curvature of $\tilde{S}$ is denoted by $\bar{\tau}(\tilde{S})$ and is defined as

$$\bar{\tau}(\tilde{S}) = \sum_{1 \leq p < q \leq m} \bar{\kappa}_{pq}, \tag{9}$$

where $\bar{\kappa}_{pq} = \bar{\kappa}(e_p \wedge e_q)$ and m is the dimension of the Riemannian manifold $\tilde{S}$. Throughout this study, we shall use the equivalent version of the above equation, which is given by

$$2\bar{\tau}(\tilde{S}) = \sum_{1 \leq p < q \leq m} \bar{\kappa}_{pq}. \tag{10}$$

In a similar way, the scalar curvature $\bar{\tau}(L_x)$ of a $L-$plane is given by

$$\bar{\tau}(L_x) = \sum_{1 \leq p < q \leq m} \bar{\kappa}_{pq}. \tag{11}$$

Let $\{e_1,\ldots,e_t\}$ be an orthonormal basis of the tangent space T_xS and if e_r belongs to the orthonormal basis $\{e_{n+1},\ldots e_m\}$ of the normal space $T^\perp S$, then we have

$$\Gamma^r_{pq} = \langle \Gamma(e_p,e_q), e_r \rangle \tag{12}$$

and

$$\|\Gamma\|^2 = \sum_{p,q=1}^{t} \langle \Gamma(e_p,e_q), \Gamma(e_p,e_q) \rangle. \tag{13}$$

Let κ_{pq} and $\bar{\kappa}_{pq}$ be the sectional curvatures of the plane sections spanned by e_p and e_q at x in the submanifold S and in the Riemannian space form $\bar{S}^m(c)$, respectively. Thus by Gauss equation, we have

$$\kappa_{pq} = \bar{\kappa}_{pq} + \sum_{r=t+1}^{m}(\Gamma^r_{pp}\Gamma^r_{qq} - (\Gamma^r_{pq})^2). \tag{14}$$

The global tensor field for orthonormal frame of vector field $\{e_1,\ldots,e_t\}$ on S is defined as

$$\bar{M}(U_1,U_2) = \sum_{i=1}^{t}\{\langle \bar{R}(e_i,U_1)U_2, e_i \rangle\}, \tag{15}$$

for all $U_1,U_2 \in T_xS$. The above tensor is called the Ricci tensor. If we fix a distinct vector e_n from $\{e_1,\ldots,e_t\}$ on S, which is governed by χ, then the Ricci curvature is defined by

$$R^S(\chi) = \sum_{\substack{p=1 \\ p\neq n}}^{t} \kappa(e_p \wedge e_n). \tag{16}$$

For a smooth function g on a Riemannian manifold S with Riemannian metric $\langle,\rangle$, the gradient of g is denoted by ∇g and is defined as

$$\langle \nabla g, U_1 \rangle = U_1 g, \tag{17}$$

for all $U_1 \in TS$.

Let the dimension of S be t and $\{e_1,e_2,\ldots,e_t\}$ be a basis of TS. Then as a result of (17), we get

$$\|\nabla g\|^2 = \sum_{i=1}^{t}(e_i(g))^2. \tag{18}$$

The Laplacian of g is defined by

$$\Delta g = \sum_{i=1}^{t}\{(\nabla_{e_i}e_i)g - e_ie_i\psi\}. \tag{19}$$

For a W-P submanifold $S_1^{t_1} \times_g S_2^{t_2}$ isometrically immersed in a Riemannian manifold $\bar{S}$, we observe the well known result, which can be described as follows [25]:

$$\sum_{p=1}^{t_1}\sum_{q=1}^{t_2}\kappa(e_p \wedge e_q) = \frac{t_2 \Delta g}{g} = t_2(\Delta \ln g - \|\nabla \ln g\|^2), \tag{20}$$

where t_1 and t_2 are the dimensions of the submanifolds $S_1^{t_1}$ and $S_2^{t_2}$ respectively.

3. Skew CR-Warped Product Submanifolds

Recently, B. Sahin [17] demonstrated the existence of SCR W-P of the type $S = S_1 \times_f S_\perp$, where S_1 is a semi-slant submanifold as defined by N. Papaghuic [4] and $S_\perp$ is a totally real submanifold. Throughout this section we consider the SCR W-P $S = S_1 \times_f S_\perp$ in a Kaehler manifold $\bar{S}$. Then it is evident that S is a proper SCR W-P of order 1. Moreover, the tangent space TS of S can be decomposed as follows.

$$TS = L^\theta \oplus L^T \oplus L^\perp, \tag{21}$$

where $L_x^\theta = L_x^{\lambda_1}$. If $L^\theta = \{0\}$, then S becomes a CR-warped product submanifold defined in [26]. If $L^T = \{0\}$, then S is reduced to a warped product hemi-slant submanifold [6]. Thus, skew CR-warped product submanifold presents a single platform to study the CR W-P submanifolds and W-P hemi-slant submanifold.

Now, we have an example of SCR W-P submanifold in an A-H manifold

Example 1. *Let S be a submanifold in R^{12} defined by $x_1 = u, x_2 = v\,sech\alpha, x_3 = k\,tanh\beta, x_4 = k\,sech\beta, x_5 = u\,sech\beta, x_6 = u\,tanh\beta, y_1 = -v, y_2 = v\,tanh\alpha, y_3 = -r\,tanh\beta, y_4 = -r\,sech\beta, y_5 = 0, y_6 = 0$. Then, we have the following basis of TS*

$$U_1 = sech\beta \frac{\partial}{\partial x_5} + tanh\beta \frac{\partial}{\partial x_6} + \frac{\partial}{\partial x_1}, \quad U_2 = sech\alpha \frac{\partial}{\partial x_2} - \frac{\partial}{\partial y_1} + tanh\alpha \frac{\partial}{\partial y_2},$$

$$U_3 = tanh\beta \frac{\partial}{\partial x_3} + sech\beta \frac{\partial}{\partial x_4}, \quad U_4 = -tanh\beta \frac{\partial}{\partial y_3} - sech\beta \frac{\partial}{\partial y_4},$$

$$U_5 = -k\,sech\beta \frac{\partial}{\partial x_3} + k\,tanh\beta \frac{\partial}{\partial x_4} + u\,tanh\beta \frac{\partial}{\partial x_5} - u\,sech\beta \frac{\partial}{\partial x_6} + r\,sech\beta \frac{\partial}{\partial y_3} - r\,tanh\beta \frac{\partial}{\partial y_4}.$$

It is straightforward to identify that $L^\theta = span\{U_1, U_2\}$ is a slant distribution with slant angle $60°$, $L = span\{U_3, U_4\}$ is a holomorphic distribution and JU_5 is orthogonal to S. Thus $L^\perp = span\{U_5\}$ is a totally real distribution. Moreover, it is easy to observe that L^θ, L and $L^\perp$ are integrable. If S_θ, S_T and $S_\perp$ are the integral manifolds of the distributions L^θ, L and $L^\perp$ respectively. Then the induced metric tensor of S is given by

$$ds^2 = \langle , \rangle_{S_\theta} + \langle , \rangle_{S_T} + (k^2 + u^2 + r^2)\langle , \rangle_{S_\perp}$$

or

$$ds^2 = \langle , \rangle_{S_1} + (k^2 + u^2 + r^2)\langle , \rangle_{S_\perp}.$$

Definition 2. *The warped product $S_1 \times_f S_2$ isometrically immersed in a Riemannian manifold $\bar{S}$ is called S_i totally geodesic if the partial second fundamental form Γ_i is zero identically. It is called S_i-minimal if the partial mean curvature vector Π^i becomes zero for $i = 1, 2$.*

Let $\{e_1, \ldots, e_p, e_{p+1} = Je_1, \ldots, e_{t_1=2p} = Je_p, e^1, \ldots, e^q, e^{q+1} = sec\theta Pe^1, \ldots, e^{(t_2=2q)} = sec\theta Pe^q, e^{t_2+1}, \ldots, e^{t_3}\}$ be a local orthonormal frame of vector fields such that $\{e_1, \ldots, e_p, e_{p+1} = Je_1, \ldots, e_{t_1=2p} = Je_p\}$ is an orthonormal basis of L, $\{e^1, \ldots, e^q, e^{q+1} = sec\theta Pe^1, \ldots, e^{(t_2=2q)} = sec\theta Pe^q\}$ is an orthonormal basis of L_θ and $\{e^{t_2+1}, \ldots, e^{t_3}\}$ is an orthonormal basis of $L^\perp$.

Throughout this paper we consider that the SCR W-P submanifold $S_1 \times_f S_\perp$ is L−minimal. Presently we have the following outcome for further applications

Lemma 1. *Let $S^t = S_1^{t_1+t_2} \times_f S_\perp^{t_3}$ be a L−minimal SCR W-P submanifold isometrically immersed in a Kaehler manifold; then*

$$\|\Pi\|^2 = \frac{1}{t^2} \sum_{r=t+1}^{m} (\Gamma_{t_1+1 t_1+1}^r + \cdots + \Gamma_{t_2 t_2}^r + \cdots + \Gamma_{tt}^r)^2, \tag{22}$$

where $\|\Pi\|^2$ represents squared mean curvature.

4. Ricci Curvature for Skew CR-Warped Product Submanifold

In this section, we investigate Ricci curvature in terms of the squared norm of mean curvature and the warping functions as follows:

Theorem 3. *Let $S^t = S_1^{t_1+t_2} \times_f S_\perp^{s_3}$ be a L−minimal SCR W-P submanifold isometrically immersed in a Complex space form $\tilde{S}^m(c)$. If the holomorphic and slant distributions L and L_θ are integrable with integral submanifolds $S_T^{t_1}$ and $S_\theta^{t_2}$ respectively, then for each orthogonal unit vector field $\chi \in T_xS$, the tangent to $S_T^{t_1}, S_\theta^{t_2}$ or $S_\perp^{t_3}$, we have that*

(1) The Ricci curvature satisfies the following expressions:

(i) If $\chi \in TS_T^{t_1}$, then

$$\frac{1}{4}t^2\|\Pi\|^2 \geq R^S(\chi) + \frac{t_3 \Delta f}{f} + \frac{c}{4}(t - t_1t_2 - t_2t_3 - t_1t_3 - \frac{1}{2}). \tag{23}$$

(ii) $\chi \in TS_\theta^{t_2}$, then

$$\frac{1}{4}t^2\|\Pi\|^2 \geq R^S(\chi) + \frac{t_3 \Delta f}{f} + \frac{c}{4}(t - t_1t_2 - t_2t_3 - t_1t_3 + 1 - \frac{3}{2}\cos^2\theta). \tag{24}$$

(iii) If $\chi \in TS_\perp^{t_2}$, then

$$\frac{1}{4}t^2\|\Pi\|^2 \geq R^S(\chi) + \frac{t_3 \Delta f}{f} + \frac{c}{4}(t - t_1t_2 - t_2t_3 - t_1t_3 + 1). \tag{25}$$

(2) If $\Gamma(x) = 0$ for each point $x \in S^t$, then there is a unit vector field χ which satisfies the equality of (1) iff S^t is mixed totally geodesic and $\chi \in \mathcal{N}_x$ at x.

(3) For the equality case we have

(a) The equality of (23) holds identically for all unit vector fields tangential to $S_T^{t_1}$ at each $x \in S^t$ iff S^t is mixed TG and L−totally geodesic SCR W-P submanifold in $\tilde{S}^m(c)$.

(b) The equality of (24) holds identically for all unit vector fields tangential to S_θ at each $x \in S^t$ iff S is mixed totally geodesic and either S^t is L_θ- totally geodesic SCR W-P submanifold or S^t is a L_θ totally umbilical in $\tilde{S}^m(c)$ with dim $L_\theta = 2$.

(c) The equality of (25) holds identically for all unit vector fields tangential to $S_\perp^{t_2}$ at each $x \in S^t$ iff S is mixed totally geodesic and either S^t is $L^\perp$- totally geodesic SCR W-P or S^t is a $L^\perp$ totally umbilical in $\tilde{S}^m(c)$ with dim $L^\perp = 2$.

(d) The equality case of (1) holds identically for all unit tangent vectors to S^t at each $x \in S^t$ iff either S^t is totally geodesic submanifold or M^t is a mixed totally geodesic totally umbilical and L totally geodesic submanifold with dim $S_\theta^{t_2} = 2$ and dim $S_\perp^{t_3} = 2$.

where t_1, t_2 and t_3 are the dimensions of $S_T^{t_1}$, $S_\theta^{t_2}$ and $S_\perp^{t_3}$ respectively.

Proof. Suppose that $S^t = S_1^{t_1+t_2} \times_f S_\perp^{t_3}$ be a SCR W-P submanifold of a CSF. From Gauss equation, we have

$$t^2\|\Pi\|^2 = 2\tau(S^t) + \|\Gamma\|^2 - 2\bar{\tau}(S^t). \tag{26}$$

Let $\{e_1,\ldots,e_{t_1},e_{t_1+1},\ldots,e_{t_2},\ldots e_t\}$ be a local orthonormal frame of vector fields on S^t such that $\{e_1,\ldots,e_{t_1}\}$ is tangential to $S_T^{t_1}$, $\{e_{t_1+1},\ldots,e_{t_2}\}$ is tangential to $S_\theta^{t_2}$ and $\{e_{t_2+1},\ldots,e_t\}$ is the tangent to $S_\perp^{t_3}$. Thus, the unit tangent vector $\chi = e_A \in \{e_1,\ldots,e_t\}$ can be expanded (26) as follows.

$$t^2\|\Pi\|^2 = 2\tau(S^t) + \tfrac{1}{2}\sum_{r=t+1}^{m}\{(\Gamma^r_{11}+\ldots\Gamma^r_{t_2t_2}+\cdots+\Gamma^r_{tt}-\Gamma^r_{AA})^2+(\Gamma^r_{AA})^2\} \\ -\sum_{r=t+1}^{m}\sum_{1\leq i\neq j\leq t}\Gamma^r_{ii}\Gamma^r_{jj}-2\bar\tau(S^t). \tag{27}$$

The above expression can be represented as

$$t^2\|\Pi\|^2 = 2\tau(S^t) + \frac{1}{2}\sum_{r=t+1}^{m}\{(\Gamma^r_{11}+\ldots\Gamma^r_{t_2t_2}+\cdots+\Gamma^r_{tt})^2 \\ + (2\Gamma^r_{AA}-(\Gamma^r_{11}+\cdots+\Gamma^r_{tt}))^2\} + 2\sum_{r=t+1}^{m}\sum_{1\leq i<j\leq t}(\Gamma^r_{ij})^2 \\ - 2\sum_{r=t+1}^{m}\sum_{1\leq i<j\leq t}\Gamma^r_{ii}\Gamma^r_{jj}-2\bar\tau(S^t).$$

In view of the assumption that SCR W-P submanifold $S_1\times_f S_\perp$ is L-minimal submanifold, the preceding expression takes the form

$$t^2\|\Pi\|^2 = 2\tau(S^t) + \frac{1}{2}\sum_{r=t+1}^{m}\{(\Gamma^r_{t_1+1t_1+1}+\ldots\Gamma^r_{t_2t_2}+\cdots+\Gamma^r_{tt})^2 \\ + \frac{1}{2}\sum_{r=t+1}^{m}(2\Gamma^r_{AA}-(\Gamma^r_{t_1+1t_1+1}+\ldots\Gamma^r_{t_2t_2}+\cdots+\Gamma^r_{tt}))^2 \\ + \sum_{r=t+1}^{m}\sum_{\substack{1\leq i<j\leq t \\ i,j\neq A}}(\Gamma^r_{ij})^2 - \sum_{r=t+1}^{m}\sum_{\substack{1\leq i<j\leq t \\ i,j\neq A}}\Gamma^r_{ii}\Gamma^r_{jj}-2\bar\tau(S^t) \\ + \sum_{r=t+1}^{m}\sum_{\substack{a=1 \\ a\neq A}}^{t}(\Gamma^r_{aA})^2 + \sum_{r=t+1}^{m}\sum_{\substack{1\leq i<j\leq t \\ i,j\neq A}}(\Gamma^r_{ij})^2 - \sum_{r=t+1}^{m}\sum_{\substack{1\leq i<j\leq t \\ i,j\neq A}}\Gamma^r_{ii}\Gamma^r_{jj}. \tag{28}$$

Equation (14) can be written as

$$\sum_{\substack{1\leq p<q\leq t \\ p,q\neq A}}\bar\kappa_{pq} - \sum_{\substack{1\leq p<q\leq t \\ p,q\neq A}}\kappa_{pq} = \sum_{r=t+1}^{m}\sum_{\substack{1\leq p<q\leq t \\ p,q\neq A}}(\Gamma^{pq}_r)^2 - \sum_{r=t+1}^{m}\sum_{\substack{1\leq p<q\leq t \\ p,q\neq A}}\Gamma^r_{pp}\Gamma^r_{qq}.$$

Substituting this value in (28), we derive

$$t^2\|\Pi\|^2 = 2\tau(S^t) + \frac{1}{2}\sum_{r=t+1}^{m}\{(\Gamma^r_{t_1+1t_1+1}+\ldots\Gamma^r_{t_2t_2}+\cdots+\Gamma^r_{tt})^2 \\ + \frac{1}{2}\sum_{r=t+1}^{m}(2\Gamma^r_{AA}-(\Gamma^r_{t_1+1t_1+1}+\ldots\Gamma^r_{t_2t_2}+\cdots+\Gamma^r_{tt}))^2 \\ + \sum_{r=t+1}^{m}\sum_{1\leq i<j\leq t}(\Gamma^r_{ij})^2 - \sum_{r=t+1}^{m}\sum_{\substack{1\leq i<j\leq t \\ i,j\neq A}}\Gamma^r_{ii}\Gamma^r_{jj}-2\bar\tau(S^t) \\ + \sum_{r=t+1}^{m}\sum_{\substack{a=1 \\ a\neq A}}^{t}(\Gamma^r_{aA})^2 + \sum_{\substack{1\leq i<j\leq t \\ i,j\neq A}}\bar\kappa_{ij} - \sum_{\substack{1\leq i<j\leq t \\ i,j\neq A}}\kappa_{ij}. \tag{29}$$

On the other hand, from (9) we have

$$\tau(S^t) = \sum_{1 \leq i < j \leq t} \kappa(e_i \wedge e_j) = \sum_{\alpha=1}^{t_1+t_2} \sum_{\beta=t_1+t_2+1}^{t} \kappa(e_\alpha \wedge e_\beta) + \sum_{1 \leq \alpha < \gamma \leq t_1} \kappa(e_\alpha \wedge e_\gamma) \\ + \sum_{t_1+1 \leq l < 0 \leq t_2} \kappa(e_l \wedge e_0) + \sum_{t_2+1 \leq u < v \leq t} \kappa(e_u \wedge e_v). \quad (30)$$

Using (9) and (20), we derive

$$\tau(S^t) = \frac{t_3 \Delta f}{f} + \bar{\tau}(S_T^{t_1}) + \bar{\tau}(S_\theta^{t_2}) + \bar{\tau}(S_\perp^{t_3}).$$

Using this in (29), we get

$$t^2 \|\Pi\|^2 = \frac{t_3 \Delta f}{f} + \frac{1}{2} \sum_{r=t+1}^{m} (\Gamma^r_{t_1+1 t_1+1} + \ldots \Gamma^r_{t_2 t_2} + \cdots + \Gamma^r_{tt})^2 \\ + \frac{1}{2} \sum_{r=t+1}^{m} (2\Gamma^r_{AA} - (\Gamma^r_{t_1+1 t_1+1} + \ldots \Gamma^r_{t_2 t_2} + \cdots + \Gamma^r_{tt}))^2 \\ + \sum_{r=t+1}^{m} \sum_{1 \leq \alpha < \beta \leq t_1} (\Gamma^r_{\alpha\alpha} \Gamma^r_{\beta\beta} - (\Gamma^r_{\alpha\beta})^2) \\ + \sum_{r=t+1}^{m} \sum_{t_1+1 \leq p < q \leq t_2} (\Gamma^r_{pp} \Gamma^r_{qq} - (\Gamma^r_{pq})^2) \\ + \sum_{r=t+1}^{m} \sum_{t_2+1 \leq s < n \leq t} (\Gamma^r_{ss} \Gamma^r_{nn} - (\Gamma^r_{sn})^2) \\ + \sum_{r=t+1}^{m} \sum_{1 \leq i < j \leq t} (\Gamma^r_{ij})^2 - \sum_{r=t+1}^{m} \sum_{\substack{1 \leq i < j \leq t \\ i,j \neq A}} (\Gamma^r_{ii} \Gamma^r_{jj}) \\ - 2\bar{\tau}(S^t) + \sum_{\substack{1 \leq i < j \leq t \\ i,j \neq A}} \bar{\kappa}_{ij} + \bar{\tau}(S_T^{t_1}) + \bar{\tau}(S_\theta^{t_2}) + \bar{\tau}(S_\perp^{t_3}). \quad (31)$$

Considering unit tangent vector $\chi = e_A$, we have three choices: χ is the tangent to the base manifold $S_T^{t_1}$ or $S_\theta^{t_2}$, or to the fiber $S_\perp^{t_3}$.

Case 1: If $\chi \in S_T^{t_1}$, then we need to choose a unit vector field from $\{e_1, \ldots, e_{t_1}\}$. Let $\chi = e_1$; then by (15) and the assumption that the submanifolds is L−minimal, we have

$$t^2 \|\Pi\|^2 \geq R^S(\chi) + \frac{1}{2} \sum_{r=t+1}^{m} (\Gamma^r_{t_1+1 t_1+1} + \ldots \Gamma^r_{t_2 t_2} + \cdots + \Gamma^r_{tt})^2 \\ + \frac{t_3 \Delta f}{f} + \frac{1}{2} \sum_{r=t+1}^{m} (2\Gamma^r_{11} - (\Gamma^r_{t_1+1 t_1+1} + \ldots \Gamma^r_{t_2 t_2} + \cdots + \Gamma^r_{tt}))^2 \\ + \sum_{r=t+1}^{m} \sum_{1 \leq \alpha < \beta \leq t_1} (\Gamma^r_{\alpha\alpha} \Gamma^r_{\beta\beta} - (\Gamma^r_{\alpha\beta})^2) \\ + \sum_{r=t+1}^{m} \sum_{t_1+1 \leq p < q \leq t_2} (\Gamma^r_{pp} \Gamma^r_{qq} - (\Gamma^r_{pq})^2) \quad (32) \\ + \sum_{r=t+1}^{m} \sum_{t_2+1 \leq s < n \leq t} (\Gamma^r_{ss} \Gamma^r_{nn} - (\Gamma^r_{sn})^2) \\ + \sum_{r=t+1}^{m} \sum_{1 \leq i < j \leq t} (\Gamma^r_{ij})^2 - \sum_{r=t+1}^{m} \sum_{2 \leq i < j \leq t} (\Gamma^r_{ii} \Gamma^r_{jj}) \\ - 2\bar{\tau}(S^t) + \sum_{2 \leq i < j \leq t} \bar{\kappa}(e_i, e_j) + \bar{\tau}(S_T^{t_1}) + \bar{\tau}(S_\theta^{t_2}) + \bar{\tau}(S_\perp^{t_3}).$$

Putting $U_1, U_3 = e_i$, $U_2, U_4 = e_j$ in the formula (3), we have

$$2\bar{\tau}(S) = \frac{c}{4}[t(t-1) + 3t_1 + 3t_2 \cos^2 \theta] \tag{33}$$

$$\sum_{2 \leq i < j \leq t} \bar{K}(e_i, e_j) = \frac{c}{8}[(t-1)(t-2) + 3(t_1-1) + 3t_2 \cos^2 \theta]$$

$$\bar{\tau}(S_T^{t_1}) = \frac{c}{8}[t_1(t_1-1) + 3t_1]$$

$$\bar{\tau}(S_\theta^{t_2}) = \frac{c}{8}[t_2(t_2-1) + 3t_2 \cos^2 \theta]$$

$$\bar{\tau}(S_\perp^{t_3}) = \frac{c}{8}[t_3(t_3-1)].$$

Using these values in (32), we get

$$t^2 \|\Pi\|^2 \geq R^S(\chi) + \frac{1}{2} t^2 \|\Pi\|^2 + \frac{1}{2} \sum_{r=t+1}^{m} (2\Gamma_{11}^r - (\Gamma_{t_1+1 t_1+1}^r + \cdots + \Gamma_{tt}^r))^2$$

$$+ \frac{t_3 \Delta f}{f} + \sum_{r=t+1}^{m} \sum_{i=1}^{t_1} \sum_{j=t_1+1}^{t_2} (\Gamma_{ij}^r)^2$$

$$+ \sum_{r=t+1}^{m} \sum_{i=1}^{t_1} \sum_{k=t_2+1}^{t} (\Gamma_{ik}^r)^2 + \sum_{r=t+1}^{m} \sum_{\beta=2}^{t_1} \Gamma_{11}^r \Gamma_{\beta\beta}^r \tag{34}$$

$$- \sum_{r=t+1}^{m} \sum_{i=2}^{t_1} \sum_{j=t_1+1}^{t_2} \Gamma_{ii}^r \Gamma_{jj}^r - \sum_{r=t+1}^{m} \sum_{i=2}^{t_1} \sum_{k=t_2+1}^{t} \Gamma_{ii}^r \Gamma_{kk}^r$$

$$+ \frac{c}{4}(t - t_1 t_2 - t_2 t_3 - t_3 t_1 - \frac{1}{2}).$$

In view of the assumption that the submanifold is $L-$minimal, then

$$\sum_{r=t+1}^{m} \sum_{\beta=2}^{t_1} \Gamma_{11}^r \Gamma_{\beta\beta}^r = \sum_{r=t+1}^{m} (\Gamma_{11}^r)^2$$

$$- \sum_{r=t+1}^{m} \sum_{i=2}^{t_1} \left[\sum_{j=t_1+1}^{t_2} \Gamma_{ii}^r \Gamma_{jj}^r + \sum_{k=t_2+1}^{t} \Gamma_{ii}^r \Gamma_{kk}^r \right] = \sum_{r=t+1}^{m} \sum_{j=t_1+1}^{t} \Gamma_{11}^r \Gamma_{jj}^r.$$

Utilizing that in (34), we have

$$t^2 \|\Pi\|^2 \geq R^S(\chi) + \frac{1}{2} t^2 \|\Pi\|^2 + \frac{1}{2} \sum_{r=t+1}^{m} (2\Gamma_{11}^r - (\Gamma_{t_1+1 t_1+1}^r + \cdots + \Gamma_{nn}^r))^2$$

$$+ \frac{t_3 \Delta f}{f} + \sum_{r=t+1}^{m} \sum_{i=1}^{t_1} \sum_{j=t_1+1}^{t_2} (\Gamma_{ij}^r)^2$$

$$+ \sum_{r=t+1}^{m} \sum_{i=1}^{t_1} \sum_{k=t_2+1}^{t} (\Gamma_{ik}^r)^2 - \sum_{r=t+1}^{m} (\Gamma_{11}^r)^2 + \sum_{i=1}^{t_1} \sum_{j=t_1+1}^{t} \Gamma_{ii}^r \Gamma_{jj}^r \tag{35}$$

$$+ \frac{c}{4}(t - t_1 t_2 - t_2 t_3 - t_3 t_1 - \frac{1}{2}).$$

The third term on the right hand side can be written as

$$\frac{1}{2}\sum_{r=t+1}^{m}(2\Gamma_{11}^r - (\Gamma_{t_1+1t_1+1}^r + \cdots + \Gamma_{t_2t_2}^r + \cdots + \Gamma_{nn}^r))^2$$
$$= 2\sum_{r=t+1}^{m}(\Gamma_{11}^r)^2 + \frac{1}{2}t^2\|\Pi\|^2 - 2\sum_{r=t+1}^{m}\left[\sum_{j=t_1+1}^{t_2}\Gamma_{11}^r\Gamma_{jj}^r\right. \tag{36}$$
$$\left.+ \sum_{k=t_2+1}^{t}\Gamma_{11}^r\Gamma_{kk}^r\right].$$

Combining above two expressions, we have

$$\frac{1}{2}t^2\|\Pi\|^2 \geq R^S(\chi) + \sum_{r=t+1}^{m}(\Gamma_{11}^r)^2 - \sum_{r=t+1}^{m}\sum_{j=t_1+1}^{t}\Gamma_{11}^r\Gamma_{jj}^r$$
$$+ \frac{1}{2}\sum_{r=t+1}^{m}(\Gamma_{t_1+1t_1+1}^r + \cdots + \Gamma_{t_2t_2}^r + \cdots + \Gamma_{nn}^r)^2 \tag{37}$$
$$+ \sum_{r=t+1}^{m}\sum_{i=1}^{t_1}\sum_{j=t_1+1}^{t}(\Gamma_{ij}^r)^2 + \frac{t_3\Delta f}{f}$$
$$+ \frac{c}{4}(t - t_1t_2 - t_2t_3 - t_3t_1 - \frac{1}{2}),$$

or equivalently

$$\frac{1}{4}t^2\|\Pi\|^2 \geq R^S(\chi) + \frac{1}{4}\sum_{r=t+1}^{m}(2\Gamma_{11}^r - (\Gamma_{t_1+1t_1+1}^r + \cdots + \Gamma_{t_2t_2}^r + \cdots + \Gamma_{nn}^r))^2$$
$$+ \sum_{r=t+1}^{m}\sum_{i=1}^{t_1}\sum_{j=t_1+1}^{t}(\Gamma_{ij}^r)^2 + \frac{t_3\Delta f}{f} \tag{38}$$
$$+ \frac{c}{4}(t - t_1t_2 - t_2t_3 - t_3t_1 - \frac{1}{2}),$$

which proves the inequality (i) of (1).

Case 2. If χ is tangential to $S_\theta^{t_2}$, we choose the unit vector from $\{e_{t_1+1},\ldots,e_{t_2}\}$. Suppose $\chi = e_{t_2}$; then from (28), we deduce

$$t^2\|\Pi\|^2 \geq R^S(\chi) + \frac{1}{2}\sum_{r=t+1}^{m}(\Gamma_{t_1+1t_1+1}^r + \ldots \Gamma_{t_2t_2}^r + \cdots + \Gamma_{tt}^r)^2$$
$$+ \frac{t_3\Delta f}{f} + \frac{1}{2}\sum_{r=t+1}^{m}((\Gamma_{t_1+1t_1+1}^r + \ldots \Gamma_{t_2t_2}^r + \cdots + \Gamma_{tt}^r) - 2\Gamma_{t_2t_2}^r)^2$$
$$+ \sum_{r=t+1}^{m}\sum_{1\leq\alpha<\beta\leq t_1}(\Gamma_{\alpha\alpha}^r\Gamma_{\beta\beta}^r - (\Gamma_{\alpha\beta}^r)^2) + \sum_{r=t+1}^{m}\sum_{t_1+1\leq s<n\leq t_2}(\Gamma_{ss}^r\Gamma_{nn}^r - (\Gamma_{sn}^r)^2) \tag{39}$$
$$+ \sum_{r=t+1}^{m}\sum_{t_2+1\leq p<q\leq t}(\Gamma_{pp}^r\Gamma_{qq}^r - (\Gamma_{pq}^r)^2) + \sum_{r=t+1}^{m}\sum_{1\leq i<j\leq t}(\Gamma_{ij}^r)^2$$
$$- \sum_{r=t+1}^{m}\sum_{\substack{1\leq i<j\leq n\\i,j\neq t_2}}(\Gamma_{ii}^r\Gamma_{jj}^r) - 2\bar{\tau}(S^t) + \sum_{\substack{1\leq i<j\leq t\\i,j\neq t_2}}\bar{R}(e_i,e_j)$$
$$+ \bar{\tau}(S_T^{t_1}) + \bar{\tau}(S_\theta^{t_2} + \bar{\tau}(S_\perp^{t_3})).$$

From (3) by putting $U_1, U_3 = e_i$, $U_2, U_3 = e_j$, one can compute

$$\sum_{\substack{1 \leq i < j \leq t \\ i,j \neq t_2}} \bar{K}(e_i, e_j) = \frac{c}{8}[(t-1)(t-2) + 3t_1 + 3t_2 \cos^2 \theta]$$

$$\bar{\tau}(S_T^{t_1}) = \frac{c}{8}[t_1(t_1 - 1) + 3t_1]$$

$$\bar{\tau}(S_\theta^{t_3}) = \frac{c}{8}[t_2(t_2 - 1) + 3t_2 \cos^2 \theta]$$

$$\bar{\tau}(S_\perp^{t_3}) = \frac{c}{8}[t_3(t_3 - 1)].$$

Using these values together with (33) in (39) and applying similar techniques as in Case 1, we obtain

$$\begin{aligned}
t^2 \|\Pi\|^2 \geq R^S(\chi) &+ \frac{1}{2} \sum_{r=t+1}^{m} ((\Gamma_{t_1+1 t_1+1}^r + \ldots \Gamma_{t_2 t_2}^r + \cdots + \Gamma_{tt}^r) - 2\Gamma_{t_2 t_2}^r))^2 \\
&+ \frac{1}{2} t^2 \|\Pi\|^2 + \frac{t_3 \Delta f}{f} + \sum_{r=t+1}^{m} \sum_{1 \leq i < j \leq n} (\Gamma_{ij}^r)^2 \\
&+ \sum_{r=t+1}^{m} \big[\sum_{n=t_1+1}^{t_2-1} \Gamma_{t_2 t_2}^r \Gamma_{nn}^r + \sum_{l=t_2+1}^{t} \Gamma_{t_2 t_2}^r \Gamma_{ll}^r \big] \\
&\sum_{r=1}^{m} \sum_{i=1}^{t_1} \big[\sum_{j=t_1+1}^{t_2-1} \Gamma_{ii}^r \Gamma_{jj}^r + \sum_{k=t_2+1}^{t} \Gamma_{ii}^r \Gamma_{kk}^r \big] \\
&+ \frac{c}{4}(t - t_1 t_2 - t_2 t_3 - t_3 t_1 + 1).
\end{aligned} \quad (40)$$

By the assumption that the submanifold S^t is L-minimal, one can conclude

$$\sum_{r=1}^{m} \sum_{i=1}^{t_1} \big[\sum_{j=t_1+1}^{t_2-1} \Gamma_{ii}^r \Gamma_{jj}^r + \sum_{k=t_2+1}^{t} \Gamma_{ii}^r \Gamma_{kk}^r \big] = 0.$$

The second and seventh terms on right hand side of (40) can be solved as follows:

$$\begin{aligned}
\frac{1}{2} \sum_{r=t+1}^{m} &((\Gamma_{t_1+1 t_1+1}^r + \cdots + \Gamma_{tt}^r) - 2\Gamma_{t_2 t_2}^r))^2 + \sum_{r=t+1}^{m} \big[\sum_{n=t_1+1}^{t_2-1} \Gamma_{t_2 t_2}^r \Gamma_{nn}^r + \sum_{l=t_2+1}^{t} \Gamma_{t_2 t_2}^r \Gamma_{ll}^r \big] \\
&= \frac{1}{2} \sum_{r=t+1}^{m} (\Gamma_{t_1+1 t_1+1}^r + \cdots + \Gamma_{nn}^r)^2 + 2 \sum_{r=t+1}^{m} (\Gamma_{t_2 t_2}^r)^2 \\
&- 2 \sum_{r=t+1}^{m} \sum_{j=t_1+1}^{t} \Gamma_{t_2 t_2}^r \Gamma_{jj}^r + \sum_{r=t+1}^{m} \sum_{n=t_1+1}^{t} \Gamma_{t_2 t_2}^r \Gamma_{nn}^r - \sum_{r=t+1}^{m} (\Gamma_{t_2 t_2}^r)^2 \\
&= \frac{1}{2} \sum_{r=t+1}^{m} (\Gamma_{t_1+1 t_1+1}^r + \cdots + \Gamma_{nn}^r)^2 + \sum_{r=t+1}^{m} (\Gamma_{t_2 t_2}^r)^2 \\
&- \sum_{r=t+1}^{m} \sum_{j=t_1+1}^{t} \Gamma_{nn}^r \Gamma_{jj}^r.
\end{aligned} \quad (41)$$

By utilizing those two values in (40), we arrive at

$$\frac{1}{2}t^2\|\Pi\|^2 \geq R^S(\chi) + \sum_{r=t+1}^{m}(\Gamma_{t_2t_2}^r)^2 - \sum_{r=t+1}^{m}\sum_{i=t_1+1}^{t}\Gamma_{nn}^r\Gamma_{jj}^r$$
$$+ \frac{1}{2}\sum_{r=t+1}^{m}(\Gamma_{t_1+1t_1+1}^r + \cdots + \Gamma_{nn}^r)^2 + \frac{1}{2}t^2\|\Pi\|^2 + \frac{t_3\Delta f}{f} \qquad (42)$$
$$+ \sum_{r=t+1}^{m}\sum_{i=1}^{t_1}\sum_{j=t_1+1}^{t}(\Gamma_{ij}^r)^2 + \frac{c}{4}(t - t_1t_2 - t_2t_3 - t_3t_1 + 1).$$

By using similar steps as in Case 1, the above inequality can be written as

$$\frac{1}{4}t^2\|\Pi\|^2 \geq R^S(\chi) + \frac{1}{4}\sum_{r=t+1}^{m}(2\Gamma_{t_2t_2}^r - (\Gamma_{t_1+1t_1+1}^r + \cdots + \Gamma_{nn}^r))^2$$
$$+ \frac{t_3\Delta f}{f} + \frac{c}{4}(t - t_1t_2 - t_2t_3 - t_1t_3 + 1). \qquad (43)$$

The last inequality leads to inequality (ii) of (1).

Case 3. If χ is tangential to $S_\perp^{t_3}$, then we choose the unit vector field from $\{e_{t_2+1}, \ldots, e_n\}$. Suppose the vector χ is e_n. Then from (28)

$$t^2\|\Pi\|^2 \geq R^S(\chi) + \frac{1}{2}\sum_{r=t+1}^{m}(\Gamma_{t_1+1t_1+1}^r + \ldots \Gamma_{t_2t_2}^r + \cdots + \Gamma_{tt}^r)^2$$
$$+ \frac{t_3\Delta f}{f} + \frac{1}{2}\sum_{r=t+1}^{m}((\Gamma_{t_1+1t_1+1}^r + \ldots \Gamma_{t_2t_2}^r + \cdots + \Gamma_{tt}^r) - 2\Gamma_{tt}^r)^2$$
$$+ \sum_{r=t+1}^{m}\sum_{1\leq\alpha<\beta\leq t_1}(\Gamma_{\alpha\alpha}^r\Gamma_{\beta\beta}^r - (\Gamma_{\alpha\beta}^r)^2) + \sum_{r=t+1}^{m}\sum_{t_1+1\leq s<n\leq t_2}(\Gamma_{ss}^r\Gamma_{nn}^r - (\Gamma_{sn}^r)^2) \qquad (44)$$
$$+ \sum_{r=t+1}^{m}\sum_{t_2+1\leq p<q\leq t}(\Gamma_{pp}^r\Gamma_{qq}^r - (\Gamma_{pq}^r)^2) + \sum_{r=t+1}^{m}\sum_{1\leq i<j\leq n}(\Gamma_{ij}^r)^2$$
$$- \sum_{r=t+1}^{m}\sum_{1\leq i<j\leq t-1}\Gamma_{ii}^r\Gamma_{jj}^r - 2\bar{\tau}(S^t) + \sum_{1\leq i<j\leq t-1}\bar{K}(e_i, e_j)$$
$$+ \bar{\tau}(S_T^{t_1}) + \bar{\tau}(S_\theta^{t_2}) + \bar{\tau}(S_\perp^{t_3}).$$

From (3), one can compute

$$\sum_{1\leq i<j\leq t-1}\bar{K}(e_i, e_j) = \frac{c}{8}[(t-1)(t-2) + 3t_1 + 3(t_2 - 1)\cos^2\theta]$$

$$\bar{\tau}(S_T^{t_1}) = \frac{c}{8}[t_1(t_1 - 1) + 3t_1]$$

$$\bar{\tau}(S_\theta^{t_2}) = \frac{c}{8}[t_2(t_2 - 1) + 3t_2\cos^2\theta]$$

$$\bar{\tau}(t_\perp^{t_3}) = \frac{c}{8}[t_3(t_3 - 1)].$$

By usage of those values together with (33) in (44), and analogously to Case 1 and Case 2, we obtain

$$
\begin{aligned}
t^2\|\Pi\|^2 \geq & R^S(\chi) + \frac{1}{2}t^2\|\Pi\|^2 + \frac{1}{2}\sum_{r=t+1}^{m}((\Gamma^r_{t_1+1t_1+1} + \ldots \Gamma^r_{t_2t_2} + \cdots + \Gamma^r_{tt}) - 2\Gamma^r_{tt})^2 \\
& + \frac{t_3\Delta f}{f} + \sum_{r=t+1}^{m}\sum_{1\leq i<j\leq t}(\Gamma^r_{ij})^2 \\
& + \sum_{r=t+1}^{m}\sum_{q=t_1+1}^{t-1}\Gamma^r_{tt}\Gamma^r_{qq} - \sum_{r=t+1}^{m}\sum_{i=1}^{t_1}\sum_{j=t_1+1}^{t-1}\Gamma^r_{ii}\Gamma^r_{jj} \\
& + \frac{c}{4}(t - t_1t_2 - t_2t_3 - t_1t_3 + 1 - \frac{3}{2}\cos^2\theta).
\end{aligned}
\tag{45}
$$

Again, using the assumption that S^t is $L-minimal$, it is easy to verify

$$\sum_{r=t+1}^{m}\sum_{i=1}^{t_1}\sum_{j=t_1+1}^{t-1}\Gamma^r_{ii}\Gamma^r_{jj} = 0. \tag{46}$$

Using in (45), we obtain

$$
\begin{aligned}
t^2\|\Pi\|^2 \geq & R^S(\chi) + \frac{1}{2}t^2\|\Pi\|^2 + \frac{1}{2}\sum_{r=t+1}^{m}((\Gamma^r_{t_1+1t_1+1} + \ldots \Gamma^r_{t_2t_2} + \cdots + \Gamma^r_{tt}) - 2\Gamma^r_{tt})^2 \\
& + \frac{t_3\Delta f}{f} + \sum_{r=t+1}^{m}\sum_{1\leq i<j\leq n}(\Gamma^r_{ij})^2 + \sum_{r=t+1}^{m}\sum_{q=t_1+1}^{t-1}\Gamma^r_{tt}\Gamma^r_{qq} \\
& + \frac{c}{4}(t - t_1t_2 - t_2t_3 - t_1t_3 + 1 - \frac{3}{2}\cos^2\theta).
\end{aligned}
\tag{47}
$$

The third and sixth terms on the right hand side of (47) in a similar way as in Case 1 and Case 2 can be simplified as

$$
\begin{aligned}
& \frac{1}{2}\sum_{r=t+1}^{m}((\Gamma^r_{t_1+1t_1+1} + \ldots \Gamma^r_{t_2t_2} + \cdots + \Gamma^r_{tt}) - 2\Gamma^r_{tt})^2 + \sum_{r=t+1}^{m}\sum_{q=t_1+1}^{t-1}\Gamma^r_{tt}\Gamma^r_{qq} \\
= & \frac{1}{2}\sum_{r=t+1}^{m}(\Gamma^r_{t_1+1t_1+1} + \ldots \Gamma^r_{t_2t_2} + \cdots + \Gamma^r_{tt})^2 + \sum_{r=t+1}^{m}(\Gamma^r_{tt})^2 \\
& - \sum_{r=t+1}^{m}\sum_{j=t_1+1}^{t}\Gamma^r_{tt}\Gamma^r_{jj}.
\end{aligned}
\tag{48}
$$

By combining (47) and (48) and using similar techniques as used in Case 1 and Case 2, we can derive

$$
\frac{1}{4}t^2\|\Pi\|^2 \geq R^S(\chi) + \frac{1}{4}\sum_{r=t+1}^{m}(2\Gamma^r_{tt} - (\Gamma^r_{t_1+1t_1+1} + \cdots + \Gamma^r_{tt}))^2 + \frac{t_3\Delta f}{f} + \frac{c}{4}(t - t_1t_2 - t_2t_3 - t_1t_3 + 1 - \frac{3}{2}\cos^2\theta).
\tag{49}
$$

The last inequality leads to inequality (iii) in (1).

Next, we explore the equality cases of (1). First, we redefine the notion of the relative null space $\mathcal{N}_x$ of the submanifold S^t in the CSF $\tilde{S}^m(c)$ at any point $x \in S^t$; the relative null space was defined by B.-Y. Chen [19], as follows:

$$\mathcal{N}_x = \{U_1 \in T_xS^t : \Gamma(U_1, U_2) = 0, \forall U_2 \in T_xS^t\}.$$

For $A \in \{1, \ldots, t\}$ a unit vector field e_A tangential to S^t at x satisfies the equality sign of (23) identically iff

$$(i)\ \sum_{p=1}^{t_1}\sum_{q=t_1+1}^{t}\Gamma^r_{pq}=0\ (ii)\ \sum_{b=1}^{t}\sum_{\substack{A=1\\b\neq A}}^{t}\Gamma^r_{bA}=0\ (iii)\ 2\Gamma^r_{AA}=\sum_{q=t_1+1}^{t}\Gamma^r_{qq},\qquad(50)$$

such that $r\in\{t+1,\ldots m\}$ the condition (i) implies that S^t is mixed totally geodesic SCR W-P submanifold. Combining statements (ii) and (iii) with the fact that S^t is L−minimal, we get that the unit vector field $\chi=e_A\in\mathcal{N}_x$. The converse is trivial; this proves statement (2).

For a SCR W-P submanifold, the equality sign of (23) holds identically for all unit tangent vector belong to $S^{t_1}_T$ at x iff

$$(i)\ \sum_{p=1}^{t_1}\sum_{q=t_1+1}^{t}\Gamma^r_{pq}=0\ (ii)\ \sum_{b=1}^{t}\sum_{\substack{A=1\\b\neq A}}^{t_1}\Gamma^r_{bA}=0\ (iii)\ 2\Gamma^r_{pp}=\sum_{q=t_1+1}^{t}\Gamma^r_{qq},\qquad(51)$$

where $p\in\{1,\ldots,t_1\}$ and $r\in\{t+1,\ldots,m\}$. Since S^t is L−minimal SCR W-P submanifold, the third condition implies that $\Gamma^r_{pp}=0$, $p\in\{1,\ldots,t_1\}$. Using this in the condition (ii), we conclude that S^t is L−totally geodesic SCR W-P submanifold in $\bar{S}^m(c)$ and totally mixed geodesicness follows from the condition (i), which proves (a) in the statement (3).

For a SCR W-P submanifold, the equality sign of (24) holds identically for all unit tangent vector fields tangential to $S^{t_2}_\theta$ at x if and only if

$$(i)\ \sum_{p=1}^{t_1}\sum_{q=t_1+1}^{t}\Gamma^r_{pq}=0\ (ii)\ \sum_{b=1}^{t}\sum_{\substack{A=t_1+1\\b\neq A}}^{t_2}\Gamma^r_{bA}=0\ (iii)\ 2\Gamma^r_{KK}=\sum_{q=t_1+1}^{t}\Gamma^r_{qq},\qquad(52)$$

such that $K\in\{t_1+1,\ldots,t_2\}$ and $r\in\{t+1,\ldots,m\}$. From the condition (iii) two cases emerge; that is,

$$\Gamma^r_{KK}=0,\ \forall K\in\{t_1+1,\ldots,t_2\}\ \text{and}\ r\in\{t+1,\ldots,m\}\ \text{or}\ dim\,S^{t_2}_\theta=2.\qquad(53)$$

If the first case of (52) is satisfied, then by virtue of condition (ii), it is easy to conclude that S^t is a D_θ− totally geodesic SCR W-P submanifold in $\bar{S}^m(c)$. This is the first case of part (b) of statement (3).

For a SCR W-P submanifold, the equality sign of (25) holds identically for all unit tangent vector fields tangent to $S^{t_3}_\perp$ at x if and only if

$$(i)\ \sum_{p=1}^{t_1}\sum_{q=t_1+1}^{t}\Gamma^r_{pq}=0\ (ii)\ \sum_{b=1}^{t}\sum_{\substack{A=t_2+1\\b\neq A}}^{t_3}\Gamma^r_{bA}=0\ (iii)\ 2\Gamma^r_{LL}=\sum_{q=t_1+1}^{t}\Gamma^r_{qq},\qquad(54)$$

such that $L\in\{t_2+1,\ldots,t\}$ and $r\in\{t+1,\ldots,m\}$. From the condition (iii) two cases arise; that is,

$$\Gamma^r_{LL}=0,\ \forall L\in\{t_2+1,\ldots,t\}\ \text{and}\ r\in\{t+1,\ldots,m\}\ \text{or}\ dim\,S^{t_3}_\perp=2.\qquad(55)$$

If the first case of (54) is satisfied, then by virtue of condition (ii), it is easy to conclude that S^t is a $L_\perp$− totally geodesic SCR W-P submanifold in $\bar{S}^m(c)$. This is the first case of part (c) of statement (3).

For the other case, assume that S^t is not $L_\perp$−totally geodesic SCR W-P submanifold and dim $S^{t_3}_\perp=2$. Then condition (ii) of (54) implies that S^t is $L_\perp$− totally umbilical SCR W-P submanifold in $\bar{S}(c)$, which is second case of this part. This verifies part (c) of (3).

To prove (d) using parts (a), (b) and (c) of (3), we combine (51), (52) and (54). For the first case of this part, assume that $dim\,S^{t_2}_\theta\neq 2$ and $dim\,S^{t_3}_\perp\neq 2$. From parts (a), (b) and (c) of statement (3) we concluded that M^t is L−totally geodesic, L_θ− is totally geodesic and $D_\perp$− is a totally geodesic submanifold in $\bar{S}^m(c)$. Hence S^t is a totally geodesic submanifold in $\bar{S}^m(c)$.

For another case, suppose that first case is not satisfied. Then parts (a), (b) and (c) provide that S^t is mixed totally geodesic and $L-$ totally geodesic submanifold of $\tilde{S}^m(c)$ with $dim S_\theta^{t_2} = 2$ and $dim S_\perp^{t_3} = 2$. From the conditions (b) and (c) it follows that S^t is $L_\theta-$ and $L_\perp-$totally umbilical SCR W-P submanifolds and from (a) it is $L-$totally geodesic, which is part (d). This proves the theorem. □

If, $S_\theta^{t_2} = \{0\}$ then the SCR W-P submanifold becomes the CR W-P submanifold. In this case we have the following corollary

Corollary 1. *Let $S^t = S_T^{t_1} \times_f S_\perp^{t_3}$ be a CR W-P submanifold isometrically immersed in a CSF $\tilde{S}^m(c)$. Then for each orthogonal unit vector field $\chi \in T_x S^t$, either tangent to $S_T^{t_1}$ or $S_\perp^{t_3}$, we have*

(1) *The Ricci curvature satisfy the following inequalities*

  (i) *If $\chi \in S_T^{t_1}$, then*
$$\frac{1}{4} t^2 \|\Pi\|^2 \geq R^S(\chi) + \frac{t_3 \Delta f}{f} + \frac{c}{4}(t - t_1 t_3 - \frac{1}{2}). \tag{56}$$

  (ii) *If $\chi \in S_\perp^{t_3}$, then*
$$\frac{1}{4} t^2 \|\Pi\|^2 \geq R^S(\chi) + \frac{t_3 \Delta f}{f} + \frac{c}{4}(t - t_1 t_3 + 1). \tag{57}$$

(2) *If $H(x) = 0$, then each point $x \in S^t$ there is a unit vector field χ which satisfies the equality case of (1) if and only if S^t is mixed totally geodesic and χ lies in the relative null space $\mathcal{N}_x$ at x.*

(3) *For the equality case we have*

  (a) *The equality case of (56) holds identically for all unit vector fields tangent to $S_T^{t_1}$ at each $x \in S^t$ iff S^t is mixed totally geodesic and $L-$totally geodesic CR W-P submanifold in $\tilde{S}^m(c)$.*

  (b) *The equality case of (57) holds identically for all unit vector fields tangent to $S_\perp^{t_3}$ at each $x \in M^t$ iff S^t is mixed totally geodesic and either S^t is $L_\perp$- totally geodesic CR-warped product or S^t is a $L_\perp$ totally umbilical in $\tilde{S}^m(c)$ with dim $L_\perp = 2$.*

  (c) *The equality case of (1) holds identically for all unit tangent vectors to S^t at each $x \in S^t$ if and only if either S^t is totally geodesic submanifold or S^t is a mixed totally geodesic totally umbilical and $L-$ totally geodesic submanifold with dim $S_\perp^{t_3} = 2$*

where t_1 and t_3 are the dimensions of $S_T^{t_1}$ and $S_\perp^{t_3}$ respectively.

In view of (20) we have the another version of the Theorem 2 as follows:

Theorem 4. *Let $S^t = S_1^{t_1+t_2} \times_f S_\perp^{t_3}$ be a $L-$minimal SCR W-P submanifold isometrically immersed in a CSF $\bar{M}(c)$. If the holomorphic and slant distributions L and L_θ are integrable with integral submanifolds $S_T^{t_1}$ and $S_\theta^{t_2}$ respectively. Then for each orthogonal unit vector field $\chi \in T_x S$, either tangent to $S_T^{t_1}$, $S_\theta^{t_2}$ or $S_\perp^{t_3}$, we have*

(1) *The Ricci curvature satisfy the following inequalities*

  (i) *If $\chi \in TS_T^{t_1}$, then*
$$\frac{1}{4} t^2 \|\Pi\|^2 \geq R^S(\chi) + t_3(\Delta \ln f - \|\nabla \ln f\|^2) + \frac{c}{4}(t - t_1 t_2 - t_2 t_3 \\ - t_1 t_3 - \frac{1}{2}). \tag{58}$$

(ii) $\chi \in TS_\theta^{t_2}$, then

$$\frac{1}{4}t^2\|\Pi\|^2 \geq R^S(\chi) + t_3(\Delta \ln f - \|\nabla \ln f\|^2) + \frac{c}{4}(t - t_1t_2 - t_2t_3 \\ - t_1t_3 + 1 - \frac{3}{2}\cos^2\theta).\quad (59)$$

(iii)   If $\chi \in TS_\perp^{t_2}$, then

$$\frac{1}{4}t^2\|\Pi\|^2 \geq R^S(\chi) + t_3(\Delta \ln f - \|\nabla \ln f\|^2) + \frac{c}{4}(t - t_1t_2 - t_2t_3 - t_1t_3 + 1). \quad (60)$$

(2) If $\Gamma(x) = 0$ for each point $x \in S^t$, then there is a unit vector field χ which satisfies the equality of (1) iff S^t is mixed totally geodesic and $\chi \in \mathcal{N}_x$ at x.

(3) For the equality case we have

   (a) The equality of (58) holds identically for all unit vector fields tangent to $S_T^{t_1}$ at each $x \in S^t$ iff S^t is mixed TG and L−totally geodesic SCR W-P submanifold in $\tilde{S}^m(c)$.

   (b) The equality of (59) holds identically for all unit vector fields tangent to S_θ at each $x \in S^t$ iff S is mixed totally geodesic and either S^t is L_θ- totally geodesic SCR W-P submanifold or S^t is a L_θ totally umbilical in $\tilde{S}^m(c)$ with dim $L_\theta = 2$.

   (c) The equality of (60) holds identically for all unit vector fields tangent to $S_\perp^{t_2}$ at each $x \in S^t$ iff S is mixed totally geodesic and either S^t is $L^\perp$- totally geodesic SCR W-P or S^t is a $L^\perp$ totally umbilical in $\tilde{S}^m(c)$ with dim $L^\perp = 2$.

   (d) The equality case of (1) holds identically for all unit tangent vectors to S^t at each $x \in S^t$ iff either S^t is totally geodesic submanifold or M^t is a mixed totally geodesic totally umbilical and L totally geodesic submanifold with dim $S_\theta^{t_2} = 2$ and dim $S_\perp^{t_3} = 2$.

Where t_1, t_2 and t_3 are the dimensions of $S_T^{t_1}$, $S_\theta^{t_2}$ and $S_\perp^{t_3}$ respectively.

5. Conclusions

In the present study we obtained some fundamental results for skew CR-warped product submanifolds in the frame of complex space forms. Further, some inequalities in terms of Ricci curvature and squared norm of mean curvature vector were derived. In particular, a Ricci curvature for CR-warped product submanifolds was also discussed. Recently, we also studied warped product submanifolds in complex space forms (see [15,16]) and obtained some inequalities in terms of squared norm of second fundamental form, slant function and the warping functions, but the results obtained in the present study are dissimilar from the previous works of the authors and were proved by using different techniques.

Author Contributions: Conceptualization, M.A.K. and I.A.; formal analysis, M.A.K.; investigation, M.A.K. and I.A.; methodology, I.A.; project administration, M.A.K.; validation, M.A.K. and I.A.; writing—original draft, M.A.K. and I.A. All authors have read and agreed to the published version of the manuscript.

Funding: This research received no external funding.

Acknowledgments: We would like to thank the anonymous reviewers for their thoughtful comments and efforts towards improving our manuscript.

Conflicts of Interest: Both the authors declares that they have no conflict of interest.

Abbreviations

W-P   Warped product
W-F   Warping function
CSF   Complex Space form

References

1. Bejancu, A. *Geometry of CR-Submanifolds*; Kluwer Academic Publishers: Dortrecht, The Netherlands, 1986.
2. Chen, B.Y. CR-submanifolds of a Kaehler manifold I. *J. Differ. Geom.* **1981**, *16*, 305–323. [CrossRef]
3. Chen, B.Y. *Geometry of Slant Submanifolds*; Katholieke Universiteit Leuven: Leuven, Belgium, 1990.
4. Papaghiuc, N. Semi-slant submanifolds of Kaehler manifold. *Analele Ştiinţifice ale Universităţii "Al. I. Cuza" din Iaşi* **1994**, *40*, 55–61.
5. Cabrerizo, J.L.; Carriazo, A.; Fernandez, L.M.; Fernandez, M. Slant submanifolds in Sasakian manifolds. *Glasg. Math. J.* **2000**, *42*, 125–138. [CrossRef]
6. Sahin, B. Warped product submanifolds of Kaehler manifolds with a slant factor. *Ann. Pol. Math.* **2009**, *95*, 207–206. [CrossRef]
7. Ronsse, G.S. Generic and skew CR-submanifolds of a Kaehler manifold. *Bull. Inst. Math. Acad. Sin.* **1990**, *18*, 127–141.
8. Bishop, R.L.; O'Neill, B. Manifolds of negative curvature. *Trans. Am. Math. Soc.* **1969**, *145*, 1–9. [CrossRef]
9. Beem, J.K.; Ehrlich, P.; Powell, T.G. *Warped Product Manifolds in Relativity, Selected Studies*; North-Holland: Amsterdam, NY, USA, 1982.
10. Hawkings, S.W.; Ellis, G.F.R. *The Large Scale Structure of Space-Time*; Cambridge University Press: Cambridge, UK, 1973.
11. O'Neill, B. *Semi-Riemannian Geometry with Application to Relativity*; Academic Press: Cambridge, MA, USA, 1983.
12. Chen, B.Y. Geometry of warped product CR-submanifolds in Kaehler manifolds, I. *Monatsh Math.* **2001**, *133*, 177–195. [CrossRef]
13. Siddiqui, A.N.; Chen, B.Y.; Bahadr, O. Statistical solitons and inequalities for statistical warped product submanifolds. *Mathematics* **2019**, *7*, 797. [CrossRef]
14. Chen, B.Y. Geometry of warped product submanifolds: A survey. *J. Adv. Math. Study* **2013**, *6*, 1–43.
15. Khan, M.A. Warped product point wise semi-slant submanifolds of the complex space forms. *Rendiconti del Circolo Matematico di Palermo Series 2* **2019**, *69*, 195–207. [CrossRef]
16. Khan, M.A.; Khan, K. Biwarped product submanifolds of complex space forms. *Int. J. Geom. Methods Mod. Phys.* **2019**, *16*, 1950072. [CrossRef]
17. Sahin, B. Skew CR-warped products of Kaehler manifolds. *Math. Commun.* **2010**, *15*, 189–204.
18. Haider, S.M.K.; Thakur, M. Warped product skew CR-submanifolds of cosymplectic manifold. *Lobachevskii J. Math.* **2012**, *33*, 262–273. [CrossRef]
19. Chen, B.Y. Relations between Ricci curvature and shape operator for submanifolds with arbitrary codimension. *Glasg. Math. J.* **1999**, *41*, 33–41. [CrossRef]
20. Cioroboiu, D.; Chen, B.Y. Inequalities for semi-slant submanifolds in Sasakian space forms. *Int. J. Math. Math. Sci.* **2003**, *27*, 1731–1738. [CrossRef]
21. Yoon, D.W. Inequality for Ricci curvature of slant submanifolds in cosymplectic space forms. *Turk. J. Math.* **2006**, *30*, 43–56.
22. Mihai, A.; Ozgur, C. Chen inequalities for submanifolds of real space forms with a semi-symmetric metric connection. *Taiwan. J. Math.* **2010**, *14*, 1465–1477. [CrossRef]
23. Tripathi, M.M. Improved Chen Ricci Inequality for curvature-like tensors and its applications. *Diff. Geom. Appl.* **2011**, *29*, 685–698. [CrossRef]
24. Ali, A.; Laurian-Ioan, P.; Al-Khalidi, A.H. Ricci curvature on warped product submanifolds in spheres with geometric applications. *J. Geom. Phys.* **2019**, *146*, 103510. [CrossRef]

25. Chen, B.Y.; Dillen, F.; Verstraelen, L.; Vrancken, L. Characterization of Riemannian space forms, Einstein spaces and conformally flate spaces. *Proc. Am. Math. Soc.* **2000**, *128*, 589–598 [CrossRef]
26. Chen, B.Y. *Differential Geometry of Warped Product Manifolds and Submanifolds*; World Scientific: Singapore, 2017.

© 2020 by the authors. Licensee MDPI, Basel, Switzerland. This article is an open access article distributed under the terms and conditions of the Creative Commons Attribution (CC BY) license (http://creativecommons.org/licenses/by/4.0/).

Article

New Refinements of the Erdös–Mordell Inequality and Barrow's Inequality

Jian Liu

East China Jiaotong University, Nanchang 330013, China; liujian999@ecjtu.jx.cn

Received: 10 July 2019; Accepted: 5 August 2019; Published: 9 August 2019

Abstract: In this paper, two new refinements of the Erdös–Mordell inequality and three new refinements of Barrow's inequality are established. Some related interesting conjectures are put forward.

Keywords: Erdös–Mordell inequality; Barrow's inequality; triangle; interior point

1. Introduction

In 1935, Erdös [1] proposed the following geometric inequality:

For any interior point P of the triangle ABC, let R_1, R_2, R_3 be the distances from P to the vertices A, B, C, respectively, and let r_1, r_2, r_3 be the distances from P to the sides BC, CA, AB, respectively. Then

$$\sum R_1 \geq 2 \sum r_1, \tag{1}$$

where $\sum$ denotes the cyclic sums (we shall use this symbol in the sequel). Equality in (1) holds if and only if the triangle ABC is equilateral and P is its center.

Two years later, Mordell and Barrow [2] first proved the inequality (1), and the latter actually obtained the following sharpness:

$$\sum R_1 \geq 2 \sum w_1, \tag{2}$$

where w_1, w_2, w_3 are the lengths of the bisectors of $\angle BPC, \angle CPA, \angle APB$, respectively.

The above two inequalities have long been famous results in the field of geometric inequalities. The former is called the Erdös–Mordell inequality, which has attracted the interest of many authors and motivated a large number of research papers (see [2–28] and the references cited therein).

In 1957, Ozeki [22] first obtained the following generalization of Barrow's inequality (2) for convex polygons: For any interior point P of the convex polygon $A_1 A_2 \cdots A_n$, it holds that

$$\sum_{i=1}^{n} R_i \geq \sec \frac{\pi}{n} \sum_{i=1}^{n} w_i, \tag{3}$$

where $R_i = PA_i$ and w_i denote the lengths of the bisectors of $\angle A_i P A_{i+1} (i = 1, 2, \cdots, n$ and $A_{n+1} = A_1)$.

Some other discussions about Barrow's inequality and (3) can be found in [4,14,19,21,23,27].

In 2012, when the author considered Oppenheim's inequality (see [24])

$$\sum R_2 R_3 \geq 2 \sum (r_3 + r_1)(r_1 + r_2), \tag{4}$$

the following sharpened version of the Erdös–Mordell inequality was found:

$$R_2 + R_3 \geq 2r_1 + \frac{(r_2 + r_3)^2}{R_1}, \tag{5}$$

with equality if and only if $\triangle ABC$ is an isosceles right triangle and P is its circumcenter. Furthermore, by using inequalities (4), (5), and other results, the author obtained a series of refinements for the Erdös–Mordell inequality in [14,16].

In this paper, we shall give two new refinements of the Erdös–Mordell inequality and three new refinements of Barrow's inequality. In addition, we shall present several interesting related conjectures in the last section.

2. Refinements of the Erdös–Mordell Inequality

In [11], the author proved the following refinement of the Erdös–Mordell inequality:

$$\sum R_1 \geq 2\sqrt{\sum h_a r_1} \geq 2 \sum r_1 \qquad (6)$$

where h_a, h_b, h_c are the corresponding altitudes of the sides BC, CA, AB of the triangle ABC.

Here, we further give the following result:

Theorem 1. *For any interior point P of the triangle ABC, it holds that*

$$\sum R_1 \geq 2\sqrt{\sum h_a w_1} \geq 2\sqrt{\sum h_a r_1} \geq 2 \sum r_1. \qquad (7)$$

Equalities in (7) all hold if and only if $\triangle ABC$ is equilateral and P is its center.

To prove Theorem 1, we first give several lemmas.

Lemma 1. *For any triangle ABC with sides a, b, c and real numbers x, y, z, it holds that*

$$\left(\sum xa\right)^2 \geq \left(2 \sum bc - \sum a^2\right) \sum yz, \qquad (8)$$

with equality if and only if $x : y : z = (b+c-a) : (c+a-b) : (a+b-c)$.

For any triangle ABC with sides a, b, c, we have $\sqrt{b} + \sqrt{c} > \sqrt{b+c} > \sqrt{a}$. Thus, $\sqrt{a}, \sqrt{b}, \sqrt{c}$ can be viewed sides of a triangle, and we see that inequality (8) can be obtained by using the following weighted Oppenheim inequality (see [19], p. 681):

$$\left(\sum xa^2\right)^2 \geq 16 S^2 \sum yz \qquad (9)$$

(where S is the area of $\triangle ABC$) and the following equivalent form of the Heron formula:

$$16 S^2 = 2 \sum b^2 c^2 - \sum a^4. \qquad (10)$$

Remark 1. *In the sixth chapter of the monograph [17], the author proved that inequality (8) is equivalent with (9) and the Wolstenholme inequality (52) below.*

In the Appendix A of my monograph [17], Theorem A3 gives an equivalent theorem for the geometric transformations, which includes the following conclusion: An inequality involving any interior point P of the triangle ABC,

$$f(a, b, c, R_1, R_2, R_3, r_1, r_2, r_3) \geq 0, \qquad (11)$$

is equivalent to

$$f\left(\frac{aR_1}{2R}, \frac{bR_2}{2R}, \frac{cR_3}{2R}, r_1, r_2, r_3, \frac{r_2 r_3}{R_1}, \frac{r_3 r_1}{R_2}, \frac{r_1 r_2}{R_3}\right) \geq 0. \qquad (12)$$

In fact, this conclusion can be extended to the following:

Lemma 2. *With above notations, the inequality*

$$f(a,b,c,R_1,R_2,R_3,r_1,r_2,r_3,w_1,w_2,w_3) \geq 0 \tag{13}$$

is equivalent to

$$f\left(\frac{aR_1}{2R}, \frac{bR_2}{2R}, \frac{cR_3}{2R}, r_1, r_2, r_3, \frac{r_2r_3}{R_1}, \frac{r_3r_1}{R_2}, \frac{r_1r_2}{R_3},\right.$$
$$\left.\frac{2r_2r_3}{r_2+r_3}\sin\frac{A}{2}, \frac{2r_3r_1}{r_3+r_1}\sin\frac{B}{2}, \frac{2r_1r_2}{r_1+r_2}\sin\frac{C}{2}\right) \geq 0. \tag{14}$$

Proof. Let DEF be the pedal triangle of P with respect to the triangle ABC (see Figure 1), and let $EF = a_p, FD = b_p, DE = c_p$, then it is easy to get

$$a_p = \frac{aR_1}{2R}, \quad b_p = \frac{bR_2}{2R}, \quad c_p = \frac{cR_3}{2R}. \tag{15}$$

Let h_1, h_2, h_3 be the distances from P to the side lines EF, FD, DE, respectively, we also easily obtain

$$h_1 = \frac{r_2r_3}{R_1}, \quad h_2 = \frac{r_3r_1}{R_2}, \quad h_3 = \frac{r_1r_2}{R_3}. \tag{16}$$

In addition, by means of the known formula in the triangle ABC

$$w_a = \frac{2bc}{b+c}\cos\frac{A}{2} \tag{17}$$

(where w_a is the bisector of $\angle BAC$) and the fact that $\angle EPF = \pi - A$, we get

$$w_1' = \frac{2r_2r_3}{r_2+r_3}\sin\frac{A}{2}, \tag{18}$$

where w_1' is the bisector of $\angle EPF$. Two similar relations hold for the bisectors w_2', w_3' of $\angle FPD, \angle DPE$, respectively.

If we apply inequality (13) to triangle DEF and point P, then

$$f(a_p, b_p, c_p, r_1, r_2, r_3, h_1, h_2, h_3, w_1', w_2', w_3') \geq 0.$$

Substituting (15), (16), and (18) into this inequality, (14) follows immediately. Conversely, we can obtain (13) from (14) by using the method of proving Theorem A3 in Appendix A of the monograph [17]. Thus, inequality (13) is equivalent with (14). The proof of Lemma 2 is completed. □

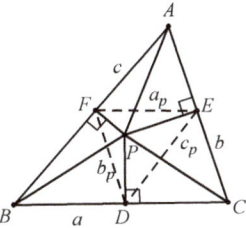

Figure 1. An inequality involving any point P inside triangle ABC is equivalent to the one involving the point P and its pedal triangle DEF with respect to ABC.

Lemma 3. *For any interior point P of the triangle ABC, it holds that*

$$r_2 + r_3 \leq 2R_1 \sin \frac{A}{2}, \qquad (19)$$

with equality if and only if $r_2 = r_3$.

Inequality (19) is well-known and is easily proved (see [29], p. 111).

Next, we prove Theorem 1.

Proof. Since $w_1 \geq r_1$ etc., the second inequality in (7) is evidently valid. In addition, the third inequality of (7) is easily obtained (see [11]).

We now prove the first inequality in (7), i.e.,

$$\left(\sum R_1\right)^2 \geq 4 \sum h_a w_1. \qquad (20)$$

By the area formula $h_a = 2S/a$ and the identity

$$\sum a r_1 = 2S, \qquad (21)$$

we see that (20) is equivalent to

$$\left(\sum R_1\right)^2 \geq 4 \sum a r_1 \sum \frac{w_1}{a}. \qquad (22)$$

According to Lemma 2 and the relations (15) and (16), we further know that inequality (22) is equivalent to

$$\left(\sum r_1\right)^2 \geq 4 \sum \frac{aR_1}{2R} \cdot \frac{r_2 r_3}{R_1} \sum \frac{2R}{aR_1} \cdot \frac{2r_2 r_3}{r_2 + r_3} \sin \frac{A}{2},$$

i.e.,

$$\left(\sum r_1\right)^2 \geq 8 \sum a r_2 r_3 \sum \frac{r_2 r_3}{a(r_2 + r_3)R_1} \sin \frac{A}{2}. \qquad (23)$$

But using $r_2 r_3 \leq (r_2 + r_3)^2/4$, Lemma 3, and the known formula

$$\sin \frac{A}{2} = \sqrt{\frac{(s-b)(s-c)}{bc}} \qquad (24)$$

(where $s = (a+b+c)/2$), we have

$$\sum \frac{r_2 r_3}{a(r_2 + r_3)R_1} \sin \frac{A}{2}$$
$$\leq \frac{1}{4} \sum \frac{r_2 + r_3}{aR_1} \sin \frac{A}{2}$$
$$\leq \frac{1}{2} \sum \frac{1}{a} \sin^2 \frac{A}{2}$$
$$= \frac{1}{2abc} \sum (s-b)(s-c).$$

Thus, in order to prove inequality (23), we only need to prove that

$$\left(\sum r_1\right)^2 \geq 4 \sum \frac{r_2 r_3}{bc} \sum (s-b)(s-c). \qquad (25)$$

Putting $x = r_1/a, y = r_2/b, z = r_3/c$ in inequality (8) of Lemma 1 and noting the fact that

$$2 \sum bc - \sum a^2 = 4 \sum (s-b)(s-c), \qquad (26)$$

we get inequality (25) immediately. Thus, inequality (20) is proved. It is easily known that the equality in (20) holds if and only if $\triangle ABC$ is equilateral and P is its center. This completes the proof of Theorem 1. □

Now we state and prove the second refinement of the Erdös–Mordell inequality.

Theorem 2. *For any interior point P of the triangle ABC, it holds that*

$$\sum R_1 \geq \sqrt{\frac{1}{2}\sum a^2 + \sum R_2 R_3 + 2\sum r_1^2}$$
$$\geq \sqrt{\frac{1}{2}\sum a^2 + \frac{3}{2}\sum (r_2 + r_3)^2} \geq 2\sum r_1. \qquad (27)$$

The first equality in (27) holds if and only if P is the circumcenter of the triangle ABC. The second and third equalities in (27) hold if and only if the triangle ABC is equilateral and P is its center.

Proof. In triangle ABC, we have the following known angle bisector formula:

$$w_a = \frac{2}{b+c}\sqrt{sbc(s-a)}. \qquad (28)$$

Noting that $\sqrt{bc} \leq (b+c)/2$ and $s = (a+b+c)/2$, we have

$$w_a \leq \frac{1}{2}\sqrt{[(b+c)^2 - a^2]}, \qquad (29)$$

with equality if and only if $b = c$. Applying this inequality to $\triangle BPC$, we get

$$\sqrt{(R_2 + R_3)^2 - a^2} \geq 2w_1. \qquad (30)$$

Hence, we have

$$\sum (R_2 + R_3)^2 \geq \sum a^2 + 4\sum w_1^2, \qquad (31)$$

that is,

$$\sum R_1^2 + \sum R_2 R_3 \geq \frac{1}{2}\sum a^2 + 2\sum w_1^2.$$

Adding $\sum R_2 R_3$ to both sides of the above inequality and then squaring root, we obtain

$$\sum R_1 \geq \sqrt{\frac{1}{2}\sum a^2 + \sum R_2 R_3 + 2\sum w_1^2}. \qquad (32)$$

Sine $w_1 \geq r_1$ etc., the first inequality in (27) obviously holds. Note that the equality in (30) holds if and only if $R_2 = R_3$, thus the equality in (31) holds if and only if $R_1 = R_2 = R_3$, which means that P is the circumcenter of the triangle ABC. Furthermore, we can conclude that the first equality in (27) holds if and only if P is the circumcenter of the triangle ABC.

Clearly, the second inequality in (27) is equivalent to

$$\sum R_2 R_3 + 2\sum r_1^2 \geq \frac{3}{2}\sum (r_2 + r_3)^2.$$

Removing $2\sum r_1^2$ to the right and arranging gives the previous Oppenheim inequality (4), which has been proved by the author in different ways (see [12,14]).

For the third inequality in (27), by squaring both sides and arranging, we know that it is equivalent to

$$2\sum r_1^2 + 10\sum r_2 r_3 \leq \sum a^2, \qquad (33)$$

which was first established by Chu in [30] and proved by the author in another way in [15]. In addition, we have known that both equalities in (4) and (33) hold if and only if $\triangle ABC$ is equilateral and P is its center. This completes the proof of Theorem 2. □

From Theorem 2, we have

Corollary 1. *For any interior point P of the triangle ABC, it holds that*

$$2\left(\sum R_1\right)^2 - 3\sum (r_2+r_3)^2 \geq \sum a^2. \tag{34}$$

Furthermore, we can easily obtain the following inequality:

Corollary 2. *For any interior point P of the triangle ABC, it holds that*

$$\left(\sum R_1\right)^2 - 2\left(\sum r_1\right)^2 \geq \frac{1}{2}\sum a^2. \tag{35}$$

3. Refinements of Barrow's Inequality

In [14], Theorem 4.3 gives the following refinement of the Erdös–Mordell inequality:

$$\sum R_1 \geq \sqrt{\sum \left[R_1^2 + 2r_1 R_1 + (r_2+r_3)^2\right]} \geq 2\sum r_1, \tag{36}$$

which is actually equivalent to

$$\sum R_1 \geq \sqrt{\sum (R_1+r_1)^2 + \left(\sum r_1\right)^2} \geq 2\sum r_1. \tag{37}$$

Now, we point out that for Barrow's inequality (2), the following similar result holds:

Theorem 3. *For any interior point P of the triangle ABC, it holds that*

$$\sum R_1 \geq \sqrt{\sum (R_1+w_1)^2 + \left(\sum w_1\right)^2} \geq 2\sum w_1. \tag{38}$$

Equalities in (38) hold if and only if $\triangle ABC$ is equilateral and P is its center.

Clearly, the first inequality in (38) is also equivalent to the following interesting form:

$$\left(\sum R_1\right)^2 - \left(\sum w_1\right)^2 \geq \sum (R_1+w_1)^2. \tag{39}$$

To prove this inequality, we first prove a strengthening of the previous inequality (5), which is posed by the author in [12] as a conjecture.

Lemma 4. *For any interior point P of the triangle ABC, it holds that*

$$R_2 + R_3 \geq 2w_1 + \frac{(w_2+w_3)^2}{R_1}, \tag{40}$$

with equality if and only if $CA = AB$ and P is the circumcenter of the triangle ABC.

Proof. We let $\angle BPC = 2\delta_1, \angle CPA = 2\delta_2, \angle APB = 2\delta_3$. By the previous formula (17), we know that inequality (40) is equivalent to

$$R_2 + R_3 \geq \left(\frac{4R_2 R_3}{R_2+R_3}\cos\delta_1 + \frac{1}{R_1}\left(\frac{2R_3 R_1}{R_3+R_1}\cos\delta_2 + \frac{2R_1 R_2}{R_1+R_2}\cos\delta_3\right)^2\right).$$

Since $R_3 + R_1 \geq 2\sqrt{R_3 R_1}$ and $R_1 + R_2 \geq 2\sqrt{R_1 R_2}$, to prove the above inequality we only need to prove that

$$R_2 + R_3 \geq \frac{4R_2 R_3}{R_2 + R_3} \cos \delta_1 + \left(\sqrt{R_3} \cos \delta_2 + \sqrt{R_2} \cos \delta_3\right)^2. \tag{41}$$

Letting $\sqrt{R_2} = y$ and $\sqrt{R_3} = z$, (41) then becomes

$$y^2 + z^2 \geq \frac{4y^2 z^2}{y^2 + z^2} \cos \delta_1 + (z \cos \delta_2 + y \cos \delta_3)^2. \tag{42}$$

Note that $\delta_1, \delta_2, \delta_3$ can be viewed angles of a non-obtuse triangle. To prove inequality (42), we only need to prove that the following inequality holds for non-obtuse triangles ABC and real numbers y, z:

$$y^2 + z^2 \geq \frac{4y^2 z^2}{y^2 + z^2} \cos A + (z \cos B + y \cos C)^2, \tag{43}$$

that is,

$$(y^2 + z^2)^2 - 4y^2 z^2 \cos A - (y^2 + z^2)(z \cos B + y \cos C)^2 \geq 0. \tag{44}$$

Multiplying both sides by $4(abc)^2$ and using the law of cosines, we can transform the proof to the following weighted inequality:

$$4(abc)^2 (y^2 + z^2)^2 - 8bca^2 y^2 z^2 (b^2 + c^2 - a^2)$$
$$- (y^2 + z^2) \left[zb(c^2 + a^2 - b^2) + yc(a^2 + b^2 - c^2)\right]^2 \geq 0. \tag{45}$$

If we denote by Q_0 the value of the left-hand side of (45), then it is easy to check the following identity:

$$Q_0 = (y^2 + z^2)(yc - zb)^2(c^2 + a^2 - b^2)(a^2 + b^2 - c^2)$$
$$+ 2a^2(b^2 + c^2 - a^2)[(y^2 c - z^2 b)^2 + y^2 z^2 (b - c)^2], \tag{46}$$

which shows that inequality $Q_0 \geq 0$ holds clearly. Moreover, from (46) we can obtain the following conclusions: (i) if $A = \pi/2$, then the equality in (43) holds if and only if $yc = zb$; (ii) if $A < \pi/2$, then the equality in (43) holds if and only if $y = z$ and $b = c$. According to this conclusion, we can further determine the equality condition of (40), just as mentioned in Lemma 4. This completes the proof of Lemma 4. □

Remark 2. *Adding R_1 to both sides of (40) and noting that*

$$R_1 + \frac{(w_2 + w_3)^2}{R_1} \geq 2(w_2 + w_3),$$

we obtain Barrow's inequality (2). Therefore, inequality (43) is actually stronger than Barrow's inequality (2).

We now prove Theorem 3.

Proof. As the proof of the first inequality (36) given in [14], we can easily prove the first inequality of (38) by using Lemma 4 (we omit the details here). By the power means inequality and Barrow's inequality (2), we have

$$\sum (R_1 + w_1)^2 \geq \frac{1}{3} \left[\sum (R_1 + w_1)\right]^2 = \frac{1}{3} \left(\sum R_1 + \sum w_1\right)^2$$
$$\geq 3 \left(\sum w_1\right)^2.$$

Hence, the second inequality of (38) follows immediately. Moreover, it is easily known that both equalities in (38) hold if and only if $\triangle ABC$ is equilateral and P is its center. The proof of Theorem 3 is completed. □

Next, we state and prove the second new refinement of Barrow's inequality (2).

Theorem 4. *For any interior point P of the triangle ABC, it holds that*

$$\sum R_1 \geq \sum \sqrt{(R_2 + R_3)^2 - a^2} \geq 2 \sum w_1. \tag{47}$$

The second equality in (47) holds if and only if P is the circumcenter of the triangle ABC.

Proof. Firstly, we prove the first of (47):

$$\sum R_1 \geq \sum \sqrt{(R_2 + R_3)^2 - a^2}. \tag{48}$$

According to Lemma 2, we only need to prove that

$$\sum r_1 \geq \sum \sqrt{(r_2 + r_3)^2 - a_p^2}. \tag{49}$$

Using the law of cosines in triangle EPF and the fact that $\angle EPF = \pi - A$ (see Figure 1), we have

$$a_p^2 = r_2^2 + r_3^2 - 2r_2r_3 \cos \angle EPF = r_2^2 + r_3^2 + 2r_2r_3 \cos A,$$

and then

$$(r_2 + r_3)^2 - a_p^2 = 4r_2r_3 \sin^2 \frac{A}{2}. \tag{50}$$

Thus, we see that inequality (49) is equivalent to

$$\sum r_1 \geq 2 \sum \sqrt{r_2 r_3} \sin \frac{A}{2}. \tag{51}$$

But, for any real numbers x, y, z and $\triangle ABC$, we have the following Wolstenholme inequality (see [19]):

$$\sum x^2 \geq 2 \sum yz \cos A, \tag{52}$$

with equality if and only if $x : y : z = \sin A : \sin B : \sin C$. Putting $x = \sqrt{r_1}, y = \sqrt{r_2}, z = \sqrt{r_3}$ in (52) and substituting $A \to (\pi - A)/2$ etc., we get inequality (51) at once. Thus, inequality (48) is proved.

The second inequality in (47) follows immediately by adding the previous inequality (30) and its two analogues. Note that the equality in (30) holds if and only if $R_2 = R_3$. We conclude that the second equality in (47) holds if and only if $R_1 = R_2 = R_3$, which means that the point P is the circumcenter of $\triangle ABC$. The proof of Theorem 4 is completed. □

Remark 3. *The author knows that the triangle ABC need not be equilateral when the first equality in (47) holds but does not know what are the barycentric coordinates of P with respect to the triangle ABC.*

Now we give an application of Theorem 4.
Squaring both sides of the first inequality of (47), we have

$$\left(\sum R_1\right)^2$$
$$\geq \sum \left[(R_2 + R_3)^2 - a^2\right] + 2 \sum \sqrt{(R_3 + R_1)^2 - b^2} \cdot \sqrt{(R_1 + R_2)^2 - c^2}.$$

Then, applying inequality (30), we further get

$$\left(\sum R_1\right)^2 \geq \sum (R_2+R_3)^2 - \sum a^2 + 8\sum w_2 w_3.$$

Expanding gives the following:

Corollary 3. *For any interior point P of the triangle ABC, it holds that*

$$\sum R_1^2 + 8\sum w_2 w_3 \leq \sum a^2. \tag{53}$$

In fact, by using the previous inequality (30), we have the following extension:

$$\sum R_1^2 + 8\sum w_2 w_3 \leq \sum a^2 \leq \sum (R_2+R_3)^2 - 4\sum w_1^2, \tag{54}$$

which implies Barrow's inequality (2).

Finally, we give the third new refinement of Barrow's inequality:

Theorem 5. *For any interior point P of the triangle ABC, it holds that*

$$\sum R_1 \geq \sqrt{\frac{1}{2}\sum a^2 + \sum R_2 R_3 + 2\sum w_1^2} \geq 2\sum w_1 \tag{55}$$

The first equality in (55) *holds if and only if P is the circumcenter of* $\triangle ABC$. *The second equality in* (55) *holds if and only if* $\triangle ABC$ *is equilateral and P is its center.*

Proof. In the proof of Theorem 2, we have proved the first inequality in (55). The second inequality in (55) is easily obtained as follows: By (53), we have

$$\frac{1}{2}\sum a^2 + \sum R_2 R_3 + 2\sum w_1^2$$
$$\geq \frac{1}{2}\sum R_1^2 + 4\sum w_2 w_3 + \sum R_2 R_3 + 2\sum w_1^2$$
$$= \frac{1}{2}\left(\sum R_1\right)^2 + 2\left(\sum w_1\right)^2$$
$$\geq \left(\sum w_1\right)^2,$$

where the last step used Barrow's inequality (2). It is not difficult to know the equality conditions of inequality chain (55). The proof of Theorem 5 is completed. □

4. Some Open Problems

In this section, we present some interesting conjectures as open problems.

For the second inequality in (27), the author guesses that the following refinement is valid.

Conjecture 1. *For any interior point P of the triangle ABC, it holds that*

$$\sqrt{\frac{1}{2}\sum a^2 + \sum R_2 R_3 + 2\sum r_1^2} \geq \frac{1}{2}\sum \sqrt{a^2 + 4r_1^2}$$
$$\geq \sqrt{\frac{1}{2}\sum a^2 + \frac{3}{2}\sum (r_2+r_3)^2}. \tag{56}$$

A similar conjecture is as follows.

Conjecture 2. *For any interior point P of the triangle ABC, it holds that*

$$\sqrt{\frac{1}{2}\sum a^2 + \sum R_2 R_3 + 2\sum w_1^2} \geq \frac{1}{2}\sum \sqrt{a^2 + 4w_1^2}$$
$$\geq \sqrt{\frac{1}{2}\sum a^2 + \frac{3}{2}\sum (w_2+w_3)^2} \geq 2\sum w_1. \tag{57}$$

Remark 4. *The last inequality of (57) is actually equivalent to*

$$2\sum w_1^2 + 10\sum w_2 w_3 \leq \sum a^2, \tag{58}$$

which is Conjecture 2 posed by the author in [15].

Next, we give a reversed inequality similar to the previous inequality (34).

Conjecture 3. *For any interior point P of the triangle ABC, it holds that*

$$\left(\sum R_1\right)^2 + 12\sum r_2 r_3 \leq 2\sum a^2. \tag{59}$$

Considering generalizations of the first inequality of (47), the author presents the following conjecture:

Conjecture 4. *Let P be an interior point of a convex polygon $A_1 A_2 \cdots A_n (n > 3)$ and $PA_i = R_i (i = 1, 2, \cdots, n)$, $R_{n+1} = R_1$, $A_i A_{i+1} = a_i (i = 1, 2, \cdots, n,$ and $A_{n+1} = A_1)$. Then*

$$2\cos\frac{\pi}{n}\sum_{i=1}^n R_i \geq \sum_{i=1}^n \sqrt{(R_i + R_{i+1})^2 - a_i^2}. \tag{60}$$

Remark 5. *By the previous inequality, (30) we know that the above inequality is stronger than inequality (3).*

We have the following refinement of the Erdös–Mordell inequality (see [10]):

$$\sum R_1 \geq \frac{1}{2}\sum \sqrt{a^2 + 4r_1^2} \geq 2\sum r_1, \tag{61}$$

in which the first inequality can easily be generalized to polygons by applying inequality (30) and $w_1 \geq r_1$. The author believes that the second inequality can also be generalized to polygons as follows:

Conjecture 5. *Let P be an interior point of convex polygon $A_1 A_2 \cdots A_n (n > 3)$, and let r_i denote the distances from P to the side lines $A_i A_{i+1} (i = 1, 2, \cdots, n$ and $A_{n+1} = A_1)$. Then*

$$\sum_{i=1}^n \sqrt{a_i^2 + 4r_i^2} \geq 2\sec\frac{\pi}{n}\sum_{i=1}^n r_i. \tag{62}$$

Similarly, we put forward the following conjecture:

Conjecture 6. *Let P be an interior point of convex polygon $A_1 A_2 \cdots A_n (n > 3)$, and let w_i denote the angle bisectors of $\angle A_i P A_{i+1} (i = 1, 2, \cdots, n$ and $A_{n+1} = A_1)$. Then*

$$\sum_{i=1}^n \sqrt{a_i^2 + 4w_i^2} \geq 2\sec\frac{\pi}{n}\sum_{i=1}^n w_i. \tag{63}$$

If the above inequality holds, then we can obtain the following refinement of inequality (3):

$$\sum_{i=1}^{n} R_i \geq \frac{1}{2} \sum_{i=1}^{n} \sqrt{a_i^2 + 4w_i^2} \geq \sec \frac{\pi}{n} \sum_{i=1}^{n} w_i, \qquad (64)$$

where $R_i = PA_i (i = 1, 2, \cdots, n)$.

Conflicts of Interest: The author declares that he has no competing interest.

References

1. Erdös, P. Problem 3740. *Am. Math. Mon.* **1935**, *42*, 396.
2. Mordell, L.J.; Barrow, D.F. Solution of problem 3740. *Am. Math. Mon.* **1937**, *44*, 252–254.
3. Avez, A. A short proof of a theorem of Erdös and Mordell. *Am. Math. Mon.* **1993**, *100*, 60–62. [CrossRef]
4. Abi-Khuzam, F.F. A trigonometric inequality and its geometric applications. *Math. Inequal. Appl.* **2000**, *3*, 437–442. [CrossRef]
5. Alsina, C.; Nelsen, R.B. A visual proof of the Erdös-Mordell inequality. *Forum Geom.* **2007**, *7*, 99–102.
6. Bankoff, L. An elementary proof of the Erdös-Mordell theorem. *Am. Math. Mon.* **1958**, *65*, 521. [CrossRef]
7. Kazarinoff, D.K. A simple proof of the Erdös-Mordell inequality for triangles. *Mich. Math. J.* **1957**, *4*, 97–98. [CrossRef]
8. Komornik, V. A short proof of the Erdös-Mordell theorem. *Am. Math. Mon.* **1997**, *104*, 57–60.
9. Lee, H. Another proof of the Erdös-Mordell theorem. *Forum Geom.* **2001**, *1*, 7–8.
10. Liu, J. A new proof of the Erdös-Mordell inequality. *Int. Electron. J. Geom.* **2011**, *4*, 114–119.
11. Liu, J. A sharpening of the Erdös-Mordell inequality and its applications. *J. Chongqing Norm. Univ.* **2005**, *2*, 12–14. (In Chinese)
12. Liu, J. On a geometric inequality of Oppenheim. *J. Sci. Arts* **2012**, *18*, 5–12.
13. Liu, J. Sharpened versions of the Erdös-Mordell inequality. *J. Inequal. Appl.* **2015**, *2015*, 206. [CrossRef]
14. Liu, J. Refinements of the Erdös-Mordell inequality, Barrow's inequality and Oppenheim's inequality. *J. Inequal. Appl.* **2016**, *2016*, 9. [CrossRef]
15. Liu, J. Three inequalities involving the distances from an interior point to the sides of a triangle. *Int. J. Geom.* **2017**, *1*, 49–60.
16. Liu, J. New refinements of the Erdös-Mordell inequality. *J. Math. Inequal.* **2018**, *12*, 63–75. [CrossRef]
17. Liu, J. *Three Sine Inequality*; Harbin Institute of Technology Press: Harbin, China, 2018.
18. Mitrinović, D.S.; Pečarić, J.E. On the Erdös Mordell's inequality for a polygon. *J. Coll. Arts Sci. Chiba Univ.* **1986**, *19*, 3–6.
19. Mitrinović, D.S.; Pečarić, J.E.; Volence, V. *Recent Advances in Geometric Inequalities*; Kluwer Academic Publishers: Dordrecht, The Netherlands; Boston, MA, USA; London, UK,1989.
20. Malešević, B.; Petrović, M.; Popkonstantinović, B. On the extension of the Erdös-Mordell type inequalities. *Math. Inequal. Appl.* **2014**, *17*, 269–281. [CrossRef]
21. Maleševic, B.; Petrović, M. Barrow's inequality and signed angle bisectors. *J. Math. Inequal.* **2014**, *8*, 537–544. [CrossRef]
22. Ozeki, N. On Paul Erdös-Mordell inequality for the triangle. *J. Coll. Arts Sci. Chiba Univ. A* **1957**, *2*, 247–250.
23. Oppenheim, A. The Erdö-Mordell inequality and other inequalities for a triangle. *Am. Math. Mon.* **1961**, *68*, 226–230. [CrossRef]
24. Oppenheim, A. New inequalities for a triangle and an internal point. *Ann. Mat. Pura Appl.* **1961**, *53*, 157–163. [CrossRef]
25. Pech, P. Erdös-Mordell inequality for space n-gons. *Math. Pannon.* **1994**, *5*, 3–6.
26. Sakurai, A. Vector analysis proof of Erdös' inequality for triangles. *Am. Math. Mon.* **2012**, *8*, 682–684. [CrossRef]
27. Wu, S.H.; Debnath, L. Generalization of the Wolstenholme cyclic inequality and its application. *Comput. Math. Appl.* **2007**, *53*, 104–114. [CrossRef]
28. Gueron, S.; Shafrir, I. A weighted Erdös-Mordell inequality for polygons. *Am. Math. Mon.* **2005**, *112*, 257–263. [CrossRef]

29. Bottema, O.; Djordjević, R.Z.; Janić, R.R.; Mitrinovicć, D.S.; Vasić, P.M. *Geometic Ineqalities*; Wolters-Noordhoff: Groningen, The Netherlands, 1969.
30. Chu, X.G. The proofs of two conjectures involving triangle inequalities. *J. Huaihua Univ.* **2007**, *5*, 71–75. (In Chinese)

© 2019 by the author. Licensee MDPI, Basel, Switzerland. This article is an open access article distributed under the terms and conditions of the Creative Commons Attribution (CC BY) license (http://creativecommons.org/licenses/by/4.0/).

Article

$\delta(2,2)$-Invariant for Lagrangian Submanifolds in Quaternionic Space Forms

Gabriel Macsim [1,†], Adela Mihai [2,†] and Ion Mihai [3,*,†]

1. Doctoral School of Mathematics, University of Bucharest, 010014 Bucharest, Romania; gabi_macsim@yahoo.com
2. Department of Mathematics and Computer Science, Technical University of Civil Engineering Bucharest, 020396 Bucharest, Romania; adela.mihai@utcb.ro
3. Department of Mathematics, University of Bucharest, 010014 Bucharest, Romania
* Correspondence: imihai@fmi.unibuc.ro
† The authors contributed equally to this work.

Received: 27 February 2020; Accepted: 23 March 2020; Published: 1 April 2020

Abstract: In the geometry of submanifolds, Chen inequalities represent one of the most important tool to find relationships between intrinsic and extrinsic invariants; the aim is to find sharp such inequalities. In this paper we establish an optimal inequality for the Chen invariant $\delta(2,2)$ on Lagrangian submanifolds in quaternionic space forms, regarded as a problem of constrained maxima.

Keywords: $\delta(2,2)$-invariant; Chen inequalities; Lagrangian submanifolds; quaternionic space forms; complex space forms

MSC: 53C40; 53C25

---

1. Preliminaries

In submanifold theory, Lagrangian submanifolds are studied not only for their special geometric properties, but also for their important roles in supersymmetric field theory and string theory. For these submanifolds in quaternionic space forms, we give an answer to one problem in submanifold theory, most precisely to find relationships between the main extrinsic invariants and intrinsic invariants.

The intrinsic characteristics of a Riemannian manifold are given by its curvature invariants. In the second section of this article, we recall the definition of δ-invariants (also known as Chen invariants) (see [1]). This theory was initiated by Chen in [2].

In Section 3 we derive an improved inequality for the Chen invariant $\delta(2,2)$ in the case of a Lagrangian submanifold in a quaternionic space form, regarded as a problem of constrained maxima, and recall the inequality which has been improved.

Let $\widetilde{M}^m$ be a complex m-dimensional Kaehler manifold endowed with an almost complex structure J and a Hermitian metric $\widetilde{g}$ and $f: M^n \to \widetilde{M}^m$ an isometric immersion of an n-dimensional manifold M^n into $\widetilde{M}^m$. The submanifold M^n is called a *totally real submanifold* if $J(T_p M^n) \subset T_p^\perp M^n$, $\forall p \in M^n$. A totally real submanifold of maximum dimension, i.e., $\dim_{\mathbb{R}} M^n = \dim_{\mathbb{C}} \widetilde{M}^n = n$, is called a *Lagrangian submanifold*.

If $\widetilde{M}^m$ has holomorphic constant sectional curvature $4c$, then it is called a *complex space form* and it is denoted by $\widetilde{M}^m(4c)$. Its Riemannian curvature tensor is given by

$$\widetilde{R}(X,Y)Z = c[g(Y,Z)X - g(X,Z)Y + g(JY,Z)JX - g(JX,Z)JY + 2g(X,JY)JZ],$$

for any vector fields X, Y, Z tangent to $\widetilde{M}^m(4c)$.

Let M^n be a Lagrangian submanifold of $\widetilde{M}^n(4c)$. One denotes by ∇ and $\widetilde{\nabla}$ the Levi-Civita connections of M^n and $\widetilde{M}^n(4c)$, respectively. The Gauss and Weingarten formulae are given respectively by

$$\widetilde{\nabla}_X Y = \nabla_X Y + h(X,Y), \tag{1}$$

$$\widetilde{\nabla}_X \xi = -A_\xi X + D_X \xi, \tag{2}$$

where X and Y are tangent vector fields, ξ is a normal vector field and D is the normal connection. The second fundamental form h and shape operator A_ξ are related by

$$g(h(X,Y),\xi) = g(A_\xi X, Y). \tag{3}$$

The mean curvature vector H of M^n is defined by

$$H = \frac{1}{n} \text{trace } h.$$

In the case of a Lagrangian submanifold in a complex space form, we have the following relations

$$D_X JY = J\nabla_X Y, \tag{4}$$

$$A_{JX} Y = -Jh(X,Y) = A_{JY} X, \tag{5}$$

and we point out that $g(h(X,Y),Z)$ is totally symmetric.

One denotes by $K(\pi)$ the sectional curvature of M^n associated with a plane section $\pi \subset T_p M^n$, $p \in M^n$ and by R the Riemannian curvature tensor of M^n. Then the Gauss equation is given by

$$\widetilde{R}(X,Y,Z,W) = R(X,Y,Z,W) - g(h(X,Z),h(Y,W)) + \tag{6}$$

$$g(h(X,W),h(Y,Z)),$$

for any vectors X,Y,Z,W tangent to M^n, where $\widetilde{R}(X,Y,Z,W) = g(\widetilde{R}(X,Y)W,Z)$ and $R(X,Y,Z,W) = g(R(X,Y)W,Z)$.

For an orthonormal basis $\{e_1, e_2, \ldots, e_n\}$ of $T_p M^n$ at a point $p \in M^n$, we put

$$h^C_{AB} = g(h(e_A, e_B), Je_C), \quad A, B, C = 1, \ldots, n.$$

Because $g(h(X,Y),Z)$ is totally symmetric, it follows that

$$h^A_{BC} = h^B_{AC} = h^C_{AB}. \tag{7}$$

On the other hand, we recall the following result for a Riemannian submanifold (M^n, g) of a Riemannian manifold $(\widetilde{M}^m, \widetilde{g})$ (of an arbitrary codimension); let consider $f \in C^\infty(\widetilde{M})$. We attach the optimum problem:

$$\min_{x \in M} f(x). \tag{8}$$

Then the following result holds (see [3]).

Theorem 1. *If $x_0 \in M^n$ is a solution of the problem (8), then*

(a) $(\text{grad})(x_0) \in T^\perp_{x_0} M^n$;
(b) *the bilinear form $\alpha : T_{x_0} M^n \times T_{x_0} M^n \to \mathbb{R}$,*

$$\alpha(X,Y) = \text{Hess}_f(X,Y) + \widetilde{g}(h(X,Y), (\text{grad})(x_0))$$

is semipositive definite, where h is the second fundamental form of the submanifold M^n in $\widetilde{M}^m$.

2. Chen Invariants

Let M^n be an n-dimensional Riemannian manifold and $K(\pi)$ the sectional curvature of M^n associated with a 2-plane section $\pi \subset T_p M^n$, $p \in M^n$.

For any orthonormal basis $\{e_1, ..., e_n\}$ of the tangent space $T_p M^n$, the scalar curvature τ at p is defined by

$$\tau(p) = \sum_{1 \leq i < j \leq n} K(e_i \wedge e_j).$$

One denotes by

$$(\inf K)(p) = \inf\{K(\pi) | \pi \subset T_p M^n, \dim \pi = 2\}.$$

The *Chen first invariant* is given by $\delta_M(p) = \tau(p) - (\inf K)(p)$.

If L is a subspace of $T_p M^n$ of dimension $r \geq 2$ and $\{e_1, ..., e_r\}$ an orthonormal basis of L, the scalar curvature $\tau(L)$ of the r-plane section L is given by

$$\tau(L) = \sum_{1 \leq \alpha < \beta \leq r} K(e_\alpha \wedge e_\beta).$$

For given integers $n \geq 3$ and $k \geq 1$, one denotes by $S(n,k)$ the finite set of all k-tuples $(n_1, ..., n_k)$ of integers satisfying $2 \leq n_1, ..., n_k < n$, $n_1 + ... + n_k \leq n$. Let $S(n) = \bigcup_{k \geq 1} S(n,k)$.

For each $(n_1, ..., n_k) \in S(n)$ and each point $p \in M^n$, B.-Y. Chen introduced a Riemannian invariant defined by

$$\delta(n_1, ..., n_k)(p) = \tau(p) - \inf\{\tau(L_1) + ... + \tau(L_k)\},$$

where $L_1, ..., L_k$ run over all k mutually orthogonal subspaces of $T_p M^n$ such that $\dim L_j = n_j$, $j = 1, ..., k$.

We recall the most important Chen inequalities for submanifolds in real space forms.

Theorem 2 ([2]). *Let M^n be an n-dimensional ($n \geq 3$) submanifold of a real space form $\tilde{M}^m(c)$ of constant sectional curvature c. Then*

$$\delta_M \leq \frac{n-2}{2}\left\{\frac{n^2}{n-1}\|H\|^2 + (n+1)c\right\}. \tag{9}$$

The equality case was characterized in terms of the shape operator.

The same inequality holds for totally real submanifolds in complex space forms. A corresponding inequality for slant submanifolds in complex space forms was obtained in [4].

However, for Lagrangian submanifolds in complex space forms the above inequality, known as *Chen first inequality*, was improved by Bolton et al. [5]. Moreover, one of the present authors improved the Chen first inequality for Kaehlerian slant submanifolds in complex space forms (see [6]).

For each $(n_1, ..., n_k) \in S(n)$, one denotes by:

$$d(n_1, ..., n_k) = \frac{n^2(n+k-1-\sum_{j=1}^k n_j)}{2(n+k-\sum_{j=1}^k n_j)},$$

$$b(n_1, ..., n_k) = \frac{1}{2}[n(n-1) - \sum_{j=1}^k n_j(n_j - 1)].$$

The following sharp inequality involving the Chen invariants and the squared mean curvature obtained in [7] plays a very important role in this topic.

Theorem 3. *For each $(n_1, ..., n_k) \in S(n)$ and each n-dimensional submanifold M^n in a Riemannian space form $\widetilde{M}^m(4c)$ of constant sectional curvature $4c$, the inequality*

$$\delta(n_1, ..., n_k) \leq d(n_1, ..., n_k) \|H\|^2 + b(n_1, ..., n_k)c \tag{10}$$

is fulfilled.

Chen also pointed-out that a similar inequality holds for totally real (in particular Lagrangian) submanifolds in a complex space form.

3. Lagrangian Submanifolds in Quaternionic Space Forms

Chen et al. established the following inequalities for Chen invariants of Lagrangian submanifolds in complex space forms, which improve the inequality (10).

Theorem 4 ([8])**.** *Let M^n be a Lagrangian submanifold of a complex space form $\widetilde{M}^n(4c)$. For a given k-tuple $(n_1, n_2, \ldots, n_k) \in S(n)$, we put $N = n_1 + n_2 + \ldots + n_k$. If $N < n$, then the inequality*

$$\delta(n_1, n_2, \ldots, n_k) \leq \frac{n^2\{n - N + 3k - 1 - 6\sum_{i=1}^{k}(2 + n_i)^{-1}\}}{2\{n - N + 3k + 2 - 6\sum_{i=1}^{k}(2 + n_i)^{-1}\}} \|H\|^2 + \tag{11}$$

$$\frac{1}{2}\left\{n(n-1) - \sum_{i=1}^{k} n_i(n_i - 1)\right\}c$$

is satisfied.

In particular, one has (see also [9]).

Theorem 5. *Let M^n be a Lagrangian submanifold of a complex space form $\widetilde{M}^n(4c)$, $n \geq 4$. Then the following inequality holds.*

$$\delta(2,2) \leq \frac{n^2}{2} \cdot \frac{n-2}{n+1} \|H\|^2 + \frac{1}{2}[n(n-1) - 4]c. \tag{12}$$

The equality sign holds at a point $p \in M^n$ if and only if there is an orthonormal basis $\{e_1, e_2, \ldots, e_n\}$ at p such that with respect to this basis the second fundamental form h satisfies the following conditions

$$h_{iA}^C = 0, \ A, C \in \{1, \ldots, n\} \setminus \{i\}, \ A < C, \ i = \overline{1,3},$$

$$h_{BC}^A = 0, \ A = \overline{1,n}, \ 4 \leq B < C \leq n, \ A \notin \{B, C\}.$$

Next, we recall some basic notions about quaternionic space forms.

Let $\widetilde{M}^{4m}$ be a differentiable manifold and we assume that there is a rank 3 subbundle σ of $\text{End}(T\widetilde{M}^{4m})$ such that a local basis $\{J_1, J_2, J_3\}$ exists on sections of σ satisfying for all $\alpha \in \{1, 2, 3\}$

$$J_\alpha^2 = -\text{Id}, \quad J_\alpha J_{\alpha+1} = -J_{\alpha+1} J_\alpha = J_{\alpha+2}, \tag{13}$$

where Id denotes the identity field of type $(1,1)$ on $\widetilde{M}^{4m}$ and the indices are taken from $\{1, 2, 3\}$ modulo 3. The bundle σ is called an *almost quaternionic structure* on $\widetilde{M}^{4m}$ and $\{J_1, J_2, J_3\}$ is called a canonical basis of σ. $(\widetilde{M}^{4m}, \sigma)$ is said to be an *almost quaternionic manifold*. It is easy to see that any almost quaternionic manifold is of dimension $4m$, $m \geq 1$.

A Riemannian metric $\widetilde{g}$ on $\widetilde{M}^{4m}$ is said to be *adapted to the almost quaternionic structure σ* if it satisfies

$$\widetilde{g}(J_\alpha X, J_\alpha Y) = \widetilde{g}(X, Y), \quad \forall \alpha \in \{1, 2, 3\}, \tag{14}$$

for all vector fields X, Y on $\widetilde{M}^{4m}$ and any canonical basis $\{J_1, J_2, J_3\}$ on σ. $(\widetilde{M}^{4m}, \sigma, \widetilde{g})$ is said to be an *almost quaternionic Hermitian manifold*.

$(\widetilde{M}^{4m}, \sigma, \widetilde{g})$ is said to be a *quaternionic Kaehler manifold* if the bundle σ is parallel with respect to the Levi-Civita connection $\widetilde{\nabla}$ of $\widetilde{g}$, i.e., locally defined 1-forms ω_1, ω_2, ω_3 exist such that we have

$$\widetilde{\nabla}_X J_\alpha = \omega_{\alpha+2}(X) J_{\alpha+1} - \omega_{\alpha+1}(X) J_{\alpha+2}, \tag{15}$$

for all $\alpha \in \{1, 2, 3\}$ and for any vector field X on $\widetilde{M}^{4m}$, where the indices are taken from $\{1, 2, 3\}$ modulo 3.

Let $(\widetilde{M}^{4m}, \sigma, \widetilde{g})$ be a quaternionic Kaehler manifold and let X be a non-null vector on $\widetilde{M}^{4m}$. The 4-plane spanned by $\{X, J_1 X, J_2 X, J_3 X\}$ is called a *quaternionic 4-plane* and is denoted by $Q(X)$. Any 2-plane in $Q(X)$ is called a *quaternionic plane*. The sectional curvature of a quaternionic plane is called a *quaternionic sectional curvature*. A quaternionic Kaehler manifold is a *quaternionic space form* if its quaternionic sectional curvature are equal to a constant, say $4c$, i.e., its curvature tensor is given by

$$\widetilde{R}(X,Y)Z = c\{\widetilde{g}(Z,Y)X - \widetilde{g}(X,Z)Y + \sum_{\alpha=1}^{3}[\widetilde{g}(Z,J_\alpha Y)J_\alpha X - \tag{16}$$

$$\widetilde{g}(Z, J_\alpha X) J_\alpha Y + 2\widetilde{g}(X, J_\alpha y) J_\alpha Z]\},$$

for all vector fields X, Y, Z on $\widetilde{M}^{4m}$ and any local basis $\{J_1, J_2, J_3\}$ on σ.

A submanifold M^n of a quaternionic space form $\widetilde{M}^{4n}(4c)$ is said to be *Lagrangian* if $J_\alpha(T_p M) \subset T_p^\perp M$, for any $p \in M$ and $\alpha = 1, 2, 3$.

On a Lagrangian submanifold M^n we can choose an orthonormal frame field in $\widetilde{M}^{4n}(4c)$

$$\{e_1, e_2, \ldots, e_n; \quad e_{\phi_1(1)} = J_1(e_1), \ldots, e_{\phi_1(n)} = J_1(e_n);$$

$$e_{\phi_2(1)} = J_2(e_1), \ldots, e_{\phi_2(n)} = J_2(e_n); \quad e_{\phi_3(1)} = J_3(e_1), \ldots, e_{\phi_3(n)} = J_3(e_n)\},$$

such that, restricted to M, $e_1, e_2, \ldots, e_n$ are tangent to M.

We set

$$h_{ij}^\xi = g(h(e_i, e_j), e_\xi), \quad \xi \in \{\phi_1(1), \ldots, \phi_1(n), \phi_2(1), \ldots, \phi_2(n), \phi_3(1), \ldots, \phi_3(n)\}$$

and then, for any $\alpha = 1, 2, 3$, we have (see (2.9) in [10])

$$h_{ij}^{\phi_\alpha(k)} = h_{ki}^{\phi_\alpha(j)} = h_{jk}^{\phi_\alpha(i)}. \tag{17}$$

We denote by $H^r = g(H, e_{\phi_1(e_r)})$, for $r = 1, \ldots, n$.

By using the method of constrained maxima, we prove the following improved Chen inequality for the invariant $\delta(2,2)$ of Lagrangian submanifolds in quaternionic space forms, the main result of this paper.

Theorem 6. *Let M^n be a Lagrangian submanifold of a quaternionic space form $\widetilde{M}^{4n}(4c)$, $n \geq 4$. Then the inequality*

$$\delta(2,2) \leq \frac{n^2}{2} \cdot \frac{n-2}{n+1} \|H\|^2 + \frac{1}{2}[n(n-1) - 4]c \tag{18}$$

is fulfilled.

The equality sign holds at a point $p \in M^n$ if and only if there is an orthonormal basis $\{e_1, e_2, \ldots, e_n\}$ at p such that with respect to this basis the second fundamental form h satisfies the following conditions:

$$h_{iA}^{\phi_\alpha(C)} = 0, \ A, C \in \{1, \ldots, n\} \setminus \{i\}, \ A < C, i = \overline{1,3}, \alpha = \overline{1,3},$$

$$h_{BC}^{\phi_\alpha(A)} = 0,\ A = \overline{1,n},\ 4 \leq B < C \leq n, A \notin \{B,C\}, \alpha = \overline{1,3}.$$

Proof. Let M^n be a Lagrangian submanifold of the quaternionic space form $\widetilde{M}^{4n}(4c)$, $p \in M^n$ and L_1 and L_2 two mutual orthogonal plane sections at p. We denote $\{e_1, e_2\} \subset L_1, \{e_3, e_4\} \subset L_2$ orthonormal bases, complete to an orthonormal basis $\{e_1, ..., e_n\} \subset T_p M^n$ and extend it to $T_p \widetilde{M}^{4n}(4c)$ as above.

Gauss equation implies

$$\tau = \sum_{\alpha=1}^{3} \sum_{A=1}^{n} \sum_{B<C} \left[h_{BB}^{\phi_\alpha(A)} h_{CC}^{\phi_\alpha(A)} - \left(h_{BC}^{\phi_\alpha(A)} \right)^2 \right] + \frac{n(n-1)}{2} c,$$

$$\tau(L_1) = \sum_{\alpha=1}^{3} \sum_{A=1}^{n} \left[h_{11}^{\phi_\alpha(A)} h_{22}^{\phi_\alpha(A)} - \left(h_{12}^{\phi_\alpha(A)} \right)^2 \right] + c,$$

$$\tau(L_2) = \sum_{\alpha=1}^{3} \sum_{A=1}^{n} \left[h_{33}^{\phi_\alpha(A)} h_{44}^{\phi_\alpha(A)} - \left(h_{34}^{\phi_\alpha(A)} \right)^2 \right] + c.$$

Then

$$\tau - \tau(L_1) - \tau(L_2) - [n(n-1) - 4]\frac{c}{2} =$$

$$\sum_{\alpha=1}^{3} \sum_{A=1}^{n} \sum_{B<C} \left[h_{BB}^{\phi_\alpha(A)} h_{CC}^{\phi_\alpha(A)} - \left(h_{BC}^{\phi_\alpha(A)} \right)^2 \right] -$$

$$\sum_{\alpha=1}^{3} \sum_{A=1}^{n} \left[h_{11}^{\phi_\alpha(A)} h_{22}^{\phi_\alpha(A)} + h_{33}^{\phi_\alpha(A)} h_{44}^{\phi_\alpha(A)} - \left(h_{12}^{\phi_\alpha(A)} \right)^2 - \left(h_{34}^{\phi_\alpha(A)} \right)^2 \right] =$$

$$\sum_{\alpha=1}^{3} \sum_{A=1}^{n} \left[\sum_{1 \leq B < C \leq n} h_{BB}^{\phi_\alpha(A)} h_{CC}^{\phi_\alpha(A)} - h_{11}^{\phi_\alpha(A)} h_{22}^{\phi_\alpha(A)} - h_{33}^{\phi_\alpha(A)} h_{44}^{\phi_\alpha(A)} \right] -$$

$$\sum_{\alpha=1}^{3} \sum_{A=1}^{n} \left[\sum_{1 \leq B < C \leq n} \left(h_{BC}^{\phi_\alpha(A)} \right)^2 - \left(h_{12}^{\phi_\alpha(A)} \right)^2 - \left(h_{34}^{\phi_\alpha(A)} \right)^2 \right].$$

Thus, we get

$$\tau - \tau(L_1) - \tau(L_2) - [n(n-1) - 4]\frac{c}{2} \leq$$

$$\sum_{\alpha=1}^{3} \sum_{A=1}^{n} \left[\left(h_{11}^{\phi_\alpha(A)} + h_{22}^{\phi_\alpha(A)} \right) \sum_{B=3}^{n} h_{BB}^{\phi_\alpha(A)} + \left(h_{33}^{\phi_\alpha(A)} + h_{44}^{\phi_\alpha(A)} \right) \sum_{B=5}^{n} h_{BB}^{\phi_\alpha(A)} + \sum_{5 \leq B < C \leq n} h_{BB}^{\phi_\alpha(A)} h_{CC}^{\phi_\alpha(A)} \right] -$$

$$\sum_{\alpha=1}^{3} \sum_{B=3}^{n} \left[\left(h_{1B}^{\phi_\alpha(1)} \right)^2 + \left(h_{1B}^{\phi_\alpha(B)} \right)^2 + \left(h_{2B}^{\phi_\alpha(2)} \right)^2 + \left(h_{2B}^{\phi_\alpha(B)} \right)^2 \right] - \sum_{\alpha=1}^{3} \sum_{B=5}^{n} \left[\left(h_{3B}^{\phi_\alpha(3)} \right)^2 + \left(h_{3B}^{\phi_\alpha(B)} \right)^2 \right] -$$

$$\sum_{\alpha=1}^{3} \sum_{4 \leq B < C \leq n} \left[\left(h_{BC}^{\phi_\alpha(B)} \right)^2 + \left(h_{BC}^{\phi_\alpha(C)} \right)^2 \right].$$

It follows that

$$\tau - \tau(L_1) - \tau(L_2) - [n(n-1) - 4]\frac{c}{2} \leq \qquad (19)$$

$$\sum_{\alpha=1}^{3} \sum_{A=1}^{n} \left[\left(h_{11}^{\phi_\alpha(A)} + h_{22}^{\phi_\alpha(A)} \right) \sum_{B=3}^{n} h_{BB}^{\phi_\alpha(A)} + \left(h_{33}^{\phi_\alpha(A)} + h_{44}^{\phi_\alpha(A)} \right) \sum_{B=5}^{n} h_{BB}^{\phi_\alpha(A)} + \sum_{5 \leq B < C \leq n} h_{BB}^{\phi_\alpha(A)} h_{CC}^{\phi_\alpha(A)} \right] -$$

$$\sum_{\alpha=1}^{3}\sum_{B=3}^{n}\left[\left(h_{11}^{\phi_\alpha(B)}\right)^2+\left(h_{BB}^{\phi_\alpha(1)}\right)^2+\left(h_{22}^{\phi_\alpha(B)}\right)^2+\left(h_{BB}^{\phi_\alpha(2)}\right)^2\right]-\sum_{\alpha=1}^{3}\sum_{B=5}^{n}\left[\left(h_{33}^{\phi_\alpha(B)}\right)^2+\left(h_{BB}^{\phi_\alpha(3)}\right)^2\right]-$$

$$\sum_{\alpha=1}^{3}\sum_{4\le B<C\le n}\left[\left(h_{BB}^{\phi_\alpha(C)}\right)^2+\left(h_{CC}^{\phi_\alpha(B)}\right)^2\right].$$

For each $\alpha \in \{1,2,3\}$, let us consider certain quadratic forms. For example, for $\alpha=1$, we will define the quadratic forms

$$f_1, f_2, \ldots, f_n : \mathbb{R}^n \to \mathbb{R}$$

by

$$f_r\left(h_{11}^{\phi_1(r)}, h_{22}^{\phi_1(r)}, \ldots, h_{nn}^{\phi_1(r)}\right) = \left(h_{11}^{\phi_1(r)}+h_{22}^{\phi_1(r)}\right)\sum_{B=3}^{n}h_{BB}^{\phi_1(r)}+\left(h_{33}^{\phi_1(r)}+h_{44}^{\phi_1(r)}\right)\sum_{B=5}^{n}h_{BB}^{\phi_1(r)}+$$

$$\sum_{5\le B<C\le n}h_{BB}^{\phi_1(r)}h_{CC}^{\phi_1(r)}-\sum_{B=3}^{n}\left(h_{BB}^{\phi_1(r)}\right)^2,$$

for $r=1,2$,

$$f_s\left(h_{11}^{\phi_1(s)}, h_{22}^{\phi_1(s)}, \ldots, h_{nn}^{\phi_1(s)}\right) = \left(h_{11}^{\phi_1(s)}+h_{22}^{\phi_1(s)}\right)\sum_{B=3}^{n}h_{BB}^{\phi_1(s)}+\left(h_{33}^{\phi_1(s)}+h_{44}^{\phi_1(s)}\right)\sum_{B=5}^{n}h_{BB}^{\phi_1(s)}+$$

$$\sum_{5\le B<C\le n}h_{BB}^{\phi_1(s)}h_{CC}^{\phi_1(s)}-\left(h_{11}^{\phi_1(s)}\right)^2-\left(h_{22}^{\phi_1(s)}\right)^2-\sum_{B=5}^{n}\left(h_{BB}^{\phi_1(s)}\right)^2,$$

for $s=3,4$,

$$f_t\left(h_{11}^{\phi_1(t)}, h_{22}^{\phi_1(t)}, \ldots, h_{nn}^{\phi_1(t)}\right) = \left(h_{11}^{\phi_1(t)}+h_{22}^{\phi_1(t)}\right)\sum_{B=3}^{n}h_{BB}^{\phi_1(t)}+\left(h_{33}^{\phi_1(t)}+h_{44}^{\phi_1(t)}\right)\sum_{B=5}^{n}h_{BB}^{\phi_1(t)}+$$

$$\sum_{5\le B<C\le n}h_{BB}^{\phi_1(t)}h_{CC}^{\phi_1(t)}-\sum_{B=1;B\ne t}^{n}\left(h_{BB}^{\phi_1(t)}\right)^2,$$

for $5 \le t \le n$.

We shall find an upper bound for f_1, subject to

$$P:\ h_{11}^{\phi_1(1)}+h_{22}^{\phi_1(1)}+\ldots+h_{nn}^{\phi_1(1)}=k^1, \tag{20}$$

where k^1 is a real number.

Let $q \in P$ an arbitrary point. The bilinear form $\gamma : T_qP \times T_qP \to \mathbb{R}$ has the expression

$$\gamma(X,Y) = \mathrm{Hess}(f_r)(X,Y) + \langle h'(X,Y), \mathrm{grad}\, f_r(q)\rangle,$$

where h' is the second fundamental form of P in $\mathbb{R}^n$ and $\langle\,,\,\rangle$ is the standard inner-product on $\mathbb{R}^n$.

The partial derivatives of the function f_1 are

$$\frac{\partial f_1}{\partial h_{11}^{\phi_1(1)}} = \sum_{B=3}^{n} h_{BB}^{\phi_1(1)},$$

$$\frac{\partial f_1}{\partial h_{22}^{\phi_1(1)}} = \sum_{B=3}^{n} h_{BB}^{\phi_1(1)},$$

$$\frac{\partial f_1}{\partial h_{rr}^{\phi_1(1)}} = h_{11}^{\phi_1(1)} + h_{22}^{\phi_1(1)} + \sum_{B=5}^{n} h_{BB}^{\phi_1(1)} - 2h_{rr}^{\phi_1(1)}, \; r = 3, 4,$$

$$\frac{\partial f_1}{\partial h_{tt}^{\phi_1(1)}} = \sum_{B=1}^{n} h_{BB}^{\phi_1(1)} - 3h_{tt}^{\phi_1(1)}, \; 5 \leq t \leq n.$$

In the standard frame of $\mathbb{R}^n$, the Hessian of f_1 has the matrix

$$\begin{pmatrix} 0 & 0 & 1 & 1 & 1 & 1 & \cdots & 1 \\ 0 & 0 & 1 & 1 & 1 & 1 & \cdots & 1 \\ 1 & 1 & -2 & 0 & 1 & 1 & \cdots & 1 \\ 1 & 1 & 0 & -2 & 1 & 1 & \cdots & 1 \\ 1 & 1 & 1 & 1 & -2 & 1 & \cdots & 1 \\ 1 & 1 & 1 & 1 & 1 & -2 & \cdots & 1 \\ & & & \cdots\cdots\cdots & & & & \\ 1 & 1 & 1 & 1 & 1 & 1 & \cdots & -2 \end{pmatrix}.$$

As P is totally geodesic in $\mathbb{R}^n$, we obtain

$$\gamma(X,X) = 2(X_1 + X_2) \sum_{i=3}^{n} X_i + 2(X_3 + X_4) \sum_{i=5}^{n} X_i + 2 \sum_{5 \leq i < j \leq n} X_i X_j - 2 \sum_{i=3}^{n} (X_i)^2 =$$

$$\left(\sum_{i=1}^{n} X_i\right)^2 - (X_1)^2 - (X_2)^2 - 3 \sum_{i=3}^{n} (X_i)^2 - 2X_1 X_2 - 2X_3 X_4 =$$

$$-(X_1 + X_2)^2 - (X_3 + X_4)^2 - 2(X_3)^2 - 2(X_4)^2 - 3 \sum_{i=5}^{n} (X_i)^2 < 0;$$

then the Hessian of f_1 is negative semidefinite.

Searching for the critical point $h_{11}^{\phi_1(1)}, h_{22}^{\phi_1(1)}, \ldots, h_{nn}^{\phi_1(1)}$ of f_1, we denote by

$$h_{33}^{\phi_1(1)} = h_{44}^{\phi_1(1)} = a^1.$$

Then,

$$4h_{33}^{\phi_1(1)} = 3h_{rr}^{\phi_1(1)}, \; r = \overline{5,n} \implies h_{rr}^{\phi_1(1)} = \frac{4a^1}{3}, \; 5 \leq r \leq n,$$

$$h_{11}^{\phi_1(1)} + h_{22}^{\phi_1(1)} = 3h_{tt}^{\phi_1(1)}, \; t = \overline{5,n} \implies h_{11}^{\phi_1(1)} + h_{22}^{\phi_1(1)} = 4a^1.$$

From (20) it follows that

$$4a^1 + 2a^1 + \frac{4a^1}{3}(n-4) = k^1 \implies a^1 = \frac{3k^1}{2(2n+1)}.$$

This implies

$$h_{11}^{\phi_1(1)} + h_{22}^{\phi_1(1)} = \frac{6k^1}{2n+1},$$

$$h_{33}^{\phi_1(1)} = h_{44}^{\phi_1(1)} = \frac{3k^1}{2(2n+1)},$$

$$h_{rr}^{\phi_1(1)} = \frac{2k^1}{2n+1}, \; 5 \leq r \leq n.$$

Thus

$$f_1 \leq \frac{6k^1}{2n+1}\left[\frac{6k^1}{4n+2} + (n-4)\frac{2k^1}{2n+1}\right] + \frac{3k^1}{2n+1}(n-4)\frac{2k^1}{2n+1} +$$

$$C_{n-4}^2\left(\frac{2k^1}{2n+1}\right)^2 - 2\left(\frac{3k^1}{4n+2}\right)^2 - (n-4)\left(\frac{2k^1}{2n+1}\right)^2 =$$

$$\frac{6k^1}{2n+1}\cdot\left[\frac{3k^1}{2n+1} + (n-4)\frac{2k^1}{2n+1}\right] + \frac{3k^1}{2(2n+1)}(n-4)\frac{2k^1}{2n+1} +$$

$$\frac{3k^1}{2(2n+1)}(n-4)\frac{2k^1}{2n+1} + C_{n-4}^2\left(\frac{2k^1}{2n+1}\right)^2 - 2\frac{(3k^1)^2}{4(2n+1)^2} - (n-4)\frac{(2k^1)^2}{(2n+1)^2} =$$

$$\frac{6k^1}{2n+1}\cdot\frac{3k^1 + 2(n-4)k^1}{2n+1} + \frac{3(n-4)(k^1)^2}{(2n+1)^2} + \frac{3(n-4)(k^1)^2}{(2n+1)^2} +$$

$$\frac{(n-4)(n-5)}{2}\cdot\frac{4(k^1)^2}{(2n+1)^2} - \frac{9(k^1)^2}{2(2n+1)^2} - \frac{4(n-4)(k^1)^2}{(2n+1)^2} =$$

$$\frac{6k^1}{2n+1}\cdot\frac{(2n-5)k^1}{(2n+1)} + \frac{6(n-4)(k^1)^2}{(2n+1)^2} + \frac{2(n-4)(n-5)(k^1)^2}{(2n+1)^2} -$$

$$\frac{9(k^1)^2}{2(2n+1)^2} - \frac{4(n-4)(k^1)^2}{(2n+1)^2} =$$

$$\frac{(k^1)^2}{(2n+1)^2}\cdot\left[6(2n-5) + 6(n-4) + 2(n-4)(n-5) - \frac{9}{2} - 4(n-4)\right] =$$

$$\frac{(k^1)^2}{(2n+1)^2}\cdot\left[12n - 30 + 6n - 24 + (2n-8)(n-5) - \frac{9}{2} - 4n + 16\right] =$$

$$\frac{(k^1)^2}{2(2n+1)^2}\cdot(24n - 60 + 12n - 48 + 4n^2 - 20n - 16n + 80 - 9 - 8n + 32) =$$

$$\frac{(k^1)^2}{2(2n+1)^2}\cdot(4n^2 - 8n - 5) = \frac{(k^1)^2}{2(2n+1)^2}\cdot(2n-5)(2n+1),$$

which implies

$$f_1 \leq \frac{(2n-5)(k^1)^2}{2(2n+1)},$$

i.e.,

$$f_1 \leq \frac{n^2}{2}\cdot\frac{2n-5}{2n+1}(H^1)^2.$$

In a similar manner, we obtain for f_2

$$f_2 \leq \frac{n^2}{2}\cdot\frac{2n-5}{2n+1}(H^2)^2.$$

Let's consider now f_3, as:

$$f_3\left(h_{11}^{\phi_1(3)}, h_{22}^{\phi_1(3)}, \ldots, h_{nn}^{\phi_1(3)}\right) = \left(h_{11}^{\phi_1(3)} + h_{22}^{\phi_1(3)}\right)\sum_{B=3}^{n} h_{BB}^{\phi_1(3)} + \left(h_{33}^{\phi_1(3)} + h_{44}^{\phi_1(3)}\right)\sum_{B=5}^{n} h_{BB}^{\phi_1(3)} +$$

$$\sum_{5\leq B<C\leq n} h_{BB}^{\phi_1(3)} h_{CC}^{\phi_1(3)} - \left(h_{11}^{\phi_1(3)}\right)^2 - \left(h_{22}^{\phi_1(3)}\right)^2 - \sum_{B=5}^{n}\left(h_{BB}^{\phi_1(3)}\right)^2.$$

The partial derivatives of the function f_3 are

$$\frac{\partial f_3}{\partial h_{rr}^{\phi_1(3)}} = \sum_{B=3}^{n} h_{BB}^{\phi_1(3)} - 2h_{rr}^{\phi_1(3)}, \; r = 1, 2,$$

$$\frac{\partial f_3}{\partial h_{ss}^{\phi_1(3)}} = h_{11}^{\phi_1(3)} + h_{22}^{\phi_1(3)} + \sum_{B=5}^{n} h_{BB}^{\phi_1(3)}, \; s = 3, 4,$$

$$\frac{\partial f_3}{\partial h_{tt}^{\phi_1(3)}} = \sum_{B=1}^{n} h_{BB}^{\phi_1(3)} - 3h_{tt}^{\phi_1(3)}, \; 5 \leq t \leq n.$$

In the standard frame of $\mathbb{R}^n$, the Hessian of f_3 has the matrix

$$\begin{pmatrix} -2 & 0 & 1 & 1 & 1 & 1 & \ldots & 1 \\ 0 & -2 & 1 & 1 & 1 & 1 & \ldots & 1 \\ 1 & 1 & 0 & 0 & 1 & 1 & \ldots & 1 \\ 1 & 1 & 0 & 0 & 1 & 1 & \ldots & 1 \\ 1 & 1 & 1 & 1 & -2 & 1 & \ldots & 1 \\ 1 & 1 & 1 & 1 & 1 & -2 & \ldots & 1 \\ & & & \ldots\ldots\ldots & & & \\ 1 & 1 & 1 & 1 & 1 & 1 & \ldots & -2 \end{pmatrix}.$$

As P is totally geodesic in $\mathbb{R}^n$, we have

$$\gamma(X, X) = -2\left[(X_1)^2 + (X_2)^2 + \sum_{i=5}^{n}(X_i)^2\right] +$$

$$2(X_1 + X_2)\sum_{i=3}^{n} X_i + 2(X_3 + X_4)\sum_{i=5}^{n} X_i + 2\sum_{5 \leq i < j \leq n} X_i X_j =$$

$$\left(\sum_{i=1}^{n} X_i\right)^2 - 2X_1 X_2 - 2X_3 X_4 - \sum_{i=1}^{n}(X_i)^2 - 2(X_1)^2 - 2(X_2)^2 - 2\sum_{i=5}^{n}(X_i)^2 =$$

$$\left(\sum_{i=1}^{n} X_i\right)^2 - (X_1 + X_2)^2 - (X_3 + X_4)^2 - 2(X_1)^2 - 2(X_2)^2 - 3\sum_{i=5}^{n}(X_i)^2 < 0,$$

and hence the Hessian of f_3 is negative semidefinite.

If we denote by $q = \left(h_{11}^{\phi_1(3)}, h_{22}^{\phi_1(3)}, \ldots, h_{nn}^{\phi_1(3)}\right)$ a solution of the extremum problem in question, then we have

$$h_{11}^{\phi_1(3)} = h_{22}^{\phi_1(3)},$$

$$h_{55}^{\phi_1(3)} = h_{66}^{\phi_1(3)} = \ldots = h_{nn}^{\phi_1(3)},$$

$$4h_{11}^{\phi_1(3)} = 3h_{55}^{\phi_1(3)},$$

$$h_{33}^{\phi_1(3)} + h_{44}^{\phi_1(3)} = 3h_{55}^{\phi_1(3)}.$$

Thus

$$h_{33}^{\phi_1(3)} + h_{44}^{\phi_1(3)} = 4h_{11}^{\phi_1(3)}.$$

Considering

$$h_{55}^{\phi_1(3)} = h_{66}^{\phi_1(3)} = \ldots = h_{nn}^{\phi_1(3)} = a^3,$$

we obtain
$$h_{11}^{\phi_1(3)} = h_{22}^{\phi_1(3)} = \frac{3a^3}{4},$$
$$h_{33}^{\phi_1(3)} + h_{44}^{\phi_1(3)} = 3a^3.$$

Since $h_{11}^{\phi_1(3)} + h_{22}^{\phi_1(3)} + \ldots + h_{nn}^{\phi_1(3)} = k^3$, then $a^3 = \frac{2k^3}{2n+1}$, which implies

$$h_{11}^{\phi_1(3)} = h_{22}^{\phi_1(3)} = \frac{3k^3}{2(2n+1)},$$

$$h_{33}^{\phi_1(3)} + h_{44}^{\phi_1(3)} = \frac{6k^3}{2n+1},$$

$$h_{55}^{\phi_1(3)} = h_{66}^{\phi_1(3)} = \ldots = h_{nn}^{\phi_1(3)} = \frac{2k^3}{2n+1}.$$

It follows that

$$f_3 \leq 2 \cdot \frac{3k^3}{2(2n+1)} \cdot \left[\frac{6k^3}{2n+1} + (n-4)\frac{2k^3}{2n+1}\right] + \frac{6k^3}{2n+1}(n-4)\frac{2k^3}{2n+1} +$$

$$C_{n-4}^2 \cdot \frac{(2k^3)^2}{(2n+1)^2} - 2\frac{(3k^3)^2}{4(2n+1)^2} - (n-4)\left(\frac{2k^3}{2n+1}\right)^2 =$$

$$\frac{3k^3}{(2n+1)} \cdot \frac{6k^3 + 2(n-4)k^3}{(2n+1)} + \frac{12(n-4)(k^3)^2}{(2n+1)^2} +$$

$$\frac{(n-4)(n-5)}{2} \cdot \frac{4(k^3)^2}{(2n+1)^2} - \frac{9(k^3)^2}{2(2n+1)^2} - \frac{4(n-4)(k^3)^2}{(2n+1)^2} =$$

$$\frac{(k^3)^2}{(2n+1)^2} \cdot \left[3(2n-2) + 12(n-4) + 2(n-4)(n-5) - \frac{9}{2} - 4(n-4)\right] =$$

$$\frac{(k^3)^2}{(2n+1)^2} \cdot \left(6n - 6 + 12n - 48 + 2n^2 - 10n - 8n + 40 - \frac{9}{2} - 4n + 16\right) =$$

$$\frac{(k^3)^2}{2(2n+1)^2} \cdot (12n - 12 + 24n - 96 + 4n^2 - 20n - 16n + 80 - 9 - 8n + 32),$$

or, equivalently,
$$f_3 \leq \frac{(k^3)^2(4n^2 - 8n - 5)}{2(2n+1)^2} = \frac{(k^3)^2(2n-5)(2n+1)}{2(2n+1)^2}.$$

Therefore,
$$f_3 \leq \frac{n^2}{2} \cdot \frac{2n-5}{2n+1}(H^3)^2.$$

In a similar manner, we prove for f_4:
$$f_4 \leq \frac{n^2}{2} \cdot \frac{2n-5}{2n+1}(H^4)^2.$$

Using the same procedure for

$$f_5\left(h_{11}^{\phi_1(5)}, h_{22}^{\phi_1(5)}, \ldots, h_{nn}^{\phi_1(5)}\right) = \left(h_{11}^{\phi_1(5)} + h_{22}^{\phi_1(5)}\right) \sum_{B=3}^{n} h_{BB}^{\phi_1(5)} + \left(h_{33}^{\phi_1(5)} + h_{44}^{\phi_1(5)}\right) \sum_{B=5}^{n} h_{BB}^{\phi_1(5)} +$$

$$\sum_{5 \leq B < C \leq n} h_{BB}^{\phi_1(5)} h_{CC}^{\phi_1(5)} - \sum_{B=1; B \neq 5}^{n} \left(h_{BB}^{\phi_1(5)}\right)^2,$$

we find the partial derivatives of f_5

$$\frac{\partial f_5}{\partial h_{rr}^{\phi_1(5)}} = \sum_{B=3}^{n} h_{BB}^{\phi_1(5)} - 2h_{rr}^{\phi_1(5)}, \ r = 1, 2,$$

$$\frac{\partial f_5}{\partial h_{ss}^{\phi_1(5)}} = h_{11}^{\phi_1(5)} + h_{22}^{\phi_1(5)} + \sum_{B=5}^{n} h_{BB}^{\phi_1(5)} - 2h_{ss}^{\phi_1(5)}, \ s = 3, 4,$$

$$\frac{\partial f_5}{\partial h_{55}^{\phi_1(5)}} = \sum_{B=1}^{n} h_{BB}^{\phi_1(5)} - h_{55}^{\phi_1(5)},$$

$$\frac{\partial f_5}{\partial h_{tt}^{\phi_1(5)}} = \sum_{B=1}^{n} h_{BB}^{\phi_1(5)} - 3h_{tt}^{\phi_1(5)}, \ 6 \leq t \leq n.$$

In the standard frame of $\mathbb{R}^n$, the Hessian of f_5 has the matrix

$$\begin{pmatrix} -2 & 0 & 1 & 1 & 1 & 1 & \ldots & 1 \\ 0 & -2 & 1 & 1 & 1 & 1 & \ldots & 1 \\ 1 & 1 & -2 & 0 & 1 & 1 & \ldots & 1 \\ 1 & 1 & 0 & -2 & 1 & 1 & \ldots & 1 \\ 1 & 1 & 1 & 1 & 0 & 1 & \ldots & 1 \\ 1 & 1 & 1 & 1 & 1 & -2 & \ldots & 1 \\ \multicolumn{8}{c}{\ldots\ldots\ldots\ldots\ldots} \\ 1 & 1 & 1 & 1 & 1 & 1 & \ldots & -2 \end{pmatrix}.$$

As P is totally geodesic in $\mathbb{R}^n$, we have

$$\gamma(X, X) = -2 \sum_{i=1; i \neq 5}^{n} (X_i)^2 +$$

$$2(X_1 + X_2) \sum_{i=3}^{n} X_i + 2(X_3 + X_4) \sum_{i=5}^{n} X_i + 2 \sum_{5 \leq i < j \leq n} X_i X_j =$$

$$\left(\sum_{i=1}^{n} X_i\right)^2 - 2X_1 X_2 - 2X_3 X_4 - \sum_{i=1}^{n} (X_i)^2 - 2 \sum_{i=1; i \neq 5}^{n} (X_i)^2 =$$

$$\left(\sum_{i=1}^{n} X_i\right)^2 - (X_1 + X_2)^2 - (X_3 + X_4)^2 - (X_5)^2 - 2\sum_{i=1}^{4}(X_i)^2 - 3\sum_{i=6}^{n}(X_i)^2 < 0,$$

and hence the Hessian of f_5 is negative semidefinite.

Using similar arguments to those in the previous problem we obtain that the solution of the associated extremum problem is

$$h_{11}^{\phi_1(5)} = h_{22}^{\phi_1(5)} = h_{33}^{\phi_1(5)} = h_{44}^{\phi_1(5)} = 3a^5,$$

$$h_{55}^{\phi_1(5)} = 12a^5,$$

$$h_{66}^{\phi_1(5)} = \ldots = h_{nn}^{\phi_1(5)} = 4a^5,$$

where a^5 is a real number.

Since $h_{11}^{\phi_1(5)} + h_{22}^{\phi_1(5)} + \ldots + h_{nn}^{\phi_1(5)} = k^5$, then $a^5 = \dfrac{k^5}{4(n+1)}$ and

$$h_1^{\phi_1(5)} = h_{22}^{\phi_1(5)} = h_{33}^{\phi_1(5)} = h_{44}^{\phi_1(5)} = \frac{3k^5}{4(n+1)},$$

$$h_{55}^{\phi_1(5)} = \frac{3k^5}{n+1},$$

$$h_{66}^{\phi_1(5)} = \ldots = h_{nn}^{\phi_1(5)} = \frac{k^5}{n+1}.$$

We have

$$f_5 \leq \frac{6k^5}{4(n+1)} \cdot \left[\frac{6k^5}{4(n+1)} + \frac{3k^5}{n+1} + (n-5)\frac{k^5}{n+1} \right] +$$

$$2 \cdot \frac{3k^5}{4(n+1)} \left[\frac{3k^5}{n+1} + (n-5)\frac{k^5}{n+1} \right] +$$

$$\frac{3k^5}{n+1}(n-5)\frac{k^5}{n+1} + C_{n-5}^2 \frac{(k^5)^2}{(n+1)^2} - 4 \cdot \frac{(3k^5)^2}{16(n+1)^2} - (n-5)\frac{(k^5)^2}{(n+1)^2} =$$

$$\frac{3k^5}{2(n+1)} \cdot \left[\frac{3k^5}{2(n+1)} + \frac{3k^5}{n+1} + \frac{(n-5)k^5}{n+1} \right] + \frac{3k^5}{2(n+1)} \cdot \left[\frac{3k^5 + (n-5)k^5}{n+1} \right] +$$

$$\frac{3(n-5)(k^5)^2}{(n+1)^2} + \frac{(n-5)(n-6)}{2} \cdot \frac{(k^5)^2}{(n+1)^2} - \frac{9(k^5)^2}{4(n+1)^2} - \frac{(n-5)(k^5)^2}{(n+1)^2} =$$

$$\frac{(k^5)^2}{4(n+1)^2} \cdot 3(9+2n-10) + \frac{(k^5)^2}{2(n+1)^2} \cdot 3(3+n-5) + \frac{3(n-5)(k^5)^2}{(n+1)^2} +$$

$$\frac{(n-5)(n-6)(k^5)^2}{2(n+1)^2} - \frac{9(k^5)^2}{4(n+1)^2} - \frac{(n-5)(k^5)^2}{(n+1)^2} =$$

$$\frac{(k^5)^2}{(n+1)^2} \left[\frac{3(2n-1)}{4} + \frac{3(n-2)}{2} + 3(n-5) + \frac{(n-5)(n-6)}{2} - \frac{9}{4} - (n-5) \right] =$$

$$\frac{(k^5)^2}{4(n+1)^2} [3(2n-1) + 6(n-2) + 12(n-5) + (2n-10)(n-6) - 9 - 4(n-5)] =$$

$$\frac{(k^5)^2}{4(n+1)^2}(6n - 3 + 6n - 12 + 12n - 60 + 2n^2 - 12n - 10n + 60 - 9 - 4n + 20) =$$

$$\frac{(k^5)^2}{4(n+1)^2}(2n^2 - 2n - 4) = \frac{(k^5)^2}{2(n+1)^2}(n^2 - n - 2) = \frac{(k^5)^2}{2(n+1)^2}(n+1)(n-2).$$

From this we get

$$f_5 \leq \frac{(k^5)^2}{2} \cdot \frac{n-2}{n+1},$$

or, equivalently,

$$f_5 \leq \frac{n^2}{2} \cdot \left(\frac{n-2}{n+1} \right) (H^5)^2.$$

In the same manner we prove for f_r, with $5 \leq r \leq n$,

$$f_r \leq \frac{n^2}{2} \cdot \left(\frac{n-2}{n+1}\right)(H^r)^2.$$

Applying this procedure for each $\alpha \in \{1,2,3\}$ and taking into account that

$$\frac{2n-5}{2n+1} < \frac{n-2}{n+1},$$

we find

$$\delta(2,2) \leq \frac{n^2}{2} \cdot \frac{n-2}{n+1}\|H\|^2 + [n(n-1)-4]\frac{c}{2},$$

which is the inequality to prove. □

Remark 1. *In [11], the first author obtained certain Chen inequalities for Lagrangian submanifolds M^n in quaternionic space forms $\widetilde{M}^{4n}(4c)$. In particular, for the Chen invariant $\delta(2,2)$ one derives the inequality*

$$\delta(2,2) \leq \frac{n^2(n-1)}{2(n+2)}\|H\|^2 + \frac{1}{2}[n(n-1)-4]c. \tag{21}$$

We want to point-out that the inequality from Theorem 6 improves the inequality (21) because $\frac{n-2}{n+1} < \frac{n-1}{n+2}$, for $n > 4$.

Author Contributions: Conceptualization, G.M.; methodology, A.M.; software, G.M.; validation, G.M., A.M. and I.M.; formal analysis, I.M.; investigation, G.M., A.M. and I.M.; writing–review and editing, G.M. and A.M.; supervision, I.M. All authors have read and agreed to the published version of the manuscript.

Funding: This research received no external funding.

Conflicts of Interest: The authors declare no conflict of interest.

References

1. Chen, B.-Y. δ-Invariants, Inequalities of Submanifolds and Their Applications. In *Topics in Differential Geometry*; Mihai, A., Mihai, I., Miron, R., Eds.; Editura Academiei Române: Bucharest, Romania, 2008; pp. 29–155.
2. Chen, B.-Y. Some pinching and classification theorems for minimal submanifolds. *Arch. Math.* **1993**, *60*, 568–578. [CrossRef]
3. Oprea, T. Optimizations on Riemannian submanifolds. *Analele Univ. Buc.* **2005**, *54*, 127–136.
4. Oiagă, A.; Mihai, I.B.Y. Chen inequalities for slant submanifolds in complex space forms. *Demonstratio Math.* **1999**, *32*, 835–846. [CrossRef]
5. Bolton, J.; Dillen, F.; Fastenakels, J.; Vrancken, L. A best possible inequality for curvature-like tensor fields. *Math. Inequal. Appl.* **2009**, *12*, 663–681. [CrossRef]
6. Mihai, A. Geometric inequalities for purely real submanifolds in complex space forms. *Results Math.* **2009**, *55*, 457–468. [CrossRef]
7. Chen, B.-Y. Some new obstructions to minimal and Lagrangian isometric immersions. *Jpn. J. Math.* **2000**, *26*, 105–127. [CrossRef]
8. Chen, B.-Y.; Dillen, F.; Van der Veken, J.; Vrancken, L. Curvature inequalities for Lagrangian submanifolds: The final solution. *Diff. Geom. Appl.* **2013**, *31*, 808–819. [CrossRef]
9. Chen, B.-Y.; Prieto, A.; Wang, X. Lagrangian submanifolds in complex space forms satisfying an improved equality involving $\delta(2,2)$. *Publ. Math. Debrecen* **2013**, *82*, 193–217. [CrossRef]

10. Chen, B.-Y.; Houh, C.S. Totally real submanifolds of a quaternion projective space. *Ann. Mat. Pura Appl.* **1979**, *120*, 185–199. [CrossRef]
11. Macsim, G. Improved Chen's inequalities for Lagrangian submanifolds in quaternionic space forms. *Romanian J. Math. Comp. Sci.* **2016**, *6*, 61–84.

 © 2020 by the authors. Licensee MDPI, Basel, Switzerland. This article is an open access article distributed under the terms and conditions of the Creative Commons Attribution (CC BY) license (http://creativecommons.org/licenses/by/4.0/).

Article

On the Extrinsic Principal Directions and Curvatures of Lagrangian Submanifolds

Marilena Moruz [1,†] **and Leopold Verstraelen** [2,*,†]

1. Department of Mathematics, KU Leuven, Celestijnenlaan 200B-Box 2400, 3001 Leuven, Belgium; marilena.moruz@kuleuven.be
2. PiT and CiT, Department of Mathematics, KU Leuven, Celestijnenlaan 200B-Box 2400, 3001 Leuven, Belgium
* Correspondence: leopold.verstraelen@kuleuven.be
† These authors contributed equally to this work.

Received: 31 July 2020; Accepted: 4 September 2020; Published: 8 September 2020

Abstract: From the basic geometry of submanifolds will be recalled what are *the extrinsic principal tangential directions*, (first studied by Camille Jordan in the 18seventies), and what are *the principal first normal directions*, (first studied by Kostadin Trenčevski in the 19nineties), and what are *their corresponding Casorati curvatures*. For reasons of simplicity of exposition only, hereafter this will merely be done explicitly in the case of arbitrary submanifolds in Euclidean spaces. Then, *for the special case of Lagrangian submanifolds* in complex Euclidean spaces, *the natural relationships between these distinguished tangential and normal directions and their corresponding curvatures will be established*.

Keywords: extrinsic principal tangential directions; principal first normal directions; Lagrangian submanifolds

---

1. The Extrinsic Tangential Principal Directions of Submanifolds

For general submanifolds M^n of dimension n (≥ 2) and of co-dimension m (≥ 1) in Euclidean spaces $\mathbb{E}^{n+m}$, Jordan [1] studied the extrinsic curvatures $c_u^T(p)$ at arbitrary points $p \in M$ in arbitrary tangential directions determined by vectors $u \in T_p M$, $\|u\| = 1$. These are the curvatures $c_u^T(p) = (d\varphi_u/ds)^2(0)$, whereby $\varphi_u(s) \in [0, \Pi/2]$ denotes *the angle in $\mathbb{E}^{n+m}$ between the tangent spaces $T_p M$ at p and $T_q M$ at a nearby point $q \in M$ in the direction u of M at p*, s being an arclength parameter of a curve γ on M from $p = \gamma(0)$ in the direction $u = \gamma'(0)$ to $q = \gamma(s)$. He defined *the tangential principal curvatures $c_1^T(p) \geq c_2^T(p) \geq \ldots \geq c_n^T(p) \geq 0$ of a submanifold M^n in $\mathbb{E}^{n+m}$ at p as the critical values of the tangential Casorati curvature function at p*, that is of the function $c^T(p) : S_p^{n-1}(1) = \{u \in T_p M | \|u\| = 1\} \to \mathbb{R}^+ : u \mapsto c_u^T(p)$, and, *he defined the tangential principal directions of a submanifold M^n in $\mathbb{E}^{n+m}$ at p as the directions in which these critical values of the curvatures $c_u^T(p)$ are attained* and proved these directions to be mutually orthogonal, say to be determined by orthonormal vectors $f_1, f_2, \ldots, f_n \in T_p M$.

In the first step of his original fundamental studies of the geometry of submanifolds, Trenčevski [2–5] re-considered this work of Jordan, and, later, Stefan Haesen and Daniel Kowalczyk and one of the authors [6] basically re-did this. In the latter paper were followed *the 1890 Casorati's views on the intuitively most natural scalar valued curvatures "as such" of surfaces M^2 in $\mathbb{E}^3$* (and in [6,7], some tangential and normal kinds of curvature of Riemannian submanifolds were named after Casorati). Accordingly, in [6], the above *tangential Casorati curvatures* rather came up as $c_u^T(p) = (d\psi_u/ds)^2(0)$, whereby $\psi_u(s)$ denotes *the angle in $\mathbb{E}^{n+m}$ between the normal spaces $T_p^\perp M$ at p and $T_q^\perp M$ at a nearby point q in the direction of u* (as was already known by Jordan, $\psi_u = \varphi_u$). As shown by Trenčevski, *the extrinsic principal unit tangential vector fields $F_1, F_2, \ldots, F_n$ of a submanifold M^n in $\mathbb{E}^{n+m}$, and their corresponding tangential Casorati principal curvature functions $c_1^T, c_2^T, \ldots, c_n^T : M \to \mathbb{R}^+ : p \mapsto c_1^T(p), c_2^T(p), \ldots, c_n^T(p)$ are essentially the orthonormal eigen vector fields, and the corresponding eigen functions of the symmetric linear Casorati operator* $A^C = \sum_\alpha A_\alpha^2$, whereby $A_\alpha = A_{\zeta_\alpha}$

are the shape operators of M^n in $\mathbb{E}^{n+m}$ for arbitrary orthonormal normal frame fields $\xi_1, \xi_2, \ldots, \xi_m$ on M^n in $\mathbb{E}^{n+m}$, such that $A^C F_i = c_i^T F_i$, $i \in \{1, 2, \ldots, n\}$, $\alpha \in \{1, 2 \ldots, m\}$; (the *intrinsic* principal tangential directions and their corresponding curvatures of a submanifold M^n in $\mathbb{E}^{n+m}$, of course, being its *Ricci* principal directions and curvatures).

From the above, in particular, one may notice that *for hypersurfaces M^n in $\mathbb{E}^{n+1}$, the extrinsic principal tangential directions are "the classical" principal directions of these hypersurfaces, whereas* $\{c_1^T, c_2^T, \ldots, c_n^T\} = \{k_1^2, k_2^2, \ldots, k_n^2\}$, $k_1, k_2, \ldots, k_n$, *being the classical principal curvatures of these hypersurfaces*, correspond to Kronecker's extension of Euler's theory of the curvature of surfaces M^2 in $\mathbb{E}^3$ to hypersurfaces M^n in $\mathbb{E}^{n+1}$ for all dimensions $n \geq 2$.

2. Felice Casorati's Study of Surfaces M^2 in $\mathbb{E}^3$

Casorati [8] defined his extrinsic scalar-valued curvature $C(p)$ of a surface M^2 in $\mathbb{E}^3$ at one of its points p as follows. On M^2, consider a small geodesic circle γ_{Δ_ρ} centered at p with radius Δ_ρ. Let q be any point on γ_{Δ_ρ}, and consider the geodesic δ from p to q parametrised by arclength, such that $p = \delta(0)$ and $q = \delta(\Delta_\rho)$; at p, this geodesic points in the tangential direction $\delta'(0) = u$ to M^2 at p. Let $\eta(p)$ and $\eta(q)$ be the unit normal vectors on the surfaces M^2 in $\mathbb{E}^3$ at p and at q, respectively, corresponding to a choice of unit normal vector field η around p on M^2 in $\mathbb{E}^3$. Then, *in Casorati's words, and according to our common sense, the angle $\Delta\psi_u$ between $\eta(p)$ and $\eta(q)$ measures well how much the surface M^2 at p curves in the direction u; the more the surface curves in the direction u, the larger this angle*. Then, joining all the points $\delta(\Delta\psi_u)$ that thus correspond to all the points q on the geodesic circle γ_{Δ_ρ} around p, associated with γ_{Δ_ρ}, one obtains on M^2 a closed curve Γ_{Δ_ρ} (which actually passes through p whenever, at p, the surface is not curved at all in some tangential directions u). Hence, according to our common sense, the bigger or the smaller the area's $A(\Gamma_{\Delta_\rho})$ enclosed on M^2 by the curves Γ_{Δ_ρ} as compared to the area's $A(\gamma_{\Delta_\rho})$ of the geodesic discs on M^2 bounded by the geodesics γ_{Δ_ρ}, the more or the less the surface M^2 "as such" in $\mathbb{E}^3$ is curved at p. It was along this line of thought that Casorati defined his curvature of a surface M in $\mathbb{E}^3$ at p as $C(p) = \lim_{\Delta_\rho \to 0}(A(\Gamma_{\Delta_\rho})/A(\gamma_{\Delta_\rho}))$, and he proved that $C(p) = \frac{1}{2} tr A^2(p) = \frac{1}{2}(k_1^2 + k_2^2)(p)$ $= \frac{1}{2}\|h\|^2(p)$, whereby k_1 and k_2 are Euler's principal curvatures, A is the shape operator of M^2 corresponding to η and h is the second fundamental form of M^2 in $\mathbb{E}^3$.

At this stage, it might be not amiss to add the following comment. *In the definition of his curvature C, Casorati followed the common basic, i.e., from the original geometrical definitions of the curvature K of Gauss and the mean curvature H of Germain via ratios of well-chosen areas related to the surfaces M^2 in $\mathbb{E}^3$.* For $K(p)$, these ratios concern regions on M^2 around p and their corresponding spherical images, and, for $H(p)$, these ratios are for discs centered at p in $T_p M$ and for the portions of the corresponding circular cylinders perpendicular to $T_p M$ in between $T_p M$ and the surface M^2 in $\mathbb{E}^3$ itself. While for the curvatures of Germain and Gauss, this lead *to the first two elementary symmetric functions of k_1 and k_2*, $H = \frac{1}{2} tr A = \frac{1}{2}(k_1 + k_2)$ and $K = det A = k_1 k_2$, Casorati's geometrical definition of his curvature yielding that $C = \frac{1}{2} tr A^2 = \frac{1}{2}(k_1^2 + k_2^2)$ lead *to the third elementary symmetric function of Euler's principal curvatures*.

3. The First Normal Principal Directions of Submanifolds

Trenčevski determined *the maximal possible dimensions of the osculating spaces of all orders for submanifolds M^n in $\mathbb{E}^{n+m}$ and, moreover, in the related succesive normal spaces, of all orders, and determined appropriate orthonormal frames of principal normal vector fields and corresponding principal normal curvatures*. For our present purpose, it may suffice here to restrict within this grand theory to what is stated in Theorem 1 of [7]: "*The first principal normal directions of a submanifold M^n in $\mathbb{E}^{n+m}$ are the normal directions of M^n in $\mathbb{E}^{n+m}$ in which the normal Casorati curvatures of M^n attain their m_1 (=dimension of the first normal space N_1) non-zero critical values.*" The first normal space N_1 of M^n in $\mathbb{E}^{n+m}$ is the subspace of the total normal space $T^\perp M$ of M^n in $\mathbb{E}^{n+m}$ given by $N_1 = Im\, h = \{h(X,Y) | X, Y \in TM\}$, whereby h is *the second fundamental form of the submanifold M*, or, N_1 is the orthogonal complement in $T^\perp M$ of the subspace consisting of *all normals* ζ with vanishing shape operators A_ζ, or, equivalently, with vanishing normal Casorati curvature $c_\zeta^\perp$; $N_1 = \{\zeta \in T^\perp M | A_\zeta = 0\}^\perp = \{\zeta \in T^\perp M | c_\zeta^\perp = 0\}^\perp$, such that *the first osculating space of M^n in $\mathbb{E}^{n+m}$ is given by $TM \oplus N_1$*.

The considerations of Casorati on surfaces M^2 in $\mathbb{E}^3$ that were recalled above can straightforwardly be taken over to general submanifolds M^n in $\mathbb{E}^{n+m}$ (and to general submanifolds M^n in ambient general Riemannian spaces $\tilde{M}^{n+m}$, for that matter) cfr. [6]. In [6], a.o. one may find that the Casorati curvature (as such) of a submanifold M^n in $\mathbb{E}^{n+m}$ equals the arithmetic mean of its tangential principal Casorati curvatures: $C = \frac{1}{n}\|h\|^2 = \frac{1}{n} tr A^C = \frac{1}{n}\sum_\alpha tr A_\alpha^2 = \frac{1}{n}\sum_i c_i^T$. Moreover, it seems not without interest to observe that $C_\zeta(p) = \frac{1}{n} tr A_\zeta^2(p)$ is the Casorati curvature (as such) at p of the projection $M_{\zeta p}^n$ of the submanifold M^n in $\mathbb{E}^{n+m}$ onto the $(n+1)D$ subspace $\mathbb{E}^{n+1}$ of $\mathbb{E}^{n+m}$, which is spanned by $T_p M = \mathbb{R}^n$ together with the normal line $[\zeta(p)]$, ζ being any unit normal vector field on M^n in $\mathbb{E}^{n+m}$, and, hence, that $C_\zeta(p) = \frac{1}{n}\sum_i c_{\zeta i}^T(p)$, i.e., $C_\zeta(p)$ is the arithmetic mean of the tangential Casorati curvatures $c_{\zeta i}^T$ of this projected hypersurface M_ζ^n at p (for some general considerations relating the contemplation and the theory of submanifolds, see [9]). The functions $c_\zeta^\perp : \mathbb{S}^{m-1}(1) = \{\zeta \in T^\perp M \mid \|\zeta\| = 1\} \to \mathbb{R}^+$: $\zeta \mapsto c_\zeta^\perp = \frac{1}{n} tr A_\zeta^2$ are called *the normal Casorati curvatures of M^n in $\mathbb{E}^{n+m}$*; more precisely, *the normal Casorati curvature of M^n in $\mathbb{E}^{n+m}$ in the direction determined by a unit normal vector field ζ is defined as $c_\zeta^\perp = \frac{1}{n} tr A_\zeta^2$*.

In the total, mD normal space $T^\perp M$ of M^n in $\mathbb{E}^{n+m}$, consider the following symmetric linear operator $a : T^\perp M \to T^\perp M : \zeta \mapsto a(\zeta) = \frac{1}{n}\|\zeta\|\sum_\alpha (tr A_\zeta A_\alpha)\zeta_\alpha$; (in [10], Bang-Yen Chen basically introduced this operator in the study of the submanifolds for which $a(\vec{H}) = \vec{0}$, $\vec{H}$ being *the mean curvature vector field of M^n in $\mathbb{E}^{n+m}$*, submanifolds which later were called Chen submanifolds; in this respect, see also [11,12]). And, by the principal axes theorem, *there exists an orthonormal frame $\eta_1, \eta_2, \ldots, \eta_{m_1}, \eta_{m_1+1}, \ldots, \eta_m$ of eigen vector fields for this operator* $a : T^\perp M \to T^\perp M$ ($m_1 = \dim N_1$), *with corresponding eigen functions* $c_1^\perp = \frac{1}{n} tr A_{\eta_1}^2 \geq c_2^\perp = \frac{1}{n} tr A_{\eta_2}^2 \geq \ldots \geq c_{m_1}^\perp = \frac{1}{n} tr A_{\eta_{m_1}}^2 > c_{m_1+1}^\perp = tr A_{\eta_{m_1+1}}^2 = \ldots = c_m^\perp = tr A_{\eta_m}^2 = 0$. The normal vector fields $\eta_1, \eta_2, \ldots, \eta_{m_1}$ span the first normal space N_1 of M^n in $\mathbb{E}^{n+m}$ and, following Trenčevski, are called *the first principal normal vector fields of the submanifold M^n in $\mathbb{E}^{n+m}$ with corresponding first-principal normal curvatures* $c_1^\perp \geq c_2^\perp \geq \ldots \geq c_{m_1}^\perp > 0$. So, with indices $\alpha_1 \in \{1, 2, \ldots, m_1\}$, $\{\eta_{\alpha_1}\}$ *is an orthonormal frame field of the first normal space N_1 for which $a(\eta_{\alpha_1}) = c_{\alpha_1}^\perp \eta_{\alpha_1}$, whereby $c_{\alpha_1}^\perp = \frac{1}{n} tr A_{\alpha_1}^2 (> 0)$ are the principal normal Casorati curvatures of M^n in $\mathbb{E}^{n+m}$*.

4. The Principal Tangent and the First Principal Normal Directions of Lagrangian Submanifolds

From *Section 16: Totally real and Lagrangian submanifolds of Kähler manifolds* of Chen's contribution on *Riemannian submanifolds* in [11], is taken the following: "The study of totally real submanifolds of a Kähler manifold from differential geometric points of views was initiated in the early 1970's (by Bang-Yen Chen and Koichi Ogiue [13]—the authors). A totally real submanifold M of a Kähler manifold $\tilde{M}$ is a submanifold such that the almost complex structure J of the ambient manifold $\tilde{M}$ carries each tangent space of M into the corresponding normal space of M, that is, $J(T_p M) \subset T_p^\perp M$ for any point $p \in M$. (...) A totally real submanifold M of a Kähler manifold $\tilde{M}$ is called Lagrangian if $\dim_\mathbb{R} M = \dim_\mathbb{C} \tilde{M}$. 1-dimensional submanifolds, that is, real curves in a Kähler manifold are always totally real. For this reason, we only consider totally real submanifolds of dimension ≥ 2.(...) For a Lagrangian submanifold M of a Kähler manifold $(\tilde{M}, g, J)$ the tangent bundle TM and the normal bundle $T^\perp M$ are isomorphic via the almost complex structure J of the ambient manifold. In particular, this implies that *the Lagrangian submanifold has flat normal connection if and only if the submanifold is a flat Riemannian manifold*".

To continue in our aim to aim for simplicity and concreteness of presentation, (although, clearly, the following matters do hold more generally), we next restrict our attention to *the real n dimensional totally real submanifolds M^n of the complex n dimensional complex Euclidean spaces $\tilde{M}^n = \mathbb{C}^n = (\mathbb{E}^{2n}, \tilde{J})$*, that is, to *the Lagrangian submanifolds M^n in $\mathbb{C}^n$*, thus having $\tilde{J}(TM) = T^\perp M$ and $\tilde{J}(T^\perp M) = TM$, $\tilde{J}$ being *the complex structure of the Kaehler manifold $\tilde{M}^n$*. On M in $\tilde{M}$, tangential vector fields will be denoted by $X, Y, Z, \ldots$ and *normal vector fields* by $\xi, \eta, \zeta, \ldots$. Further, let $\tilde{g}$ and $\tilde{\nabla}$, respectively, g and ∇, be *the metrics and the corresponding Riemannian connections* on $\tilde{M}$ and M, respectively. The *equations of Gauss and of Weingarten* are given by

$$\tilde{\nabla}_X Y = \nabla_X Y + h(X, Y), \tag{1}$$

$$\tilde{\nabla}_X \xi = -A_\xi(X) + \nabla_X^\perp \xi, \tag{2}$$

whereby $\nabla^\perp$ is *the normal connection* and h *the second fundamental form* and A_ξ *the shape operator with respect to* ξ of the submanifold M in $\tilde{M}$, so that

$$\tilde{g}(h(X, Y), \xi) = g(A_\xi(X), Y). \tag{3}$$

Applying the complex structure $\tilde{J}$ to (1), it follows that

$$\tilde{J}(\tilde{\nabla}_X Y) = \tilde{J}(\nabla_X Y) + \tilde{J}(h(X, Y)), \tag{4}$$

while writing (2) out for $\xi = \tilde{J}Y$, it follows that

$$\tilde{\nabla}_X(\tilde{J}Y) = -A_{\tilde{J}Y}(X) + \nabla_X^\perp(\tilde{J}Y). \tag{5}$$

By the parallelity of $\tilde{J}$, $\tilde{\nabla}\tilde{J} = 0$, or, still, $\tilde{J}(\tilde{\nabla}_X Y) = \tilde{\nabla}_X(\tilde{J}Y)$, the left-hand sides in (4) and (5) are equal, and, hence, in particular, the tangential components of the right-hand sides in (4) and (5) are also equal:

$$\tilde{J}(h(X, Y)) = -A_{\tilde{J}Y}(X). \tag{6}$$

Writing out (3) for $\xi = \tilde{J}Z$, it follows that

$$\tilde{g}(h(X, Y), \tilde{J}Z) = g(A_{\tilde{J}Z}(X), Y), \tag{7}$$

which, by (6), leads to

$$\tilde{g}(h(X, Y), \tilde{J}Z) = g(-\tilde{J}(h(X, Z)), Y). \tag{8}$$

Since $\tilde{J}$ is almost complex, $\tilde{J}^2 = -I$, and since $\tilde{g}$ is *Hermitian*, so that $\tilde{g}(\tilde{J}\tilde{V}, \tilde{J}\tilde{W}) = \tilde{g}(\tilde{V}, \tilde{W})$ for all vector fields $\tilde{V}$ and $\tilde{W}$, and, hence, in particular, for $\tilde{V} = -\tilde{J}(h(X, Z))$ and $\tilde{W} = Y$, (8) becomes

$$\begin{aligned}\tilde{g}(h(X, Y), \tilde{J}Z) &= g(-\tilde{J}^2(h(X, Z)), \tilde{J}Y)\\ &= g(h(X, Z), \tilde{J}Y).\end{aligned} \tag{9}$$

In view of its crucial importance in what comes next, we have worked out in detail this property from [13], as obtained in (9), which may be stated as follows. For all tangential vector fields X, Y, Z on a Lagrangian submanifold $M^n \subset \mathbb{C}^n(\tilde{M}^n)$:

$$\tilde{g}(h(X, Y), \tilde{J}Z) = \tilde{g}(h(X, Z), \tilde{J}Y) = \tilde{g}(h(Y, Z), \tilde{J}X). \tag{10}$$

For any tangential orthonormal frame field $\mathcal{F} = \{E_1, E_2, \ldots, E_n\}$ on a Lagrangian submanifold M^n, $\tilde{\mathcal{F}} = \{E_1, E_2, \ldots, E_n, \xi_1 = \tilde{J}E_1, \xi_2 = \tilde{J}E_2, \ldots, \xi_n = \tilde{J}E_n\} = \{E_i, \xi_i = \tilde{J}E_i\}$, $(i, j, k, \alpha, \beta \in \{1, 2, \ldots, n\})$ *is a corresponding adapted orthonormal frame field of* $\mathbb{C}^n(\tilde{M}^n)$ along M^n. The local coordinates of the operator $A^C : TM \to TM$ of Casorati and of the operator $a : T^\perp M \to T^\perp M$ of Trenčevski with respect to such frame fields $\tilde{\mathcal{F}}$ are given by

$$A_{ik}^C = (\sum_\alpha A_\alpha^2)_{ik} = \sum_\alpha (A_\alpha^2)_{ik}$$
$$= \sum_\alpha \sum_j h_{ij}^\alpha h_{jk}^\alpha \qquad (11)$$

and

$$a_{\alpha\beta} = tr(A_\alpha A_\beta)$$
$$= \sum_i \sum_j h_{ij}^\alpha h_{ji}^\beta, \qquad (12)$$

whereby h_{ij}^β are the local coordinates of the symmetric second fundamental form $h: TM \times TM \to T^\perp M$. Therefore, E_i determines a Casorati principal tangential vector field on M^n in $\mathbb{C}^n$ with corresponding principal tangential Casorati curvature c_i^T if and only if

$$\forall k \neq i \,:\, A_{ik}^C = \sum_\alpha \sum_j h_{ij}^\alpha h_{jk}^\alpha = 0, \qquad (13)$$

whereby

$$c_i^T = A_{ii}^C = \sum_\alpha \sum_j (h_{ij}^\alpha)^2, \qquad (14)$$

and, $\zeta_\alpha = \tilde{J} E_\alpha$ determines a first principal normal vector field, or, first Casorati principal normal vector field (as these vector fields later on also might be termed), on M^n in $\mathbb{C}^n$ with corresponding principal normal Casorati curvature c_α^T if and only if

$$\forall \beta \neq \alpha \,:\, a_{\alpha\beta} = \sum_i \sum_j h_{ij}^\alpha h_{ji}^\beta = 0, \qquad (15)$$

whereby then

$$c_\alpha^\perp = a_{\alpha\alpha} = \sum_i \sum_j (h_{ij}^\alpha)^2. \qquad (16)$$

Written in local coordinates, *the above property* (10) *amounts to*

$$\forall i,j,k \,:\, h_{ij}^k = h_{ik}^j = h_{jk}^i, \qquad (17)$$

so that, from (13) and (15), and, from (14) and (16), in particular, we may conclude the following.

Theorem 1. *Let M^n be a Lagrangian submanifold of the complex Euclidean space $\mathbb{C}^n$ (or of any Kaehler manifold $\tilde{M}^n$). Then, a tangential vector field T is a tangential principal Casorati vector field with corresponding tangential Casorati principal curvature $c^T (> 0)$ if and only if $N = \tilde{J} T$ is a normal principal Casorati vector field—whereby $\tilde{J}$ is the complex structure of $\mathbb{C}^n$ (or, of the ambient Kaehler space, $\tilde{M}^n$)—with corresponding normal Casorati principal curvature $c^\perp = c^T (> 0)$.*

Theorem 2. *Let M^n be a Lagrangian submanifold of the complex Euclidean space $\mathbb{C}^n$ (or, of any ambient Kaehler manifold $\tilde{M}^n$) with first normal space of maximal dimension ($m_1 = dim N_1 = n = co - dim M$). Then, M^n admits an adapted orthonormal frame field $\tilde{\mathcal{F}} = \{F_1, F_2, \ldots, F_n, \eta_1 = \tilde{J} F_1, \eta_2 = \tilde{J} F_2, \ldots, \eta_n = \tilde{J} F_n\}$ in $\mathbb{C}^n$ ($\tilde{M}^n$), of which the n tangential vector fields are the principal Casorati tangential vector fields and of which the n normal vector fields are the principal Casorati normal vector fields of M^n in $\mathbb{C}^n$ ($\tilde{M}^n$), and the corresponding tangential and normal principal curvatures are equal, ($\forall i \,:\, c_i^T = c_i^\perp$).*

Author Contributions: Investigation, M.M. and L.V. The authors contributed equally to this work. All authors have read and agreed to the published version of the manuscript.

Funding: M.Moruz is a postdoctoral fellow of the Research Foundation—Flanders (FWO).

Conflicts of Interest: The authors declare no conflict of interest.

References

1. Jordan, C. Généralization du théorème d'Euler sur la courbure des surfaces. *C. R. Acad. Sci. Paris* **1874**, *79*, 909–912.
2. Trenčevski, K. Principal directions for submanifolds imbedded in Euclidean spaces of arbitrary codimension. *Proc. Third Intern. Workshop on Diff. Geom. Appl. First Ger. Rom. Semin. Geom. Gen. Math.* **1997**, *5*, 385–392.
3. Trenčevski, K. New approach for submanifolds of the Euclidean space. *Balkan J. Geom. Appl.* **1997**, *2*, 117–127.
4. Trenčevski, K. Geometrical interpretation of the principal directions and principal curvatures of submanifolds. *Diff. Geom. Dyn. Syst.* **2000**, *2*, 50–58.
5. Trenčevski, K. On the osculating spaces of submanifolds in Euclidean spaces. *Kragujevic J. Math.* **2012**, *36*, 45–49.
6. Haesen, S.; Kowalczyk, D.; Verstraelen, L. On the extrinsic principal directions of Riemannian submanifolds. *Note Mat.* **2009**, *29*, 41–51.
7. Verstraelen, L. Geometry of submanifolds I. The first Casorati curvature indicatrices. *Kragujevic J. Math.* **2013**, *37*, 5–23.
8. Casorati, F. Mesure de la courbure des surfaces suivant l'idée commune. *Acta Math.* **1890**, *14*, 95–110. [CrossRef]
9. Verstraelen, L. *Submanifolds Theory—A Contemplation of Submanifolds, "Frontmatter" in Geometry of Submanifolds, AMS Contemporary Mathematics*; volume in honor of Professor Bang-Yen Chen; Van der Veken, J., Cariazo, A., Suceava, B.D., Oh, Y.M., Vrancken, L., Eds.; American Mathematical Society: Providence, RI, USA, 2020.
10. Chen, B.Y. *Geometry of Submanifolds*; Marcel Dekker Publ. Co.: New York, NY, USA, 1973.
11. Chen, B.Y. Riemannian submanifolds. In *Handbook of Differential Geometry*; Dillen, F.J.E., Ed.; Elsevier: Amsterdam, The Netherlands, 2000; Volume 1, Chapter 3, pp. 187–418.
12. Rouxel, B. Sur quelques propriétés anallagmatiques de l'espace euclidien $\mathbb{E}^4$. In *Mémmoire Couronné*; Acad. Royale Belge: Brussels, Belgium, 1982; 128p.
13. Chen, B.Y.; Ogiue, K. On totally real submanifolds. *Trans. Am. Math. Soc.* **1974**, *193*, 257–266. [CrossRef]

© 2020 by the authors. Licensee MDPI, Basel, Switzerland. This article is an open access article distributed under the terms and conditions of the Creative Commons Attribution (CC BY) license (http://creativecommons.org/licenses/by/4.0/).

Article

On the Betti and Tachibana Numbers of Compact Einstein Manifolds

Vladimir Rovenski [1,*], Sergey Stepanov [2] and Irina Tsyganok [3]

[1] Department of Mathematics, University of Haifa, Mount Carmel, Haifa 31905, Israel
[2] Department of Mathematics, Russian Institute for Scientific and Technical Information of the Russian Academy of Sciences, 20, Usievicha Street, 125190 Moscow, Russia; s.e.stepanov@mail.ru
[3] Department of Data Analysis and Financial Technologies, Finance University, 49-55, Leningradsky Prospect, 125468 Moscow, Russia; i.i.tsyganok@mail.ru
* Correspondence: vrovenski@univ.haifa.ac.il

Received: 23 November 2019; Accepted: 7 December 2019; Published: 9 December 2019

Abstract: Throughout the history of the study of Einstein manifolds, researchers have sought relationships between the curvature and topology of such manifolds. In this paper, first, we prove that a compact Einstein manifold (M, g) with an Einstein constant $\alpha > 0$ is a homological sphere when the minimum of its sectional curvatures $> \alpha/(n+2)$; in particular, (M, g) is a spherical space form when the minimum of its sectional curvatures $> \alpha/n$. Second, we prove two propositions (similar to the above ones) for Tachibana numbers of a compact Einstein manifold with $\alpha < 0$.

Keywords: Einstein manifold; sectional curvature; Betti number; Tachibana number; spherical space form

MSC: 53C20; 53C43; 53C44

1. Introduction

The study of Einstein manifolds has a long history in Riemannian geometry. Throughout the history of the study of Einstein manifolds, researchers have sought relationships between curvature and topology of such manifolds. A. Besse [1] summarized the results. We present here some interesting facts related to the classification of all compact Einstein manifolds satisfying a suitable curvature inequality, which is one of the subjects of our research.

Recall that an n-dimensional $(n \geq 2)$ connected manifold M with a Riemannian metric g is said to be an *Einstein manifold* with *Einstein constant* α if its Ricci tensor satisfies $\mathrm{Ric} = \alpha\, g$; moreover, we have $\alpha = s/n$ for its scalar curvature s. Therefore, any Einstein manifold of dimensions two and three is a space form (i.e., has constant sectional curvature). The study of Einstein manifolds is more complicated in dimension four and higher (see [1] (p. 44)).

An important problem in differential geometry is to determine whether a smooth manifold admits an Einstein metric. When $\alpha > 0$, the example are symmetric spaces, which include the sphere $\mathbb{S}^n(1)$ with $\alpha = n-1$ and the sectional curvature $\sec = 1$, the product of two spheres $\mathbb{S}^n(1) \times \mathbb{S}^n(1)$ with $\alpha = n-1$ and $0 \leq \sec \leq 1$, and the complex projective space $\mathbb{C}P^m = \mathbb{S}^{2m+1}/\mathbb{S}^1$ with the Fubini–Study metric, $\alpha = 2m+2$ and $1 \leq \sec \leq 4$ (see [2] (pp. 86, 118, 149–150)). Recall that if (M,g) is a compact Einstein manifold with curvature bounds of the type $3n/(7n-4) < \sec \leq 1$, then (M,g) is isometric to a spherical space form. This might be not the best estimate: for $n = 4$ the sharp bound is $1/4$ (see [1] (p. 6)). In both these cases, the manifolds are real *homology spheres* (see [3] (p. XVI)). Therefore, any such manifold has the homology groups of an n-sphere; in particular, its Betti numbers are $b_1(M) = \ldots = b_{n-1}(M) = 0$.

One of the basic problems in Riemannian geometry was to classify Einstein four-manifolds with positive or nonnegative sectional curvature in the categories of either topology, diffeomorphism, or isometry (see, for example, [4–7]). It was conjectured that an Einstein four-manifold with $\alpha > 0$ and non-negative sectional curvature must be either $\mathbb{S}^4$, $\mathbb{CP}^2$, $\mathbb{S}^2(1) \times \mathbb{S}^2(1)$ or a quotient. For example, if the maximum of the sectional curvatures of a compact Einstein four-manifold is bounded above by $(2/3)\alpha$, or if $\alpha = 1$ and the minimum of the sectional curvatures $\geq (1/6)(2 - \sqrt{2})$, then the manifold is isometric to $\mathbb{S}^4$, $\mathbb{RP}^4$ or $\mathbb{CP}^2$ (see [6]). Classification of four-dimensional complete Einstein manifolds with $\alpha > 0$ and pinched sectional curvature was obtained in [7].

Here, we consider this problem from another side. Given a Riemannian manifold (M, g), the notion of symmetric *curvature operator* $\bar{R}$, acting on the space $\Lambda^2 M$ of 2-forms, is an important invariant of a Riemannian metric (see [2] (p. 83); [8,9]). The Tachibana Theorem (see [10]) asserts that a compact Einstein manifold (M, g) with $\bar{R} > 0$ is a spherical space form. Later on, it was proved that compact manifolds with $\bar{R} > 0$ are spherical space forms (see [11]).

Denote by $\overset{\circ}{R}$ the symmetric *curvature operator of the second kind*, acting on the space $S_0^2 M$ of traceless symmetric two-tensors (see [1] (p. 52); [9,12]). Kashiwada (see [9]) proved that a compact Einstein manifold with $\overset{\circ}{R} > 0$ is a spherical space form. This statement is an analogue of the theorem of Tachibana in [10]. In contrast, if a complete Riemannian manifold (M, g) satisfies sec $\geq \delta > 0$, then M is compact with $\text{diam}(M, g) \leq \pi/\sqrt{\delta}$ (see [2] (p. 251)).

Remark 1 (By [2] (Theorem 10.3.7)). *There are manifolds with metrics of positive or nonnegative sectional curvature but not admitting any metric with $\bar{R} \geq 0$ (see also [2] (p. 352)). In particular, for three-dimensional manifolds the inequality* sec > 0 *is equivalent to the inequality* $\bar{R} > 0$ *(see [9]).*

Using Kashiwada's theorem from [9] we can prove the following.

Theorem 1. *Let (M, g) be a compact Einstein manifold with Einstein constant $\alpha > 0$, and let δ be the minimum of its positive sectional curvature. If $\delta > \alpha/n$, then (M, g) is a spherical space form.*

We can present a generalization of above result in the following form.

Theorem 2. *Let (M, g) be a compact Einstein manifold with Einstein constant $\alpha > 0$ and let δ be the minimum of its positive sectional curvature. If $\delta > \alpha/(n+2)$, then (M, g) is a homological sphere.*

Obviously, $\mathbb{S}^n(1) \times \mathbb{S}^n(1)$ is not an example for Theorem 1 because the minimum of its sectional curvature is zero and $\alpha = n - 1$. On the other hand, the complex projective space $\mathbb{CP}^m$ is an Einstein manifold with $\alpha = 2m + 2$ and sectional curvature bounded below by $\delta = 1$. Then the inequality $\alpha < (n+2)\delta$ can be rewritten in the form $\delta > 1$ because $n = 2m$. Therefore, $\mathbb{CP}^m$ is not an example for Theorem 1. Moreover, all even dimensional Riemannian manifolds with positive sectional curvature have vanishing odd-dimensional homology groups. Thus, Theorem 1 complements this statement (see [2] (p. 328)).

Let (M, g) be an n-dimensional compact connected Riemannian manifold. Denote by $\Delta^{(p)}$ the *Hodge Laplacian* acting on differential p-forms on M for $p = 1, \ldots, n - 1$. The spectrum of $\Delta^{(p)}$ consists of an unbounded sequence of nonnegative eigenvalues which starts from zero if and only if the p-th Betti number $b_p(M)$ of (M, g) does not vanish (see [13]). The sequence of positive eigenvalues of $\Delta^{(p)}$ is denoted by

$$0 < \lambda_1^{(p)} < \ldots < \lambda_m^{(p)} < \ldots \to \infty.$$

In addition, if $F_p(\omega) \geq \sigma > 0$ (see Equation (4) of F_p) at every point of M, then $\lambda_1^{(p)} \geq \sigma$ (see [13] (p. 342)). Using this and Theorem 1, we get the following.

Corollary 1. *Let (M,g) be a compact Einstein manifold with positive Einstein constant α and sectional curvature bounded below by a constant $\delta > 0$ such that $\delta > \alpha/(n+2)$. Then the first eigenvalue $\lambda_1^{(p)}$ of the Hodge Laplacian $\Delta^{(p)}$ satisfies the inequality $\lambda_1^{(p)} \geq (1/3)\left((n+2)\delta - \alpha\right)(n-p)$.*

Remark 2. *In particular, if (M,g) is a Riemannian manifold with curvature operator of the second kind bounded below by a positive constant $\rho > 0$, then using the main theorem from [14], we conclude that $\lambda_1^{(p)} \geq \rho(n-p)$.*

Conformal Killing p-forms ($p = 1, \ldots, n-1$) were defined on Riemannian manifolds more than fifty years ago by S. Tachibana and T. Kashiwada (see [15,16]) as a natural generalization of conformal Killing vector fields.

The vector space of conformal Killing p-forms on a compact Riemannian manifold (M,g) has finite dimension $t_p(M)$ named the *Tachibana number* (see e.g., [17–19]). Tachibana numbers $t_1(M), \ldots, t_{n-1}(M)$ are conformal scalar invariants of (M,g) satisfying the duality condition $t_p(M) = t_{n-p}(M)$. The condition is an analog of the *Poincaré duality* for Betti numbers. Moreover, Tachibana numbers $t_1(M), \ldots, t_{n-1}(M)$ are equal to zero on a compact Riemannian manifold with negative curvature operator or negative curvature operator of the second kind (see [18,19]).

We obtain the following theorem, which is an analog of Theorem 1.

Theorem 3. *Let (M,g) be an Einstein manifold with sectional curvature bounded above by a negative constant $-\delta$ such that $\delta > -\alpha/(n+2)$ for the Einstein constant α. Then Tachibana numbers $t_1(M), \ldots, t_{n-1}(M)$ are zero.*

2. Proof of Results

Let (M,g) be an n-dimensional ($n \geq 2$) Riemannian manifold and let R_{ijkl} and R_{ij} be, respectively, the components of the Riemannian curvature tensor and the Ricci tensor in orthonormal basis $\{e_1, \ldots, e_n\}$ of T_xM at any point $x \in M$. We consider an arbitrary symmetric two-tensor φ on (M,g). At any point $x \in M$, we can diagonalize φ with respect to g, using orthonormal basis $\{e_1, \ldots, e_n\}$ of T_xM. In this case, the components of φ have the form $\varphi_{ij} = \lambda_i \delta_{ij}$. Let $\sec(e_i, e_j)$ be the sectional curvature of the plane of T_xM generated by e_i and e_j. We can express $\sec(e_i, e_j)$ in the following form (see [1] (p. 436); [20]):

$$\frac{1}{2} \sum_{i \neq j} \sec(e_i, e_j)(\lambda_i - \lambda_j)^2 = R_{ijlk}\varphi^{ik}\varphi^{jl} + R_{ij}\varphi^{ik}\varphi_k^j \qquad (1)$$

If (M,g) is an Einstein manifold and its sectional curvature satisfies the inequality $\sec \geq \delta$ for a positive constant δ, then from Equation (1) we obtain the inequality

$$R_{ijlk}\varphi^{ik}\varphi^{jl} + \frac{s}{n}\varphi^{ik}\varphi_{ik} \geq (\delta/2) \sum_{i \neq j}(\lambda_i - \lambda_j)^2. \qquad (2)$$

If $\text{trace}_g \varphi = \sum_i \lambda_i = 0$, then the identity holds $\sum_i (\lambda_i)^2 = -2 \sum_{i<j} \lambda_i \lambda_j$. In this case, the following identities are true:

$$\frac{1}{2} \sum_{i \neq j}(\lambda_i - \lambda_j)^2 = (n-1) \sum_i (\lambda_i)^2 - 2 \sum_{i<j} \lambda_i \lambda_j = n \sum_i (\lambda_i)^2 = n\|\varphi\|^2.$$

Then the inequality in Equation (2) can be rewritten in the form

$$R_{ijlk}\varphi^{ik}\varphi^{jl} + \frac{s}{n}\varphi^{ik}\varphi_{ik} \geq n\delta\|\varphi\|^2. \qquad (3)$$

From Equation (3) we obtain the inequality

$$R_{ijlk}\varphi^{ik}\varphi^{jl} \geq (n\delta - \alpha)\|\varphi\|^2.$$

Then $\overset{\circ}{R} > 0$ for the case when $\alpha < n\,\delta$, where $\alpha = s/n$ is the Einstein constant of (M,g). If (M,g) is compact then it is a spherical space form (see [9]). Theorem 1 is proven.

Define the quadratic form

$$F_p(\omega) = R_{ij}\,\omega^{i\,i_2\ldots i_p}\,\omega^{j}_{i_2\ldots i_p} - \frac{p-1}{2}\,R_{ijkl}\,\omega^{ij\,i_3\ldots i_p}\,\omega^{kl}_{i_2\ldots i_p} \qquad (4)$$

for the components $\omega_{i_1\ldots i_p} = \omega(e_{i_1},\ldots,e_{i_p})$ of an arbitrary differential p-form ω. If the quadratic form $F_p(\omega)$ is positive definite on a compact Riemannian manifold (M,g), then the p-th Betti number of the manifold vanishes (see [21] (p. 61); [3] (p. 88)). At the same time, in [22] the following inequality

$$F_p(\omega) \geq p\,(n-p)\,\varepsilon\,\|\omega\|^2 > 0$$

was proved for any nonzero p-form ω on a Riemannian manifold with $\bar{R} \geq \varepsilon > 0$. On the other hand, in [14] the inequality

$$F_p(\omega) \geq p(n-p)\,\delta\|\omega\|^2 > 0$$

was proved for any nonzero p-form ω on a Riemannian manifold with $\overset{\circ}{R} \geq \delta > 0$. In these cases, $b_1(M),\ldots,b_{n-1}(M)$ are zero (see [21]). We can improve these results for the case of Einstein manifolds. First, we will prove the following.

Lemma 1. *Let (M,g) be an Einstein manifold with Einstein constant α and sectional curvature bounded below by a constant $\delta > 0$. If $\alpha < (n+2)\delta$ then*

$$F_p(\omega) \geq (1/3)((n+2)\,\delta - \alpha)(n-p)\,\|\omega\|^2 > 0$$

for any nonzero p-form ω and an arbitrary $1 \leq p \leq n-1$.

Proof. Let $p \leq [n/2]$, then we can define the symmetric traceless two-tensor $\varphi^{(i_1 i_2 \ldots i_p)}$ with components (see [14])

$$\varphi^{(i_1 i_2 \ldots i_p)}_{jk} = \sum_{a=1}^{p}\left(\omega_{i_1\ldots i_{a-1}j i_{a+1}\ldots i_p}g_{ki_a} + \omega_{i_1\ldots i_{a-1}k i_{a+1}\ldots i_p}g_{ji_a}\right) - \frac{2p}{n}\,g_{jk}\,\omega_{i_1\ldots i_p}$$

for each set of values of indices $(i_1\,i_2\ldots i_p)$ such that $1 \leq i_1 < i_2 < \ldots < i_p \leq n$. After long but simple calculations we obtain the identities (see also [14]),

$$R_{ijkl}\,\varphi^{il\,(i_1\ldots i_p)}\,\varphi^{jk}_{(i_1\ldots i_p)} = p\left(\frac{2(n+4p)}{n}\,R_{ij}\,\omega^{i\,i_2\ldots i_p}\,\omega^{j}_{i_2\ldots i_p}\right.$$
$$\left. -3\,(p-1)\,R_{ijkl}\,\omega^{ij\,i_3\ldots i_p}\,\omega^{kl}_{i_3\ldots i_p} - \frac{4p}{n^2}s\,\|\omega\|^2\right); \qquad (5)$$

$$\|\bar{\varphi}\|^2 = \frac{2p(n+2)(n-p)}{n}\,\|\omega\|^2, \qquad (6)$$

where

$$\|\bar{\varphi}\|^2 = g^{ik}g^{jl}g_{i_1 j_1}\ldots g_{i_p j_p}\,\varphi^{(i_1\ldots i_p)}_{ij}\,\varphi^{(j_1\ldots j_p)}_{kl},$$
$$\|\omega\|^2 = \omega^{i_1 i_2 \ldots i_p}\omega_{i_1 i_2 \ldots i_p} = g^{i_1 j_1}\ldots g^{i_p j_p}\omega_{i_1\ldots i_p}\,\omega_{j_1\ldots j_p}$$

for $g^{ij} = (g^{-1})_{ij}$. If (M,g) is an Einstein manifold, then Equations (4) and (5) can be rewritten in the form

$$F_p(\omega) = \frac{s}{n}\,\|\omega\|^2 - \frac{p-1}{2}\,R_{ijkl}\,\omega^{ij\,i_3\ldots i_p}\,\omega^{kl}_{i_3\ldots i_p},$$

$$R_{ijkl}\,\varphi^{il(i_1...i_p)}\,\varphi^{jk}_{(i_1...i_p)} = p\left(\frac{2n+4p}{n^2}\,s\,\|\omega\|^2 - 3(p-1)R_{ijkl}\,\omega^{ij\,i_3...i_p}\,\omega^{kl}_{i_3...i_p}\right). \quad (7)$$

On the other hand, for a fixed set of values of indices $(i_1, i_2, \ldots, i_p)$ such that $1 \leq i_1 < i_2 < \ldots < i_p \leq n$, the equality in Equation (3) can be rewritten in the form

$$R_{ijkl}\,\varphi^{il\,(i_1...i_p)}\,\varphi^{jk(i_1...i_p)} + \frac{s}{n}\,\varphi^{ik\,(i_1...i_p)}\,\varphi^{(i_1...i_p)}_{ik} \geq n\,\delta\,\varphi^{kl\,(i_1...i_p)}\,\varphi^{(i_1...i_p)}_{kl}. \quad (8)$$

Then from Equation (8) we obtain the inequality

$$R_{ijkl}\,\varphi^{il\,(i_1...i_p)}\,\varphi^{jk}_{(i_1...i_p)} \geq \left(n\delta - \frac{s}{n}\right)\|\bar{\varphi}\|^2. \quad (9)$$

Using Equation (9) we deduce from Equation (7) the following inequality:

$$6p\,F_p(\omega) \geq \left(n\,\delta - \frac{s}{n+2}\right)\|\bar{\varphi}\|^2. \quad (10)$$

Thus, using Equation (6) we can rewrite Equation (10) in the following form:

$$F_p(\omega) \geq (1/3)((n+2)\,\delta - \alpha)\,(n-p)\|\omega\|^2. \quad (11)$$

It is obvious that if the sectional curvature of an Einstein manifold (M, g) satisfies the inequality $\sec \geq \delta$ for a positive constant δ, then the scalar curvature of (M, g) satisfies the inequality $s \geq n(n-1)\,\delta > 0$. In this case, if $(n-1)\,\delta \leq \alpha < (n+2)\,\delta$, then from Equation (11) we deduce that the quadratic form $F_p(\omega)$ is positive definite for any $p \leq [n/2]$. It is known [23] that $F_p(\omega) = F_{n-p}(*\,\omega)$ and $\|\omega\|^2 = \|*\omega\|^2$ for any p-form ω with $1 \leq p \leq n-1$ and the Hodge star operator $* : \Lambda^p M \to \Lambda^{n-p} M$ acting on the space of p-forms $\Lambda^p M$. Therefore, the inequality in Equation (11) holds for any $p = 1, \ldots, n-1$. □

Recall that if on an n-dimensional compact Riemannian manifold (M, g) the quadratic form $F_p(\omega)$ is positive definite for any smooth p-form ω with $p = 1, \ldots, n-1$, then the Betti numbers $b_1(M), \ldots, b_{n-1}(M)$ vanish (see [3] (p. 88); [13] (pp. 336–337)). In this case, Theorem 2 directly follows from Lemma 1.

If the curvature of an Einstein manifold (M, g) satisfies $\sec \leq -\delta < 0$ for a positive constant δ, then the Einstein constant of (M, g) satisfies the the obvious inequality $\alpha \leq -(n-1)\,\delta < 0$. On the other hand, from Equation (1) we deduce the inequality $R_{ijlk}\varphi^{ik}\varphi^{jl} \leq -(n\,\delta + \alpha)\,\|\varphi\|^2$. Therefore, if $\delta > -\alpha/n$, then $\overset{\circ}{R} < 0$. In this case, the Tachibana numbers $t_1(M), \ldots, t_{n-1}(M)$ are equal to zero (see [19]). We proved the following.

Proposition 1. *Let (M^n, g) be an Einstein manifold with sectional curvature bounded above by a negative constant $-\delta$ such that $\delta > -\alpha/n$ for the Einstein constant α. Then the Tachibana numbers $t_1(M), \ldots, t_{n-1}(M)$ are zero.*

We can complete this result. If an Einstein manifold (M^n, g) satisfies the curvature inequality $\sec \leq -\delta < 0$ for a positive constant δ, then from Equations (3) and (7) we deduce the inequality $F_p(\omega) \leq -\frac{1}{3}((n+2)\,\delta + \alpha)(n-p)\|\omega\|^2$ for any $p = 1, \ldots, n-1$. Therefore, the Tachibana numbers $t_1(M), \ldots, t_{n-1}(M)$ of a compact Einstein manifold with sectional curvature bounded above by a negative constant $-\delta$ such that $\delta \geq -\alpha/(n+2)$ are zero.

Author Contributions: Investigation, V.R., S.S. and I.T.

Funding: This research received no external funding.

Conflicts of Interest: The authors declare no conflict of interest.

References

1. Besse, A. *Einstein Manifolds*; Springer: Berlin, Germany, 1987.
2. Petersen, P. *Riemannian Geometry*; Springer Science: New York, NY, USA, 2016.
3. Goldberg, S.I. *Curvature and Homology*; Dover Publications Inc.: Mineola, NY, USA, 1998.
4. Yang, D. Rigidity of Einstein 4-manifolds with positive curvature. *Invent. Math.* **2000**, *142*, 435–450. [CrossRef]
5. Wu, P. Curvature decompositions on Einstein four-manifolds. *N. Y. J. Math.* **2017**, *23*, 1739–1749.
6. Ézio, A.C. On Einstein four-manifolds. *J. Geom. Phys.* **2004**, *51*, 244–255.
7. Cao, X.; Tran, H. Einstein four-manifolds of pinched sectional curvature. *Adv. Math.* **2018**, *335*, 322–342. [CrossRef]
8. Bourguignon, J.P.; Karcher, H. Curvature operators: Pinching estimates and geometric examples. *Ann. Sci. l'École Normale Supérieure* **1978**, *11*, 71–92. [CrossRef]
9. Kashiwada, T. *On the Curvature Operator of the Second Kind*; Natural Science Report; Ochanomozu University: Tokyo, Japan, 1998; Volume 44, pp. 69–73.
10. Tachibana, S. A theorem on Riemannian manifolds with positive curvature operator. *Proc. Jpn. Acad.* **1974**, *50*, 301–302. [CrossRef]
11. Böhm, C.; Wilking, B. Manifolds with positive curvature operators are space forms. *Ann. Math.* **2008**, *167*, 1079–1097. [CrossRef]
12. Nishikawa, S. On deformation of Riemannian metrics and manifolds with positive curvature operator. In *Curvature and Topology of Riemannian Manifolds*; Springer: London, UK, 1986; pp. 202–211.
13. Chavel, I. *Eigenvalues in Riemannian Geometry*; Academic Press Inc.: Orlando, FL, USA, 1984.
14. Tachibana, S.; Ogiue, K. Les variétés riemanniennes dont l'opérateur de coubure restreint est positif sont des sphères d'homologie réelle. *CR Acad. Sci. Paris* **1979**, *289*, 29–30.
15. Kashiwada, T. *On Conformal Killing Tensor*; Natural Science Report; Ochanomizu University: Tokyo, Japan, 1968; Volume 19, pp. 67–74.
16. Tachibana, S. On conformal Killing tensor in a Riemannian space. *Tohoku Math. J.* **1969**, *21*, 56–64. [CrossRef]
17. Stepanov, S.E.; Mikeš, J. Betti and Tachibana numbers of compact Riemannian manifolds. *Differ. Geom. Its Appl.* **2013**, *31*, 486–495. [CrossRef]
18. Stepanov, S.E. Curvature and Tachibana numbers. *J. Math. Sci.* **2011**, *202*, 135–146.
19. Stepanov, S.E.; Tsyganok, I.I. Theorems of existence and non-existence of conformal Killing forms. *Russ. Math. (Iz. VUZ)* **2014**, *58*, 46–51. [CrossRef]
20. Berger, M.; Ebin, D. Some decompositions of the space of symmetric tensors on a Riemannian manifold. *J. Differ. Geom.* **1969**, *3*, 379–392. [CrossRef]
21. Bochner, S.; Yano, K. *Curvature and Betti Numbers*; Princeton Univ. Press: Princeton, NJ, USA, 1953.
22. Meyer, D. Sur les variétés riemanniennes á operateur de courbure positif. *CR Acad. Sci. Paris* **1971**, *272*, 482–485.
23. Kora, M. On conformal Killing forms and the proper space of Δ for p-forms. *Math. J. Okayama Univ.* **1980**, *22*, 195–204.

© 2019 by the authors. Licensee MDPI, Basel, Switzerland. This article is an open access article distributed under the terms and conditions of the Creative Commons Attribution (CC BY) license (http://creativecommons.org/licenses/by/4.0/).

Article

Statistical Solitons and Inequalities for Statistical Warped Product Submanifolds

Aliya Naaz Siddiqui [1], Bang-Yen Chen [2,*] and Oguzhan Bahadir [3]

1. Department of Mathematics, Faculty of Natural Sciences, Jamia Millia Islamia, New Delhi 110025, India
2. Department of Mathematics, Michigan State University, 619 Red Cedar Road, East Lansing, MI 48824-1027, USA
3. Department of Mathematics, Faculty of Science and Letters, Kahramanmaras Sutcu Imam University, Kahrmanmaras 46100, Turkey
* Correspondence: chenb@msu.edu

Received: 25 July 2019; Accepted: 26 August 2019; Published: 1 September 2019

Abstract: Warped products play crucial roles in differential geometry, as well as in mathematical physics, especially in general relativity. In this article, first we define and study statistical solitons on Ricci-symmetric statistical warped products $\mathbb{R} \times_f N_2$ and $N_1 \times_f \mathbb{R}$. Second, we study statistical warped products as submanifolds of statistical manifolds. For statistical warped products statistically immersed in a statistical manifold of constant curvature, we prove Chen's inequality involving scalar curvature, the squared mean curvature, and the Laplacian of warping function (with respect to the Levi–Civita connection). At the end, we establish a relationship between the scalar curvature and the Casorati curvatures in terms of the Laplacian of the warping function for statistical warped product submanifolds in the same ambient space.

Keywords: statistical warped product submanifold; statistical manifold; B.Y.Chen inequality; Casorati curvatures; statistical soliton

MSC: 53B30; 53C15; 53C25

1. Introduction

Statistical manifolds were introduced in 1985 by S. Amari [1] in terms of information geometry, and they were applied by Lauritzen in [2]. Such manifolds have an important role in statistics as the statistical model often forms a geometrical manifold.

Let $\tilde{\nabla}$ be an affine connection on a (pseudo-)Riemannian manifold $(\tilde{N}, \tilde{g})$. The affine connection $\tilde{\nabla}^*$ on $\tilde{N}$ satisfying:

$$E\tilde{g}(F,G) = \tilde{g}(\tilde{\nabla}_E F, G) + \tilde{g}(F, \tilde{\nabla}_E^* G), \quad \forall E, F, G \in \Gamma(T\tilde{N}),$$

is called a *dual connection* of $\tilde{\nabla}$ with respect to $\tilde{g}$.

The triplet $(\tilde{N}, \tilde{\nabla}, \tilde{g})$ is called a *statistical manifold* if:

(a) the Codazzi equation $(\tilde{\nabla}_E \tilde{g})(F,G) = (\tilde{\nabla}_F \tilde{g})(E,G)$ holds, for any $E, F, G \in \Gamma(T\tilde{N})$;
(b) the torsion tensor field of $\tilde{\nabla}$ vanishes.

If $(\tilde{\nabla}, \tilde{g})$ is a statistical structure on $\tilde{N}$, then $(\tilde{\nabla}^*, \tilde{g})$ is also a statistical structure. The connections $\tilde{\nabla}$ and $\tilde{\nabla}^*$ satisfy $(\tilde{\nabla}^*)^* = \tilde{\nabla}$. On the other hand, we have $\tilde{\nabla}^0 = \frac{1}{2}(\tilde{\nabla} + \tilde{\nabla}^*)$, where $\tilde{\nabla}^0$ is the Levi–Civita connection of $\tilde{N}$.

One of the most fruitful generalizations of Riemannian products is the warped product defined in [3]. The notion of warped products plays very important roles in differential geometry and in mathematical physics, especially in general relativity. For instance, space-time models in general relativity are usually expressed in terms of warped products (cf., e.g., [4,5]).

In 2006, L. Todjihounde [6] defined a suitable dualistic structure on warped product manifolds. Furthermore, Furuhata et al. [7] defined Kenmotsu statistical manifolds and studied how to construct such structures on the warped product of a holomorphic statistical manifold [8] and a line. In [9], H. Aytimur and C. Ozgur studied Einstein statistical warped product manifolds. Further, C. Murathan and B. Sahin [10] studied and obtained the Wintgen-like inequality for statistical submanifolds of statistical warped product manifolds.

The Ricci solitons are special solutions of the Ricci flow of the Hamilton. In Section 4, we define statistical solitons and study the problem under what conditions the base manifold or fiber manifold of a statistical warped product manifold is a statistical soliton.

Curvature invariants play the most fundamental and natural roles in Riemannian geometry. A fundamental problem in the theory of Riemannian submanifolds is (cf. [11]):

Problem A. *"Establish simple optimal relationships between the main intrinsic invariants and the main extrinsic invariants of a submanifold."*

The first solutions of this problem for warped product submanifolds were given in [11,12]. In Section 5, we study this fundamental problem for statistical warped product submanifolds in any statistical manifolds of constant curvature. Our solution to this problem given in this section is derived via the fundamental equations of statistical submanifolds.

An extrinsic curvature of a Riemannian submanifold was defined by Casorati in [13], as the normalized square of the length of the second fundamental form. Casorati curvature has nice applications in computer vision. It was preferred by Casorati over the traditional curvature since it corresponds better to the common intuition of curvature.

Several sharp inequalities between extrinsic and intrinsic curvatures for different submanifolds in real, complex, and quaternionic space forms endowed with various connections have been obtained (e.g., [14–21]). Such inequalities with a pair of conjugate affine connections involving the normalized scalar curvature of statistical submanifolds in different ambient spaces were obtained in [22–26].

Inspired by historical development on the classifications of Casorati curvatures and Ricci curvatures, we establish in Section 6 an inequality for statistical warped product submanifolds in a statistical manifold of constant curvature. In the last section, we provide two examples of statistical warped product submanifolds in the same environment.

2. Preliminaries

Let $(\tilde{N}, \tilde{\nabla}, \tilde{g})$ be a statistical manifold and N be a submanifold of $\tilde{N}$. Then, (N, ∇, g) is also a statistical manifold with the statistical structure (∇, g) on N induced from $(\tilde{\nabla}, \tilde{g})$, and we call (N, ∇, g) a statistical submanifold.

The fundamental equations in the geometry of Riemannian submanifolds are the Gauss and Weingarten formulae and the equations of Gauss, Codazzi, and Ricci (cf. [4,5,27]). In the statistical setting, the Gauss and Weingarten formulae are defined respectively by [28]:

$$\left. \begin{array}{ll} \tilde{\nabla}_E F = \nabla_E F + h(E,F), & \tilde{\nabla}^*_E F = \nabla^*_E F + h^*(E,F), \\ \tilde{\nabla}_E \xi = -A_\xi(E) + \nabla^\perp_E \xi, & \tilde{\nabla}^*_E \xi = -A^*_\xi(E) + \nabla^{\perp *}_E \xi, \end{array} \right\} \quad (1)$$

for any $E, F \in \Gamma(TN)$ and $\xi \in \Gamma(T^\perp N)$, where $\tilde{\nabla}$ and $\tilde{\nabla}^*$ (resp., ∇ and ∇^*) are the dual connections on $\tilde{N}$ (resp., on N).

The symmetric and bilinear imbedding curvature tensor of N in $\tilde{N}$ with respect to $\tilde{\nabla}$ and $\tilde{\nabla}^*$ is denoted as h and h^*, respectively. The relation between h (resp. h^*) and A_ζ (resp. A^*_ζ) is defined by [28]:

$$\left.\begin{array}{l} \tilde{g}(h(E,F),\zeta) = g(A^*_\zeta E, F), \\ \tilde{g}(h^*(E,F),\zeta) = g(A_\zeta E, F), \end{array}\right\} \quad (2)$$

for any $E, F \in \Gamma(TN)$ and $\zeta \in \Gamma(T^\perp N)$.

Let $\tilde{R}$ and R be the curvature tensor fields of $\tilde{\nabla}$ and ∇, respectively. The corresponding Gauss, Codazzi, and Ricci equations are given by [28]:

$$\begin{aligned} \tilde{g}(\tilde{R}(E,F)G, H) &= g(R(E,F)G, H) + \tilde{g}(h(E,G), h^*(F,H)) \\ &\quad - \tilde{g}(h^*(E,H), h(F,G)), \end{aligned} \quad (3)$$

$$\begin{aligned} (\tilde{R}(E,F)G)^\perp &= \nabla^\perp_E h(F,G) - h(\nabla_E F, G) - h(F, \nabla_E G) \\ &\quad - \{\nabla^\perp_F h(E,G) - h(\nabla_F E, G) - h(E, \nabla_F G)\}, \end{aligned} \quad (4)$$

$$\tilde{g}(\tilde{R}^\perp(E,F)\zeta, \eta) = \tilde{g}(R(E,F)\zeta, \eta) + g([A^*_\zeta, A_\eta]E, F), \quad (5)$$

for any $E, F, G, H \in \Gamma(TN)$ and $\zeta, \eta \in \Gamma(T^\perp N)$, where $R^\perp$ is the Riemannian curvature tensor on $T^\perp N$.

Similarly, $\tilde{R}^*$ and R^* are respectively the curvature tensor fields with respect to $\tilde{\nabla}^*$ and ∇^*. We can obtain the duals of all Equations (3)–(5) with respect to $\tilde{\nabla}^*$ and ∇^*. Furthermore,

$$\tilde{S} = \frac{1}{2}(\tilde{R} + \tilde{R}^*) \quad \text{and} \quad S = \frac{1}{2}(R + R^*) \quad (6)$$

are respectively the curvature tensor fields of $\tilde{N}$ and N given by [7]. Thus, the sectional curvature $\mathbb{K}^{\nabla,\nabla^*}$ on N of $\tilde{N}$ is defined by [29,30]:

$$\begin{aligned} \mathbb{K}^{\nabla,\nabla^*}(E \wedge F) &= g(S(E,F)F, E) \\ &= \frac{1}{2}(g(R(E,F)F, E) + g(R^*(E,F)F, E)), \end{aligned} \quad (7)$$

for any orthonormal vectors $E, F \in T_p N$, $p \in N$.

Suppose that $\dim(N) = m$ and $\dim(\tilde{N}) = n$. Let $\{e_1, \ldots, e_m\}$ and $\{e_{m+1}, \ldots, e_n\}$ be respectively the orthonormal basis of $T_p N$ and $T^\perp_p N$ for $p \in N$. Then, the scalar curvature σ^{∇,∇^*} of N is given by:

$$\sigma^{\nabla,\nabla^*} = \sum_{1 \leq i < j \leq m} \mathbb{K}^{\nabla,\nabla^*}(e_i \wedge e_j). \quad (8)$$

The normalized scalar curvature ρ of N is defined as:

$$\rho^{\nabla,\nabla^*} = \frac{2\sigma^{\nabla,\nabla^*}}{m(m-1)}.$$

The mean curvature vectors $\mathcal{H}$ and $\mathcal{H}^*$ of N in $\tilde{N}$ are:

$$\mathcal{H} = \frac{1}{m}\sum_{i=1}^m h(e_i, e_i), \quad \mathcal{H}^* = \frac{1}{m}\sum_{i=1}^m h^*(e_i, e_i).$$

Furthermore, we set:

$$h_{ij}^a = \tilde{g}(h(e_i, e_j), e_a), \quad h_{ij}^{*a} = \tilde{g}(h^*(e_i, e_j), e_a),$$

for $i, j \in \{1, \ldots, m\}$, $a \in \{m+1, \ldots, n\}$.

A statistical manifold $(\tilde{N}, \tilde{\nabla}, \tilde{g})$ is said to be *of constant curvature* $\tilde{c} \in \mathbb{R}$, denoted by $\tilde{N}(\tilde{c})$, if the following curvature equation holds:

$$\tilde{S}(E, F)G = \tilde{c}(g(F, G)E - g(E, G)F), \quad \forall E, F, G \in \Gamma(T\tilde{N}). \tag{9}$$

3. Basics on Statistical Warped Product Manifolds

Definition 1. *[3] Let (N_1, g_1) and (N_2, g_2) be two (pseudo)-Riemannian manifolds and $\mathfrak{f} > 0$ be a differentiable function on N_1. Consider the natural projections $\pi : N_1 \times N_2 \to N_1$ and $\pi' : N_1 \times N_2 \to N_2$. Then, the warped product $N = N_1 \times_\mathfrak{f} N_2$ with warping function $\mathfrak{f}$ is the product manifold $N_1 \times N_2$ equipped with the Riemannian structure such that:*

$$\tilde{g}(E, F) = g_1(\pi_* E, \pi_* F) + \mathfrak{f}^2(u) g_2(\pi'_* E, \pi'_* F), \tag{10}$$

for $E, F \in \Gamma(T_{(u,v)} N)$, $u \in N_1$, and $v \in N_2$, where $*$ denotes the tangent map.

Let $\chi(N_1)$ and $\chi(N_2)$ be the set of all vector fields on $N_1 \times N_2$, which is the horizontal lift of a vector field on N_1 and the vertical lift of a vector field on N_2, respectively. We have $T(N_1 \times N_2) = \chi(N_1) \oplus \chi(N_2)$. Thus, it can be seen that $\pi_*(\chi(N_1)) = \Gamma(TN_1)$ and $\pi'_*(\chi(N_2)) = \Gamma(TN_2)$. Therefore, $\pi_*(X) = E_1 \in \Gamma(TN_1)$, $\pi_*(Y) = F_1 \in \Gamma(TN_1)$, $\pi'_*(U) = E_2 \in \Gamma(TN_2)$ and $\pi'_*(V) = F_2 \in \Gamma(TN_2)$, for any $X, Y \in \chi(N_1)$ and $U, V \in \chi(N_2)$.

Recall the following general result from [6] for a dualistic structure on the warped product manifold $N_1 \times_\mathfrak{f} N_2$.

Proposition 1. *Let $(g_1, \nabla^{N_1}, \nabla^{N_1*})$ and $(g_2, \nabla^{N_2}, \nabla^{N_2*})$ be dualistic structures on N_1 and N_2, respectively. For $X, Y \in \chi(N_1)$ and $U, V \in \chi(N_2)$, D, D^* on $N_1 \times N_2$ satisfy:*

(a) $D_X Y = \nabla^{N_1}_{E_1} F_1$,
(b) $D_X U = D_U X = \frac{E_1 \mathfrak{f}}{\mathfrak{f}} E_2$,
(c) $D_U V = \nabla^{N_2}_{E_2} F_2 - \frac{\tilde{g}(U,V)}{\mathfrak{f}} \mathrm{grad}\ \mathfrak{f}$,
(d) $D^*_X Y = \nabla^{N_1*}_{E_1} F_1$,
(e) $D^*_X U = D^*_U X = \frac{E_1 \mathfrak{f}}{\mathfrak{f}} E_2$,
(f) $D^*_U V = \nabla^{N_2*}_{E_2} F_2 - \frac{\tilde{g}(U,V)}{\mathfrak{f}} \mathrm{grad}\ \mathfrak{f}$,

where $\nabla^{N_1}_{E_1} F_1 = \pi_*(D_X Y)$, $\nabla^{N_1*}_{E_1} F_1 = \pi_*(D^*_X Y)$, $\nabla^{N_2}_{E_2} F_2 = \pi'_*(D_U V)$, and $\nabla^{N_2*}_{E_2} F_2 = \pi'_*(D^*_U V)$. Then, $(\tilde{g}, D, D^*)$ is a dualistic structure on $N_1 \times N_2$.

Furthermore, Todjihounde [6] derived the curvature of the statistical warped product $\tilde{N} = N_1 \times_\mathfrak{f} N_2$ in terms of the curvature tensors R_1 and R_2 of N_1 and N_2, respectively, and its warping function $\mathfrak{f}$.

Lemma 1. *Let $(\tilde{N} = N_1 \times_\mathfrak{f} N_2, D, D^*, \tilde{g})$ be a statistical warped product manifold. For $X, Y, Z \in \chi(N_1)$ and $U, V, W \in \chi(N_2)$, we have:*

(a) $\tilde{R}(X, Y)Z = R_1(E_1, F_1) G_1$,
(b) $\tilde{R}(U, Y)Z = -\mathfrak{f}^{-1} \mathrm{Hess}_\mathfrak{f}(Y, Z) U$,
(c) $\tilde{R}(X, Y)W = 0$,
(d) $\tilde{R}(U, V)Z = 0$,

(e) $\tilde{R}(X,V)W = -\mathfrak{f}^{-1}\tilde{g}(V,W)D_X(\text{grad }\mathfrak{f})$,
(f) $\tilde{R}(U,V)W = R_2(E_2,F_2)G_2 + ||\text{grad }\mathfrak{f}||^2[g_2(U,W)V - g_2(V,W)U]$,

where $\tilde{R}$ denotes the curvature tensor field of $(\tilde{N} = N_1 \times_{\mathfrak{f}} N_2, D, D^*, \tilde{g})$ and $\text{Hess}_{\mathfrak{f}}(X,Y) = X(Y\mathfrak{f}) - (\nabla_X^{N_1} Y)\mathfrak{f}$ is the Hessian function of $\mathfrak{f}$ with respect to ∇^{N_1}.

The next result from [9] provides the Ricci tensor $\tilde{Ric}$ of the statistical warped product manifold.

Lemma 2. Let $(\tilde{N} = N_1 \times_{\mathfrak{f}} N_2, D, D^*, \tilde{g})$ be a statistical warped product manifold. For $X, Y \in \chi(N_1)$ and $U, V \in \chi(N_2)$, we have:

(a) $\tilde{Ric}(X,Y) = Ric_1(X,Y) - dim(N_2)\mathfrak{f}^{-1}\text{Hess}_{\mathfrak{f}}(X,Y)$,
(b) $\tilde{Ric}(X,V) = 0$,
(c) $\tilde{Ric}(U,V) = Ric_2(U,V) - [\mathfrak{f}(\Delta\mathfrak{f}) + (dim(N_2) - 1)||\text{grad }\mathfrak{f}||^2]g_2(U,V)$,

where Ric_1 and Ric_2 are the Ricci tensors of N_1 and N_2, respectively, and $\Delta\mathfrak{f} = div(\text{grad }\mathfrak{f})$ is the Laplacian of $\mathfrak{f}$ with respect to D.

We recall the following result from [31]. This result is useful in some Riemannian problems like the study of the distance between two manifolds, of the extremes of sectional curvature and is applied successfully in the demonstration of the Chen inequality.

Let (N, g) be a Riemannian submanifold of a Riemannian manifold $(\tilde{N}, \tilde{g})$, and let $f : \tilde{N} \to \mathbb{R}$ be a differentiable function. Let:

$$\min_{x_0 \in N} f(x_0) \tag{11}$$

be the constrained extremum problem.

Theorem 1. If $x \in N$ is the solution of the problem (11), then:

(a) $(\text{grad } f)(x) \in T_x^{\perp} N$,
(b) the bilinear form $\Theta : T_x N \times T_x N \to \mathbb{R}$,

$$\Theta(E, F) = \text{Hess}_f(E, F) + \tilde{g}(h'(E, F), (\text{grad } f)(x))$$

is positive semi-definite, where h' is the second fundamental form of N in $\tilde{N}$ and grad f denotes the gradient of f.

4. Statistical Solitons on Statistical Warped Product Manifolds

The Ricci solitons model the formation of singularities in the Ricci flow, and they correspond to self-similar solutions. R. Hamilton [32] introduced the study of Ricci solitons as fixed or stationary points of the Ricci flow in the space of the metrics on Riemannian manifolds modulo diffeomorphisms and scaling. Since then, many researchers studied Ricci solitons for different reasons and in different ambient spaces (for example [33–35]). A complete Riemannian manifold $(\tilde{N}, \tilde{g})$ is called a *Ricci soliton* $(\tilde{N}, \tilde{g}, \zeta, \lambda)$ if there exists a smooth vector field ζ and a constant $\lambda \in \mathbb{R}$ such that:

$$2\tilde{Ric} = 2\lambda\tilde{g} - \mathcal{L}_{\zeta}\tilde{g},$$

where $\mathcal{L}_{\zeta}$ denotes the Lie derivative along ζ and $\tilde{Ric}$ is the Ricci tensor of $\tilde{g}$.

A generalization of Ricci solitons in the framework of manifolds endowed with an arbitrary linear connection $\tilde{\nabla}$, different from the Levi–Civita connection of $\tilde{g}$, is defined in [36] as follows:

Let $(\tilde{N}, \tilde{\nabla})$ be a manifold and $\zeta \in \chi(\tilde{N})$. A triple $(\tilde{g}, \zeta, \lambda)$ is called a $\tilde{\nabla}$-Ricci soliton if $\tilde{\nabla}\zeta + \tilde{Q} + \lambda I = 0$ holds, where $\tilde{Q}$ is the Ricci operator of $\tilde{N}$ defined by $\tilde{g}(\tilde{Q}E, F) = \tilde{Ric}(E, F)$, for vector fields E, F on $\tilde{N}$.

The statistical manifold $(\tilde{N}, \tilde{\nabla}, \tilde{g})$ is called *Ricci-symmetric* if the Ricci operator $\tilde{Q}$ with respect to $\tilde{\nabla}$ (equivalently, the dual operator $\tilde{Q}^*$ with respect to $\tilde{\nabla}^*$) is symmetric (cf. [36,37]).

Based on these, we have the following.

Definition 2. *A pair (ζ, λ) is called a statistical soliton on a Ricci-symmetric statistical manifold $(\tilde{N}, \tilde{\nabla}, \tilde{g})$ if the triple $(\tilde{g}, \zeta, \lambda)$ is $\tilde{\nabla}$-Ricci and $\tilde{\nabla}^*$-Ricci solitons, i.e., we have:*

$$\tilde{\nabla}\zeta + \tilde{Q} + \lambda I = 0, \tag{12}$$

and:

$$\tilde{\nabla}^*\zeta + \tilde{Q}^* + \lambda I = 0, \tag{13}$$

where $\tilde{g}(\tilde{Q}E, F) = \tilde{Ric}(E, F)$ and $\tilde{g}(\tilde{Q}^*E, F) = \tilde{Ric}^*(E, F)$, for all vector fields on $\tilde{N}$, and $\tilde{Ric}$ and $\tilde{Ric}^*$ denote the Ricci tensor fields with respect to $\tilde{\nabla}$ and $\tilde{\nabla}^*$, respectively.

The main purpose of this section is to study the problem: *under what conditions does the base manifold or fiber manifold of the statistical warped product manifold become a statistical soliton?*

Let $(N_1, \nabla^{N_1}, \nabla^{N_1 *}, g_1)$ and $(N_2, \nabla^{N_2}, \nabla^{N_2 *}, g_2)$ be the Ricci-symmetric statistical manifolds. Denote the Ricci-symmetric statistical warped product manifold by $(\tilde{N} = N_1 \times_f N_2, D, D^*, \tilde{g} = g_1 + f^2 g_2)$. Let $\zeta = (\zeta_1, \zeta_2) \in \chi(\tilde{N})$ be a vector field on $\tilde{N}$. Then, the pair (ζ, λ) on $(\tilde{N}, \tilde{\nabla}, \tilde{g})$ is called a *statistical soliton* if the triple $(\tilde{g}, \zeta, \lambda)$ is both D-Ricci and D^*-Ricci solitons, given by (12) and (13).

It follows from Lemma 2 that the Ricci tensor of $\tilde{N}$ is given as below:

$$\tilde{Ric} = Ric_1 - f^{-1}\dim(N_2) Hess_f + Ric_2 \\ - [f(\Delta f) + (\dim(N_2) - 1)||grad\, f||^2] g_2. \tag{14}$$

Thus, (12) and (13) can be rewritten as:

$$\nabla^{N_1}\zeta_1 + \nabla^{N_2}\zeta_2 + Ric_1 - f^{-1}\dim(N_2) Hess_f + Ric_2 \\ - [f(\Delta f) + (\dim(N_2) - 1)||grad\, f||^2] g_2 + \lambda g_1 + \lambda f^2 g_2 = 0, \tag{15}$$

and:

$$\nabla^{N_1 *}\zeta_1 + \nabla^{N_2 *}\zeta_2 + Ric_1^* - f^{-1}\dim(N_2) Hess_f^{D^*} + Ric_2^* \\ - [f(\Delta^{D^*} f) + (\dim(N_2) - 1)||grad\, f||^2] g_2 + \lambda g_1 + \lambda f^2 g_2 = 0, \tag{16}$$

respectively.

Throughout this section, we use the statistical warped products as Ricci-symmetric. We give the following results by applying Lemma 2:

Lemma 3. *Let $(\tilde{N} = \mathbb{R} \times_f N_2, D, D^*, \tilde{g})$ be a statistical warped product manifold, where $(\mathbb{R}, \nabla^{\mathbb{R}}, dz^2)$ is a trivial statistical manifold of dimension one and $\dim(N_2) = k$. Then, for $U, V \in \chi(N_2)$, we have:*

(a) $\tilde{Ric}(\partial z, \partial z) = -k f^{-1} \ddot{f}$,
(b) $\tilde{Ric}(\partial z, V) = 0$,
(c) $\tilde{Ric}(U, V) = Ric_2(U, V) - [f\ddot{f} + (k-1)\dot{f}^2] g_2(U, V)$.

Proposition 2. Let (ζ, λ) be a statistical soliton on statistical warped product manifold $(\tilde{N} = \mathbb{R} \times_\mathfrak{f} N_2, D, D^*, \tilde{g} = dz^2 + \mathfrak{f}^2 g_2)$ with $\dim(\mathbb{R}) = 1$ and $\dim(N_2) = k$. Then:

$$\mathrm{Hess}_\mathfrak{f} = \frac{\mathfrak{f}\lambda}{k}.$$

Proof. Since $\tilde{N}$ is a statistical soliton, then from (6), we have:

$$\tilde{g}(\tilde{\nabla}_{\partial z}\zeta, \partial z) + \tilde{R}ic(\partial z, \partial z) + \lambda \tilde{g}(\partial z, \partial z) = 0.$$

By taking into account Lemma 3 and $Ric_1(\partial z, \partial z) = 0$, we get:

$$-\tilde{g}(\zeta, \tilde{\nabla}^*_{\partial z}\partial z) - k\mathfrak{f}^{-1}\mathrm{Hess}_\mathfrak{f}(\partial z, \partial z) + \lambda \tilde{g}(\partial z, \partial z) = 0,$$

which gives $\mathrm{Hess}_\mathfrak{f}(\partial z, \partial z) = (\frac{\mathfrak{f}\lambda}{k})\tilde{g}(\partial z, \partial z)$. □

Theorem 2. Let $\zeta = (\partial z, \zeta_2) \in \chi(\tilde{N})$ be a vector field on statistical warped product manifold $(\tilde{N} = \mathbb{R} \times_\mathfrak{f} N_2, D, D^*, \tilde{g} = dz^2 + \mathfrak{f}^2 g_2)$ with $\dim(\mathbb{R}) = 1$ and $\dim(N_2) = k$. If (ζ, λ) is a statistical soliton on $\tilde{N}$, then:

(a) $(N_2, g_2, \zeta_2, \lambda_2)$ is a statistical soliton on $(N_2, \nabla^{N_2}, \nabla^{N_2*}, g_2)$, where $\lambda_2 = (k-1)[\mathfrak{f}\ddot{\mathfrak{f}} - \dot{\mathfrak{f}}^2]$,
(b) $\mathfrak{f}(z) = az + b$ if $\lambda = 0$,
(c) $\mathfrak{f}(z) = \cosh(az + b)$ if $\lambda \neq 0$,

where $a, b \in \mathbb{R}$.

Proof. From Equation (15) and Lemma 3, we have:

$$\nabla^{N_1}\partial z + \nabla^{N_2}\zeta_2 + Ric_1 - k\mathfrak{f}^{-1}\ddot{\mathfrak{f}} + Ric_2$$
$$- (\mathfrak{f}\ddot{\mathfrak{f}} + (k-1)\dot{\mathfrak{f}}^2)g_2 + \lambda g_1 + \lambda \mathfrak{f}^2 g_2 = 0.$$

Note $g_1(\nabla^{N_1}_{\partial z}\partial z, \partial z) = 0$ and $Ric_1(\partial z, \partial z) = 0$. Thus, the above equation becomes:

$$\nabla^{N_2}\zeta_2 - k\mathfrak{f}^{-1}\ddot{\mathfrak{f}} + Ric_2 - (\mathfrak{f}\ddot{\mathfrak{f}} + (k-1)\dot{\mathfrak{f}}^2)g_2 + \lambda g_1 + \lambda \mathfrak{f}^2 g_2 = 0,$$

from which we get:

$$\lambda = k\mathfrak{f}^{-1}\ddot{\mathfrak{f}}, \tag{17}$$

$$\nabla^{N_2}\zeta_2 + Ric_2 + [\lambda \mathfrak{f}^2 - (\mathfrak{f}\ddot{\mathfrak{f}} + (k-1)\dot{\mathfrak{f}}^2)]g_2 = 0. \tag{18}$$

Putting (17) into the Equation (18), we arrive at:

$$\nabla^{N_2}\zeta_2 + Ric_2 + (k-1)[\mathfrak{f}\ddot{\mathfrak{f}} - \dot{\mathfrak{f}}^2]g_2 = 0.$$

Similarly, by using (16), we derive:

$$\nabla^{N_2*}\zeta_2 + Ric_2^* + (k-1)[\mathfrak{f}\ddot{\mathfrak{f}} - \dot{\mathfrak{f}}^2]g_2 = 0.$$

Thus, $(N_2, g_2, \zeta_2, (k-1)[\mathfrak{f}\ddot{\mathfrak{f}} - \dot{\mathfrak{f}}^2])$ is a statistical soliton provided that $(k-1)[\mathfrak{f}\ddot{\mathfrak{f}} - \dot{\mathfrak{f}}^2]$ is constant. On the other hand, by using (17), we have the following cases:

(a) if $\lambda = 0$, then $\mathfrak{f}(z) = az + b$, and
(b) if $\lambda \neq 0$, then $\mathfrak{f}(z) = \cosh(az + b)$ [9],

where a, b are real constants. □

Before proving the next result, we state the following:

Lemma 4. *Let $(\tilde{N} = N_1 \times_\mathfrak{f} \mathbb{R}, D, D^*, \tilde{g})$ be a statistical warped product manifold, where $(\mathbb{R}, \nabla^\mathbb{R}, dz^2)$ is a trivial statistical manifold of dimension one and $\dim(N_1) = k$. For $X, Y \in \chi(N_1)$, we have:*

(a) $\tilde{Ric}(X, Y) = Ric_1(X, Y) - \mathfrak{f}^{-1} Hess_\mathfrak{f}(X, Y),$
(b) $\tilde{Ric}(X, \partial z) = 0,$
(c) $\tilde{Ric}(\partial z, \partial z) = -\mathfrak{f}(\Delta \mathfrak{f}) g_2(\partial z, \partial z).$

Theorem 3. *Let $\zeta = (\zeta_1, \partial z) \in \chi(\tilde{N})$ be a vector field on statistical warped product manifold $(\tilde{N} = N_1 \times_\mathfrak{f} \mathbb{R}, D, D^*, \tilde{g} = g_1 + \mathfrak{f}^2 dz^2)$ with $\dim(\mathbb{R}) = 1$ and $\dim(N_1) = k$. Suppose that $Hess_\mathfrak{f} = 0$. Then, (ζ, λ) is a statistical soliton on $\tilde{N}$ if and only if $(\zeta_1, \lambda = \mathfrak{f}^{-1}(\Delta \mathfrak{f}))$ is a statistical soliton on N_1.*

Proof. Since $g_2(\nabla^{N_1}_{\partial z} \partial z, \partial z) = 0$ and $Ric_2(\partial z, \partial z) = 0$, then by using Equation (15) and Lemma 4, we get:

$$\nabla^{N_1} \zeta_1 + Ric_1 - \mathfrak{f}(\Delta \mathfrak{f}) g_2 + \lambda g_1 + \lambda \mathfrak{f}^2 g_2 = 0.$$

Therefore, we have:

$$\nabla^{N_1} \zeta_1 + Ric_1 + \lambda g_1 = 0. \tag{19}$$

Furthermore, $\mathfrak{f}^{-1}(\Delta \mathfrak{f}) = \lambda = $ constant. Putting this into (19), we get:

$$\nabla^{N_1} \zeta_1 + Ric_1 + \mathfrak{f}^{-1}(\Delta \mathfrak{f}) g_1 = 0.$$

Similarly, by using (16), we obtain:

$$\nabla^{N_1 *} \zeta_1 + Ric_1^* + \mathfrak{f}^{-1}(\Delta^* \mathfrak{f}) g_1 = 0.$$

Since $\mathfrak{f}^{-1}(\Delta \mathfrak{f})$ is constant, $(N_1, g_1, \zeta_1, \lambda = \mathfrak{f}^{-1}(\Delta \mathfrak{f}))$ is a statistical soliton. Conversely, if $(\zeta_1, \lambda = \mathfrak{f}^{-1}(\Delta \mathfrak{f}))$ is a statistical soliton on N_1, then:

$$\nabla^{N_1} \zeta_1 + \nabla^{N_2} \partial z + Ric_1 - \mathfrak{f}^{-1} k_2 Hess_\mathfrak{f} + Ric_2$$
$$- [\mathfrak{f}(\Delta \mathfrak{f}) + (k_2 - 1) \|grad\, \mathfrak{f}\|^2] g_2$$
$$= \nabla^{N_1} \zeta_1 + Ric_1 + \mathfrak{f}^{-1}(\Delta \mathfrak{f}) g_1 - \mathfrak{f}^{-1}(\Delta \mathfrak{f}) g_1 - \mathfrak{f}(\Delta \mathfrak{f}) g_2$$
$$= -\mathfrak{f}^{-1}(\Delta \mathfrak{f}) g_1 - \mathfrak{f}(\Delta \mathfrak{f}) g_2 = -\mathfrak{f}^{-1}(\Delta \mathfrak{f})(g_1 + g_2)$$
$$= -\lambda \tilde{g}.$$

Thus, $D\zeta + \tilde{Q} + \lambda I = 0$. Similarly, $D^* \zeta + \tilde{Q}^* + \lambda I = 0$. Hence, (ζ, λ) is a statistical soliton on $\tilde{N}$. □

An immediate consequence of Theorem 3 is as follows:

Corollary 1. *Let $(\tilde{N}, \tilde{g}, \zeta, \lambda)$ be a Statistical soliton on statistical manifold $(\tilde{N} = N_1 \times_\mathfrak{f} \mathbb{R}, D, D^*, \tilde{g} = g_1 + \mathfrak{f}^2 dz^2)$ with $\dim(\mathbb{R}) = 1$ and $\dim(N_1) = k$. If $Hess_\mathfrak{f} = \varrho g_1$, $\varrho \in C^\infty(N_1)$, then $(N_1, g_1, \zeta_1, \mathfrak{f}^{-1}(\Delta \mathfrak{f}) - \mathfrak{f}^{-1} \varrho)$ is a statistical soliton.*

5. B.Y. Chen Inequality

A universal sharp inequality for submanifolds in a Riemannian manifold of constant sectional curvature was established in [38], known as the first Chen inequality. The main purpose of this section is to establish the corresponding inequality for statistical warped product manifolds statistically immersed in a statistical manifold of constant curvature.

Let $\varphi : N = N_1 \times_{\mathfrak{f}} N_2 \to \tilde{N}(\tilde{c})$ be an isometric statistical immersion of a warped product $N_1 \times_{\mathfrak{f}} N_2$ into a statistical manifold of constant sectional curvature $\tilde{c}$. We denote by r, k, and $m = r + k$ the dimensions of N_1, N_2, and $N_1 \times N_2$, respectively. Since $N_1 \times_{\mathfrak{f}} N_2$ is a statistical warped product, we have:

$$\nabla_{E_1} E_2 = \nabla_{E_2} E_1 = (E_1 \ln \mathfrak{f}) E_2,$$

for unit vector fields E_1 and E_2 tangent to N_1 and N_2, respectively. Hence, we derive:

$$\mathbb{K}(E_1 \wedge E_2) = \frac{1}{\mathfrak{f}}\{(\nabla_{E_1} E_1)\mathfrak{f} - E_1^2 \mathfrak{f}\}. \qquad (20)$$

If we choose a local orthonormal frame $\{e_1, \ldots, e_m\}$ such that $\{e_1, \ldots, e_r\}$ are tangent to N_1 and $\{e_{r+1}, \ldots, e_{r+k} = e_m\}$ are tangent to N_2, then we have:

$$\frac{\Delta \mathfrak{f}}{\mathfrak{f}} = \sum_{i=1}^{r} \mathbb{K}(e_i \wedge e_j), \qquad (21)$$

for each $j = r + 1, \ldots, m$.

On the other hand, let E_1 and E_2 be two unit local vector fields tangent to N_1 and N_2, respectively, such that $e_1 = E_1$ and $e_{r+1} = E_2$. By taking into account Equations (3), (6), and (9), we derive (7) as follows:

$$\begin{aligned}
\mathbb{K}^{\nabla, \nabla^*}(e_1 \wedge e_{r+1}) &= \frac{\tilde{c}}{2}\{2g(e_{r+1}, e_{r+1})g(e_1, e_1) - 2g(e_1, e_{r+1})g(e_{r+1}, e_1)\} \\
&+ \frac{1}{2}\{g(h^*(e_1, e_1), h(e_{r+1}, e_{r+1})) \\
&+ g(h(e_1, e_1), h^*(e_{r+1}, e_{r+1})) - 2g(h(e_1, e_{r+1}), h^*(e_1, e_{r+1}))\} \\
&= \tilde{c} + \frac{1}{2}\sum_{a=m+1}^{n}\{h_{11}^{*a} h_{r+1,r+1}^{a} + h_{11}^{a} h_{r+1,r+1}^{*a} - 2h_{1,r+1}^{a} h_{1,r+1}^{*a}\}.
\end{aligned}$$

We rewrite the terms of the RHS of the previous equation as:

$$\begin{aligned}
\mathbb{K}^{\nabla, \nabla^*}(e_1 \wedge e_{r+1}) = \tilde{c} + \frac{1}{2}\sum_{a=m+1}^{n} &\{(h_{11}^{a} + h_{11}^{*a})(h_{r+1,r+1}^{a} + h_{r+1,r+1}^{*a}) \\
&- (h_{1,r+1}^{a} + h_{1,r+1}^{*a})^2 + (h_{1,r+1}^{a})^2 + (h_{1,r+1}^{*a})^2 \\
&- h_{11}^{a} h_{r+1,r+1}^{a} - h_{11}^{*a} h_{r+1,r+1}^{*a}\}.
\end{aligned}$$

Since, $2h^0 = h + h^*$, we get:

$$\begin{aligned}
\mathbb{K}^{\nabla, \nabla^*}(e_1 \wedge e_{r+1}) = \tilde{c} + \frac{1}{2}\sum_{a=m+1}^{n} &\{4 h_{11}^{0a} h_{r+1,r+1}^{0a} \\
&- (h_{11}^{a} h_{r+1,r+1}^{a} - (h_{1,r+1}^{a})^2) \\
&- (h_{11}^{*a} h_{r+1,r+1}^{*a} - (h_{1,r+1}^{*a})^2) - 4(h_{1,r+1}^{0a})^2\}.
\end{aligned}$$

Thus, we have:

$$\mathbb{K}^{\nabla,\nabla^*}(e_1 \wedge e_{r+1}) = \tilde{c} + \sum_{a=m+1}^{n} \{2(h_{11}^{0a}h_{r+1,r+1}^{0a} - (h_{1,r+1}^{0a})^2)$$
$$- \frac{1}{2}(h_{11}^a h_{r+1,r+1}^a - (h_{1,r+1}^a)^2) - \frac{1}{2}(h_{11}^{*a}h_{r+1,r+1}^{*a} - (h_{1,r+1}^{*a})^2)\}. \tag{22}$$

Using the Gauss equation for the Levi–Civita connection, we arrive at:

$$\mathbb{K}^0(e_1 \wedge e_{r+1}) = \tilde{c} - \sum_{a=m+1}^{n} \{(h_{1,r+1}^{0a})^2 - h_{11}^{0a}h_{r+1,r+1}^{0a}\},$$

which can be rewritten as:

$$\sum_{a=m+1}^{n} \{(h_{1,r+1}^{0a})^2 - h_{11}^{0a}h_{r+1,r+1}^{0a}\} = \mathbb{K}^0(e_1 \wedge e_{r+1}) - \tilde{c}. \tag{23}$$

Substituting (23) into (22), we get:

$$\mathbb{K}^{\nabla,\nabla^*}(e_1 \wedge e_{r+1}) = 2\mathbb{K}^0(e_1 \wedge e_{r+1}) - \tilde{c} - \frac{1}{2}\sum_{a=m+1}^{n} \{h_{11}^a h_{r+1,r+1}^a$$
$$- (h_{1,r+1}^a)^2 + h_{11}^{*a}h_{r+1,r+1}^{*a} - (h_{1,r+1}^{*a})^2\}. \tag{24}$$

Furthermore, we derive (8) as:

$$\sigma^{\nabla,\nabla^*} = \frac{m(m-1)\tilde{c}}{2} + \frac{1}{2}\sum_{a=m+1}^{n}\sum_{i<j}\{h_{ii}^{*a}h_{jj}^a + h_{ii}^a h_{jj}^{*a} - 2h_{ij}^a h_{ij}^{*a}\}$$
$$= \frac{m(m-1)\tilde{c}}{2} + \frac{1}{2}\sum_{a=m+1}^{n}\sum_{i<j}\{(h_{ii}^a + h_{ii}^{*a})(h_{jj}^a + h_{jj}^{*a})$$
$$- h_{ii}^a h_{jj}^a - h_{ii}^{*a}h_{jj}^{*a} - (h_{ij}^a + h_{ij}^{*a})^2 + (h_{ij}^a)^2 + (h_{ij}^{*a})^2\}.$$

By a similar argument as above, we deduce that:

$$\sigma^{\nabla,\nabla^*} = \frac{m(m-1)\tilde{c}}{2} + \frac{1}{2}\sum_{a=m+1}^{n}\sum_{i<j}\{2(h_{ii}^{0a}h_{jj}^{0a} - (h_{ij}^{0a})^2)$$
$$- \frac{1}{2}(h_{ii}^a h_{jj}^a - (h_{ij}^a)^2) - \frac{1}{2}(h_{ii}^{*a}h_{jj}^{*a} - (h_{ij}^{*a})^2)\}. \tag{25}$$

Again by the Gauss equation for the Levi–Civita connection, we find that:

$$\sigma^0 = \frac{m(m-1)\tilde{c}}{2} + \sum_{a=m+1}^{n}\sum_{i<j}\{h_{ii}^{0a}h_{jj}^{0a} - (h_{ij}^{0a})^2\},$$

or:

$$\sum_{a=m+1}^{n}\sum_{i<j}\{h_{ii}^{0a}h_{jj}^{0a} - (h_{ij}^{0a})^2\} = \sigma^0 - \frac{m(m-1)\tilde{c}}{2}. \tag{26}$$

Inserting (26) into (25), we have:

$$\begin{aligned}\sigma^{\nabla,\nabla^*} &= 2\sigma^0 - \frac{m(m-1)\tilde{c}}{2} - \frac{1}{2}\sum_{a=m+1}^{n}\sum_{i<j}\{h_{ii}^a h_{jj}^a - (h_{ij}^a)^2 \\ &\quad + h_{ii}^{*a}h_{jj}^{*a} - (h_{ij}^{*a})^2\}. \end{aligned} \qquad (27)$$

By subtracting (24) from (27), we can state the following result:

Lemma 5. *Let $N = N_1 \times_f N_2$ be an m-dimensional statistical warped product submanifold immersed into an n-dimensional statistical manifold of constant sectional curvature $\tilde{c}$. Then:*

$$\begin{aligned}\sigma^{\nabla,\nabla^*} - \mathbb{K}^{\nabla,\nabla^*}(e_1 \wedge e_{r+1}) &= 2(\sigma^0 - \mathbb{K}^0(e_1 \wedge e_{r+1})) - \frac{(m-2)(m+1)\tilde{c}}{2} \\ &\quad - \frac{1}{2}\sum_{a=m+1}^{n}\sum_{i<j}\{h_{ii}^a h_{jj}^a - (h_{ij}^a)^2 + h_{ii}^{*a}h_{jj}^{*a} \\ &\quad - (h_{ij}^{*a})^2\} + \frac{1}{2}\sum_{a=m+1}^{n}\{h_{11}^a h_{r+1,r+1}^a - (h_{1,r+1}^a)^2 \\ &\quad + h_{11}^{*a}h_{r+1,r+1}^{*a} - (h_{1,r+1}^{*a})^2\}.\end{aligned}$$

Further, we have:

$$\begin{aligned}\sigma^{\nabla,\nabla^*} - \mathbb{K}^{\nabla,\nabla^*}(e_1 \wedge e_{r+1}) &\geq 2(\sigma^0 - \mathbb{K}^0(e_1 \wedge e_{r+1})) - \frac{(m-2)(m+1)\tilde{c}}{2} \\ &\quad - \frac{1}{2}\sum_{a=m+1}^{n}\sum_{i<j}\{h_{ii}^a h_{jj}^a + h_{ii}^{*a}h_{jj}^{*a}\} \\ &\quad + \frac{1}{2}\sum_{a=m+1}^{n}\{h_{11}^a h_{r+1,r+1}^a + h_{11}^{*a}h_{r+1,r+1}^{*a}\},\end{aligned}$$

or we write it as:

$$\begin{aligned}2(\sigma^0 - \mathbb{K}^0(e_1 \wedge e_{r+1})) &\leq \sigma^{\nabla,\nabla^*} - \mathbb{K}^{\nabla,\nabla^*}(e_1 \wedge e_{r+1}) + \frac{(m-2)(m+1)\tilde{c}}{2} \\ &\quad + \frac{1}{2}\sum_{a=m+1}^{n}\{\sum_{i<j}\{h_{ii}^a h_{jj}^a\} - h_{11}^a h_{r+1,r+1}^a\} \\ &\quad + \frac{1}{2}\sum_{a=m+1}^{n}\{\sum_{i<j}\{h_{ii}^{*a} h_{jj}^{*a}\} - h_{11}^{*a} h_{r+1,r+1}^{*a}\}.\end{aligned} \qquad (28)$$

We use an optimization technique: For $a \in [m+1, n]$, we consider the quadratic forms:

$$\phi_a : \mathbb{R}^m \to \mathbb{R}, \quad \phi_a^* : \mathbb{R}^m \to \mathbb{R}$$

given by:

$$\phi_a(h_{11}^a, \ldots, h_{mm}^a) = \sum_{i<j}\{h_{ii}^a h_{jj}^a\} - h_{11}^a h_{r+1,r+1}^a, \qquad (29)$$

and:

$$\phi_a^*(h_{11}^{*a}, \ldots, h_{mm}^{*a}) = \sum_{i<j}\{h_{ii}^{*a} h_{jj}^{*a}\} - h_{11}^{*a} h_{r+1,r+1}^{*a}. \qquad (30)$$

The constrained extremum problem is max ϕ_a subject to:

$$Q: h_{11}^a + \cdots + h_{mm}^a = t^a, \quad (t^a \text{ is any constant}).$$

The partial derivatives of ϕ_a are:

$$\frac{\partial \phi_a}{\partial h_{11}^a} = \sum_{i=2}^m h_{ii}^a - h_{r+1,r+1}^a,$$

$$\frac{\partial \phi_a}{\partial h_{r+1,r+1}^a} = \sum_{i \in \overline{1,m}\ r+1} h_{ii}^a - h_{11}^a,$$

$$\frac{\partial \phi_a}{\partial h_{ll}^a} = \sum_{i \in \overline{1,m}\ \{l\}} h_{ii}^a, \quad l \in [r+2, m].$$

For an optimal solution $(h_{11}^a, \ldots, h_{mm}^a)$ of the above problem and grad (ϕ_a) normal at Q, we obtain:

$$(h_{11}^a, h_{22}^a, \ldots, h_{mm}^a) = (0, \alpha^a, \ldots, \alpha^a). \tag{31}$$

As $t^a = \sum_{i=1}^m h_{ii}^a = (m-1)\alpha^a$, then we have:

$$\alpha^a = \frac{t^a}{m-1}. \tag{32}$$

As ϕ_a is obtained from the similar function studied in [39] by subtracting some square terms, $\phi_a|Q$ will have the Hessian semi-negative definite. Consequently, the point in (31), together with (32) is a global maximum point, and hence, we calculate:

$$\phi_a \leq \frac{(m-1)(m-2)(\alpha^a)^2}{2}$$

$$= \frac{(m-2)(t^a)^2}{2(m-1)} = \frac{m^2(m-2)}{2(m-1)}(\mathcal{H}^a)^2.$$

Similarly, one gets:

$$\phi_a^* \leq \frac{m^2(m-2)}{2(m-1)}(\mathcal{H}^{*a})^2,$$

by considering (30) and the constrained extremum problem max ϕ_a^* subject to:

$$Q^*: h_{11}^{*a} + \cdots + h_{mm}^{*a} = t^{*a}, \quad (t^{*a} \text{ is any constant}).$$

Thus, (28) becomes:

$$2(\sigma^0 - \mathbb{K}^0(e_1 \wedge e_{r+1})) \leq \sigma^{\nabla,\nabla^*} - \mathbb{K}^{\nabla,\nabla^*}(e_1 \wedge e_{r+1}) + \frac{(m-2)(m+1)\tilde{c}}{2}$$

$$+ \frac{m^2(m-2)}{4(m-1)}(||\mathcal{H}||^2 + ||\mathcal{H}^*||^2).$$

By summarizing, we state the following:

Theorem 4. Let $N = N_1 \times_{\mathfrak{f}} N_2$ be an m-dimensional statistical warped product submanifold immersed into an n-dimensional statistical manifold of constant sectional curvature $\tilde{c}$. Then:

$$\sigma^{\nabla,\nabla^*} - \mathbb{K}^{\nabla,\nabla^*}(e_1 \wedge e_{r+1}) \geq 2(\sigma^0 - \mathbb{K}^0(e_1 \wedge e_{r+1})) - \frac{(m-2)(m+1)\tilde{c}}{2}$$
$$- \frac{m^2(m-2)}{4(m-1)}(||\mathcal{H}||^2 + ||\mathcal{H}^*||^2).$$

By using (20), we obtain:

$$\mathbb{K}^{\nabla,\nabla^*}(e_1 \wedge e_{r+1}) = \frac{1}{2}(\mathbb{K}(e_1 \wedge e_{r+1}) + \mathbb{K}^*(e_1 \wedge e_{r+1}))$$
$$= \frac{1}{2\mathfrak{f}}\{(\nabla_{e_1} e_1)\mathfrak{f} - e_1^2\mathfrak{f} + (\nabla^*_{e_1} e_1)\mathfrak{f} - e_1^2\mathfrak{f}\}.$$

For $b = 1, 2, \ldots, r$, we also have:

$$\mathbb{K}^{\nabla,\nabla^*}(e_b \wedge e_{r+1}) = \frac{1}{2\mathfrak{f}}\{(\nabla_{e_b} e_b)\mathfrak{f} - e_b^2\mathfrak{f} + (\nabla^*_{e_b} e_b)\mathfrak{f} - e_b^2\mathfrak{f}\}.$$

By summing up b from one to r, we find that:

$$\sum_{b=1}^{r} \frac{1}{2\mathfrak{f}}\{(\nabla_{e_b} e_b)\mathfrak{f} - e_b^2\mathfrak{f} + (\nabla^*_{e_b} e_b)\mathfrak{f} - e_b^2\mathfrak{f}\} = \frac{1}{2}\left(\frac{\Delta^{N_1}\mathfrak{f}}{\mathfrak{f}} + \frac{\Delta^{N_1*}\mathfrak{f}}{\mathfrak{f}}\right) = \frac{\Delta^{N_1 0}\mathfrak{f}}{\mathfrak{f}},$$

where Δ^{N_1} and Δ^{N_1*} are dual Laplacians of N_1 and $\Delta^{N_1 0}$ denotes the Laplacian operator of N_1 for the Levi–Civita connection [37]. Thus, we have:

Theorem 5. Let $N = N_1 \times_{\mathfrak{f}} N_2$ be an m-dimensional statistical warped product submanifold immersed into an n-dimensional statistical manifold of constant sectional curvature $\tilde{c}$. Then, the scalar curvature σ^{∇,∇^*} of N satisfies:

$$\sigma^{\nabla,\nabla^*} \geq 2\sigma^0 - \frac{\Delta^{N_1 0}\mathfrak{f}}{r\mathfrak{f}} - \frac{(m-2)(m+1)\tilde{c}}{2}$$
$$- \frac{m^2(m-2)}{4(m-1)}(||\mathcal{H}||^2 + ||\mathcal{H}^*||^2).$$

6. Optimal Casorati Inequality

Let $\{e_1, \ldots, e_m\}$ and $\{e_{m+1}, \ldots, e_n\}$ be respectively the orthonormal basis of $T_p N$ and $T_p^\perp N$, $p \in N$. Then, the squared norm of second fundamental forms h and h^* is denoted by $\mathcal{C}$ and $\mathcal{C}^*$, respectively, called the Casorati curvatures of N in $\tilde{N}$. Therefore, we have:

$$\mathcal{C} = \frac{1}{m}||h||^2, \quad \mathcal{C}^* = \frac{1}{m}||h^*||^2, \tag{33}$$

where:

$$||h||^2 = \sum_{a=m+1}^{n} \sum_{i,j=1}^{m} (h_{ij}^a)^2, \quad ||h^*||^2 = \sum_{a=m+1}^{n} \sum_{i,j=1}^{m} (h_{ij}^{*a})^2.$$

If W is a q-dimensional subspace of TN, $q \geq 2$, and $\{e_1, \ldots, e_q\}$ an orthonormal basis of W. Then, the scalar curvature of the q-plane section W is:

$$\sigma^{\nabla,\nabla^*}(W) = \sum_{1 \leq i < j \leq q} S(e_i, e_j, e_j, e_i),$$

and the Casorati curvatures of the subspace W are as follows:

$$\mathcal{C}(W) = \frac{1}{q}\sum_{a=m+1}^{n}\sum_{i,j=1}^{q}(h_{ij}^a)^2, \quad \mathcal{C}^*(W) = \frac{1}{q}\sum_{a=m+1}^{n}\sum_{i,j=1}^{q}(h_{ij}^{*a})^2.$$

(1) The normalized Casorati curvatures $\delta_\mathcal{C}(m-1)$ and $\delta_\mathcal{C}^*(m-1)$ are defined as:

$$[\delta_\mathcal{C}(m-1)]_p = \frac{1}{2}\mathcal{C}_p + (\frac{m+1}{2m})\inf\{\mathcal{C}(W) | W : \text{a hyperplane of } T_pN\},$$

and $[\delta_\mathcal{C}^*(m-1)]_p = \frac{1}{2}\mathcal{C}_p^* + (\frac{m+1}{2m})\inf\{\mathcal{C}^*(W) | W : \text{a hyperplane of } T_pN\}.$

(2) The normalized Casorati curvatures $\widehat{\delta}_\mathcal{C}(m-1)$ and $\widehat{\delta}_\mathcal{C}^*(m-1)$ are defined as:

$$[\widehat{\delta}_\mathcal{C}(m-1)]_p = 2\mathcal{C}_p - (\frac{2m-1}{2m})\sup\{\mathcal{C}(W) | W : \text{a hyperplane of } T_pN\},$$

and $[\widehat{\delta}_\mathcal{C}^*(m-1)]_p = 2\mathcal{C}_p^* - (\frac{2m-1}{2m})\sup\{\mathcal{C}^*(W) | W : \text{a hyperplane of } T_pN\}.$

Let $\varphi : N = N_1 \times_f N_2 \to \widetilde{N}(\tilde{c})$ be an isometric statistical immersion of a warped product $N_1 \times_f N_2$ into a statistical manifold of constant sectional curvature $\tilde{c}$. If we chose a local orthonormal frame $\{e_1, \ldots, e_m\}$ such that $\{e_1, \ldots, e_r\}$ are tangent to N_1 and $\{e_{r+1}, \ldots, e_{r+k} = e_m\}$ are tangent to N_2, then the two partial mean curvature vectors $\mathcal{H}_1$ (resp. $\mathcal{H}_1^*$) and $\mathcal{H}_2$ (resp. $\mathcal{H}_2^*$) of N are given by:

$$\mathcal{H}_1 = \frac{1}{r}\sum_{i=1}^{r}h(e_i, e_i), \quad \mathcal{H}_1^* = \frac{1}{r}\sum_{i=1}^{r}h^*(e_i, e_i),$$

and:

$$\mathcal{H}_2 = \frac{1}{k}\sum_{j=1}^{k}h(e_{r+j}, e_{r+j}), \quad \mathcal{H}_2^* = \frac{1}{k}\sum_{j=1}^{k}h^*(e_{r+j}, e_{r+j}).$$

Furthermore, the Casorati curvatures are:

$$\mathcal{C}_1 = \frac{1}{r}\sum_{a=m+1}^{n}\sum_{i,j=1}^{r}(h_{ij}^a)^2, \quad \mathcal{C}_1^* = \frac{1}{r}\sum_{a=m+1}^{n}\sum_{i,j=1}^{r}(h_{ij}^{*a})^2, \qquad (34)$$

and:

$$\mathcal{C}_2 = \frac{1}{k}\sum_{a=m+1}^{n}\sum_{i,j=1}^{k}(h_{r+i r+j}^a)^2, \quad \mathcal{C}_2^* = \frac{1}{k}\sum_{a=m+1}^{n}\sum_{i,j=1}^{k}(h_{r+i r+j}^{*a})^2. \qquad (35)$$

Equation (21) implies:

$$\frac{k\Delta^{N_1 0}f}{f} = \sigma^{\nabla,\nabla^*} - \sum_{1\leq i\leq j\leq r}\mathbb{K}^{\nabla,\nabla^*}(e_i \wedge e_j) - \sum_{r+1\leq l\leq s\leq m}\mathbb{K}^{\nabla,\nabla^*}(e_l \wedge e_s).$$

By using (8), the previous equation becomes:

$$2\sigma^{\nabla,\nabla^*} = \frac{k\Delta^{N_1 0}\mathfrak{f}}{\mathfrak{f}} + r(r-1)\tilde{c} + k(k-1)\tilde{c} + 2r^2||\mathcal{H}_1^0||^2$$
$$- \frac{r^2}{2}(||\mathcal{H}_1||^2 + ||\mathcal{H}_1^*||^2) - \frac{k^2}{2}(||\mathcal{H}_2||^2 + ||\mathcal{H}_2^*||^2)$$
$$+ 2k^2||\mathcal{H}_2^0||^2 - 2r\mathcal{C}_1^0 + \frac{r}{2}(\mathcal{C}_1 + \mathcal{C}_1^*)$$
$$- 2k\mathcal{C}_2^0 + \frac{k}{2}(\mathcal{C}_2 + \mathcal{C}_2^*). \tag{36}$$

We define a polynomial P in terms of the components of the second fundamental form h^0 (with respect to the Levi–Civita connection) of N.

$$P = 2r(r-1)\mathcal{C}_1^0 + (r^2-1)\mathcal{C}_1^0(W_1) + \frac{r}{2}(\mathcal{C}_1 + \mathcal{C}_1^*)$$
$$+ 2k(k-1)\mathcal{C}_2^0 + (k^2-1)\mathcal{C}_2^0(W_2) + \frac{k}{2}(\mathcal{C}_2 + \mathcal{C}_2^*)$$
$$+ \frac{k\Delta^{N_1 0}\mathfrak{f}}{\mathfrak{f}} + r(r-1)\tilde{c} + k(k-1)\tilde{c} - \frac{r^2}{2}(||\mathcal{H}_1||^2 + ||\mathcal{H}_1^*||^2)$$
$$- \frac{k^2}{2}(||\mathcal{H}_2||^2 + ||\mathcal{H}_2^*||^2) - 2\sigma^{\nabla,\nabla^*}. \tag{37}$$

Without loss of generality, we assume that W_1 and W_2 are respectively spanned by $\{e_1,\ldots,e_{r-1}\}$ and $\{e_{r+1},\ldots,e_{r+k-1}\}$. Then, by (36) and (37), we derive:

$$P = \sum_{a=m+1}^{n} \left\{ \sum_{i,j=1}^{r} \frac{r+3}{2}(h_{ij}^{0a})^2 + \frac{r+1}{2}\sum_{i,j=1}^{r-1}(h_{ij}^{0a})^2 - 2(\sum_{i=1}^{r-1} h_{ii}^{0a})^2 \right\}$$
$$+ \sum_{a=m+1}^{n} \left\{ \sum_{l,s=1}^{k} \frac{k+3}{2}(h_{ls}^{0a})^2 + \frac{k+1}{2}\sum_{l,s=1}^{k-1}(h_{ls}^{0a})^2 - 2(\sum_{l=1}^{k-1} h_{ll}^{0a})^2 \right\}$$
$$= \sum_{a=m+1}^{n} \left\{ 2(r+2) \sum_{1\leq i<j\leq r-1}(h_{ij}^{0a})^2 + (r+3)\sum_{i=1}^{r-1}(h_{ir}^{0a})^2 \right.$$
$$+ r\sum_{i=1}^{r-1}(h_{ii}^{0a})^2 - 4 \sum_{1\leq i<j\leq r}(h_{ii}^{0a}h_{jj}^{0a}) + \frac{r-1}{2}(h_{rr}^{0a})^2 \right\}$$
$$+ \sum_{a=m+1}^{n} \left\{ 2(k+2) \sum_{1\leq l<s\leq k-1}(h_{ls}^{0a})^2 + (k+3)\sum_{l=1}^{k-1}(h_{lk}^{0a})^2 \right.$$
$$+ k\sum_{l=1}^{k-1}(h_{ll}^{0a})^2 - 4 \sum_{1\leq l<s\leq k}(h_{ll}^{0a}h_{ss}^{0a}) + \frac{k-1}{2}(h_{kk}^{0a})^2 \right\}$$
$$\geq \sum_{a=m+1}^{n} \left\{ \sum_{i=1}^{r-1} r(h_{ii}^{0a})^2 + \frac{r-1}{2}(h_{rr}^{0a})^2 - 4\sum_{1\leq i<j\leq r} h_{ii}^{0a} h_{jj}^{0a} \right\}$$
$$+ \sum_{a=m+1}^{n} \left\{ \sum_{l=1}^{k-1} k(h_{ll}^{0a})^2 + \frac{k-1}{2}(h_{kk}^{0a})^2 - 4\sum_{1\leq l<s\leq k} h_{ll}^{0a} h_{ss}^{0a} \right\}.$$

For any $a \in \{m+1, \ldots, n\}$, we define two quadratic forms $\phi_a : \mathbb{R}^r \to \mathbb{R}$ and $\varphi_a : \mathbb{R}^k \to \mathbb{R}$ by:

$$\phi_a(h_{11}^{0a}, h_{22}^{0a}, \ldots, h_{r-1,r-1}^{0a}, h_{rr}^{0a})$$
$$= \sum_{i=1}^{r-1} r(h_{ii}^{0a})^2 + \frac{r-1}{2}(h_{rr}^{0a})^2 - 4 \sum_{1 \le i < j \le r} h_{ii}^{0a} h_{jj}^{0a}, \qquad (38)$$

and:

$$\varphi_a(h_{11}^{0a}, h_{22}^{0a}, \ldots, h_{k-1,k-1}^{0a}, h_{kk}^{0a})$$
$$= \sum_{l=1}^{k-1} k(h_{ll}^{0a})^2 + \frac{k-1}{2}(h_{kk}^{0a})^2 - 4 \sum_{1 \le l < s \le k} h_{ll}^{0a} h_{ss}^{0a}. \qquad (39)$$

First, we consider the constrained extremum problem $\min \phi_a$ subject to:

$$Q : h_{11}^{0a} + \cdots + h_{rr}^{0a} = t^a, \quad (t^a \text{ is any constant}).$$

From (38), we find that the critical points

$$h^{0c} = (h_{11}^{0a}, h_{22}^{0a}, \ldots, h_{r-1,r-1}^{0a}, h_{rr}^{0a})$$

of Q are the solutions of the following system of linear homogeneous equations.

$$\left.\begin{array}{l} \dfrac{\partial \phi_a}{\partial h_{ii}^{0a}} = 2(r+2)(h_{ii}^{0a}) - 4 \sum_{j=1}^{r} h_{jj}^{0a} = 0, \\[2mm] \dfrac{\partial \phi_a}{\partial h_{rr}^{0a}} = (r-1)h_{rr}^{0a} - 4 \sum_{j=1}^{r-1} h_{jj}^{0a} = 0, \end{array}\right\} \qquad (40)$$

for $i \in \{1, 2, \ldots, r-1\}$ and $a \in \{m+1, \ldots, n\}$. Hence, every solution h^{0c} has:

$$h_{ii}^{0a} = \frac{1}{r+1} t^a, \quad h_{rr}^{0a} = \frac{4}{r+3} t^a,$$

for $i \in \{1, 2, \ldots, r-1\}$ and $a \in \{m+1, \ldots, n\}$.

Now, we fix $x \in Q$. The bilinear form $\Theta : T_x Q \times T_x Q \to \mathbb{R}$ has the following expression (cf. Theorem 1):

$$\Theta(E, F) = \mathrm{Hess}_{\phi_a}(E, F) + \langle h'(E, F), \mathrm{grad}(\phi_a)(x) \rangle,$$

where h' denotes the second fundamental form of Q in $\mathbb{R}^r$ and $\langle \cdot, \cdot \rangle$ denotes the standard inner product on $\mathbb{R}^r$. The Hessian matrix of ϕ_a is given by:

$$\mathrm{Hess}_{\phi_a} = \begin{pmatrix} 2(r+2) & -4 & \cdots & -4 & -4 \\ -4 & 2(r+2) & \cdots & -4 & -4 \\ \vdots & \vdots & \ddots & \vdots & \vdots \\ -4 & -4 & \cdots & 2(r+2) & -4 \\ -4 & -4 & \cdots & -4 & (r-1) \end{pmatrix}.$$

Take a vector $E \in T_x Q$, which satisfies a relation $\sum_{i=1}^{r} E_i = 0$. As the hyperplane is totally geodesic, i.e., $h' = 0$ in $\mathbb{R}^r$, we get:

$$\Theta(E, E) = \text{Hess}_{\phi_a}(E, E)$$

$$= 2(r+2) \sum_{i=1}^{r-1} E_i^2 + (r-1) E_r^2 - 8 \sum_{i \neq j=1}^{r} E_i E_j$$

$$= 2(r+2) \sum_{i=1}^{r-1} E_i^2 + (r-1) E_r^2 - 4 \left\{ (\sum_{i=1}^{r} E_i)^2 - \sum_{i=1}^{r} E_i^2 \right\}$$

$$= 2(r+4) \sum_{i=1}^{r-1} E_i^2 + (r+3) E_r^2$$

$$\geq 0.$$

However, the point h^{0c} is the only optimal solution, i.e., the global minimum point of problem, and reaches a minimum $Q(h^{0c}) = 0$ by considering (39) and the constrained extremum problem min φ_a subject to:

$$Q' : h_{11}^{0a} + \cdots + h_{kk}^{0a} = \alpha^a, \quad (\alpha^a \text{ is any constant}).$$

Thus, we have:

$$2\sigma^{\nabla, \nabla^*} \leq r(r-1)\mathcal{C}_1^0 + (r^2 - 1)\mathcal{C}_1^0(W_1) + \frac{r}{2}(\mathcal{C}_1 + \mathcal{C}_1^*)$$

$$+ k(k-1)\mathcal{C}_2^0 + (k^2 - 1)\mathcal{C}_2^0(W_2) + \frac{k}{2}(\mathcal{C}_2 + \mathcal{C}_2^*)$$

$$+ \frac{k \Delta^{N_1 0} f}{f} + r(r-1)\tilde{c} + k(k-1)\tilde{c}$$

$$- \frac{r^2}{2}(||\mathcal{H}_1||^2 + ||\mathcal{H}_1^*||^2) - \frac{k^2}{2}(||\mathcal{H}_2||^2 + ||\mathcal{H}_2^*||^2).$$

Consequently, we get immediately the following theorem from the above relation:

Theorem 6. *Let $N = N_1 \times_f N_2$ be an m-dimensional statistical warped product submanifold immersed into an n-dimensional statistical manifold of constant sectional curvature $\tilde{c}$. Then, the Casorati curvatures satisfy:*

$$2\sigma^{\nabla, \nabla^*} \leq r(r-1)\mathcal{C}_1^0 + (r^2 - 1)\mathcal{C}_1^0(W_1) + r\mathcal{C}_1^0$$

$$+ k(k-1)\mathcal{C}_2^0 + (k^2 - 1)\mathcal{C}_2^0(W_2) + k\mathcal{C}_2^0$$

$$+ \frac{k \Delta^{N_1 0} f}{f} + r(r-1)\tilde{c} + k(k-1)\tilde{c}$$

$$- \frac{r^2}{2}(||\mathcal{H}_1||^2 + ||\mathcal{H}_1^*||^2) - \frac{k^2}{2}(||\mathcal{H}_2||^2 + ||\mathcal{H}_2^*||^2),$$

where W_1 and W_2 are respectively the hyperplanes of $T_p N_1$ and $T_p N_2$, $\mathcal{C}_1^0 = \frac{1}{2}(\mathcal{C}_1 + \mathcal{C}_1^*)$, and $\mathcal{C}_2^0 = \frac{1}{2}(\mathcal{C}_2 + \mathcal{C}_2^*)$.

7. Examples

We provide examples of statistical warped product submanifolds as follows:

Example 1. *By generalizing Example 2.7 of [10] to higher dimensions, we see that:*

$$\left(\mathbb{R} \times_{e^z} \mathbb{R}^n, \tilde{g} = dz^2 + e^{2z}(dx_1^2 + \cdots + dx_n^2), \nabla, \nabla^* \right)$$

is a statistical warped product manifold. Furthermore, the hyperbolic space:

$$\mathbb{H}^{n+1}(-1) = \left(\{(x_0, \ldots, x_{n+1}) \in \mathbb{R}^{n+1} | x_0 > 0\}, \tilde{g} = \frac{dx_0^2 + \cdots + dx_{n+1}^2}{x_0^2}, \tilde{\nabla}, \tilde{\nabla}^* \right)$$

is the statistical manifold of constant sectional curvature -1. Thus, with respect to the Levi–Civita connection, $\mathbb{R} \times_{e^z} \mathbb{R}^{n-1}$ admits an isometric minimal immersion into $\mathbb{H}^{n+1}(-1)$.

Example 2. $(\mathbb{R} \times_z \mathbb{R}^n, \tilde{g} = dt^2 + t^2(dx_1^2 + \cdots + dx_n^2), \nabla, \nabla^*)$ is a statistical warped product manifold, and it is isometric to the Euclidean $(n+1)$-space $\mathbb{E}^{n+1}$. Let N be a minimal submanifold of the unit hypersphere $S^n(1) \subset \mathbb{E}^{n+1}$ center at the origin $o \in \mathbb{E}^{n+1}$, and let $C(N)$ be the cone over N with the vertex at o.

The metric of $C(N)$ is the warped product metric $g_{C(N)} = dt^2 + t^2 g_N$, where g_N denotes the metric of N. Any open submanifold M of $C(N)$ is a warped product manifold, which admits an isometric minimal immersion into the statistical manifold $\mathbb{E}^{n+1}$ of constant sectional curvature zero.

Author Contributions: Conceptualization, A.N.S., B.-Y.C. and O.B.; Methodology, A.N.S., B.-Y.C. and O.B.; Software, A.N.S. and B.-Y.C.; Validation, A.N.S., B.-Y.C. and O.B.; Formal Analysis, A.N.S.; Investigation, A.N.S., B.-Y.C. and O.B.; Resources, A.N.S.; Data curation, A.N.S. and B.-Y.C.; Writing-Original Draft Preparation, A.N.S.; Writing-Review and Editing, A.N.S. and B.-Y.C.; Visualization, A.N.S.; Supervision, B.-Y.C.; Project Administration, A.N.S., B.-Y.C. and O.B.; Funding Acquisition, A.N.S., B.-Y.C. and O.B.

Funding: This research received no external funding.

Acknowledgments: The authors thank the referees for many valuable and helpful suggestions to improve the presentation of this article.

Conflicts of Interest: The authors declare no conflict of interest.

References

1. Amari, S. *Differential-Geometrical Methods in Statistics*; Lecture Notes in Statistics; Springer: New York, NY, USA, 1985; Volume 28.
2. Lauritzen, S. Statistical manifolds. In *Differential Geometry in Statistical Inference*; Amari, S.I., Barndorff-Nielsen, O.E., Kass, R.E., Lauritzen, S.L., Rao, C.R., Eds.; IMS Lecture Notes Institute of Mathematical Statistics: Hayward, CA, USA, 1987; Volume 10, pp. 163–216.
3. Bishop, R.L.; O'Neill, B. Manifolds of negative curvature. *Trans. Am. Math. Soc.* **1969**, *145*, 1–49. [CrossRef]
4. Chen, B.-Y. *Pseudo-Riemannian Geometry, δ-Invariants and Applications*; Worlds Scientific: Hackensack, NJ, USA, 2011.
5. Chen, B.-Y. *Differential Geometry of Warped Product Manifolds and Submanifolds*; Worlds Scientific: Hackensack, NJ, USA, 2017.
6. Todjihounde, L. Dualistic structures on warped product manifolds. *Differ. Geom.-Dyn. Syst.* **2006**, *8*, 278–284.
7. Furuhata, H.; Hasegawa, I.; Okuyama, Y.; Sato, K. Kenmotsu statistical manifolds and warped product. *J. Geom.* **2017**, *108*, 1175–1191. [CrossRef]
8. Furuhata, H. Hypersurfaces in statistical manifolds. *Differ. Geom. Appl.* **2009**, *27*, 420–429. [CrossRef]
9. Aytimur, H.; Ozgur, C. Einstein statistical warped product manifolds. *Filomat* **2018**, *32*, 3891–3897. [CrossRef]
10. Murathan, C.; Sahin, B. A study of Wintgen like inequality for submanifolds in statistical warped product manifolds. *J. Geom.* **2018**, *109*, 30. [CrossRef]
11. Chen, B.-Y. On isometric minimal immersions from warped products into real space forms. *Proc. Edinb. Math. Soc.* **2002**, *45*, 579–587. [CrossRef]
12. Chen, B.-Y. Warped products in real space forms. *Rocky Mt. J. Math.* **2004**, *34*, 551–563. [CrossRef]
13. Casorati, F. Mesure de la courbure des surfaces suivant l'idée commune. *Acta Math.* **1890**, *14*, 95–110. [CrossRef]
14. Lee, C.W.; Lee, J.W.; Vilcu, G.E.; Yoon, D.W. Optimal inequalities for the Casorati curvatures of the submanifolds of generalized space form endowed with semi-symmetric metric connections. *Bull. Korean Math. Soc.* **2015**, *52*, 1631–1647. [CrossRef]

15. Lee, C.W.; Vilcu, G.E. Inequalities for generalized normalized Casorati curvatures of slant submanifolds in quaternion space forms. *Taiwan. J. Math.* **2015**, *19*, 691–702. [CrossRef]
16. Lee, C.W.; Lee, J.W.; Vilcu, G.E. Optimal inequalities for the normalized δ−Casorati curvatures of submanifolds in Kenmotsu space forms. *Adv. Geom.* **2017**, *17*, 355–362. [CrossRef]
17. Shahid, M.H.; Siddiqui, A.N. Optimizations on totally real submanifolds of LCS-manifolds using Casorati curvatures. *Commun. Korean Math. Soc.* **2019**, *34*, 603–614.
18. Siddiqui, A.N.; Shahid, M.H. A lower bound of normalized scalar curvature for bi-slant submanifolds in generalized Sasakian space forms using Casorati curvatures. *Acta Math. Univ. Comen.* **2018**, *87*, 127–140.
19. Siddiqui, A.N. Upper bound inequalities for δ-Casorati curvatures of submanifolds in generalized Sasakian space forms admitting a semi-Symmetric metric connection. *Int. Electron. J. Geom.* **2018**, *11*, 57–67.
20. Siddiqui, A.N. Optimal Casorati inequalities on bi-slant submanifolds of generalized Sasakian space forms. *Tamkang J. Math.* **2018**, *49*, 245–255. [CrossRef]
21. Slesar, V.; Sahin, B.; Vilcu, G.E. Inequalities for the Casorati curvatures of slant submanifolds in quaternionic space forms. *J. Inequal. Appl.* **2014**, *2014*, 123. [CrossRef]
22. Aydin, M. E.; Mihai, I. Wintgen inequality for statistical surfaces. *Math. Inequal. Appl.* **2019**, *22*, 123–132. [CrossRef]
23. Alkhaldi, A.H.; Aquib, M.; Siddiqui, A.N.; Shahid, M.H. Pinching theorems for statistical submanifolds in Sasaki-like statistical space forms. *Entropy* **2018**, *20*, 690. [CrossRef]
24. Decu, S.; Haesen, S.; Verstraelen, L.; Vilcu, G.E. Curvature invariants of statistical submanifolds in Kenmotsu statistical manifolds of constant ϕ-sectional curvature. *Entropy* **2018**, *20*, 529. [CrossRef]
25. Lee, C.W.; Yoon, D.W.; Lee, J.W. A pinching theorem for statistical manifolds with Casorati curvatures. *J. Nonlinear Sci. Appl.* **2017**, *10*, 4908–4914. [CrossRef]
26. Siddiqui, A.N.; Shahid, M.H. Optimizations on statistical hypersurfaces with Casorati curvatures. *Kragujevac J. Math.* **2021**, *45*, 449–463.
27. Yano, K.; Kon, M. *Structures on Manifolds*; Worlds Scientific: Singapore, 1984.
28. Vos, P.W. Fundamental equations for statistical submanifolds with applications to the Bartlett correction. *Ann. Inst. Stat. Math.* **1989**, *41*, 429–450. [CrossRef]
29. Opozda, B. Bochner's technique for statistical structures. *Ann. Glob. Anal. Geom.* **2015**, *48*, 357–395. [CrossRef]
30. Opozda, B. A sectional curvature for statistical structures. *Linear Algebra Appl.* **2016**, *497*, 134–161. [CrossRef]
31. Oprea, T. Optimization methods on Riemannian submanifolds. *Analele Univ. Buc.* **2005**, *54*, 127–136.
32. Hamilton, R.S. The Ricci flow on surfaces. *Contemp. Math.* **1988**, *71*, 237–261.
33. Chen, B.-Y. Ricci solitons on Riemannian submanifolds. In *Riemannian Geometry and Applications-Proceedings RIGA*; University of Bucharest Press: Bucharest, Romania, 2014; pp. 30–45.
34. Mantica, C.A.; Shenawy, S.; Unal, B. Ricci solitons on singly warped product manifolds and applications. *arXiv* **2019**, arXiv:1508.02794v2.
35. Meric, S.E.; Kilic, E. Some inequalities for Ricci solitons. In Proceedings of International Conference on Mathematics and Mathematics Education (ICMME 2018), Ordu, Turkey, 27–30 June 2018; Volume 10, pp. 160–164.
36. Crasmareanu, M. A new approach to gradient Ricci solitons and generalizations. *Filomat* **2018**, *32*, 3337–3346. [CrossRef]
37. Calin, O.; Udriste, C. *Geometric Modeling in Probability and Statistics*; Springer: Cham, Switzerland, 2014.
38. Chen, B.-Y. Some pinching and classification theorems for minimal submanifolds. *Arch. Math.* **1993**, *60*, 568–578. [CrossRef]
39. Oprea, T. On a Riemannian invariant of Chen type. *Rocky Mountain J. Math.* **2008**, *38*, 567–581. [CrossRef]

© 2019 by the authors. Licensee MDPI, Basel, Switzerland. This article is an open access article distributed under the terms and conditions of the Creative Commons Attribution (CC BY) license (http://creativecommons.org/licenses/by/4.0/).

MDPI
St. Alban-Anlage 66
4052 Basel
Switzerland
Tel. +41 61 683 77 34
Fax +41 61 302 89 18
www.mdpi.com

Mathematics Editorial Office
E-mail: mathematics@mdpi.com
www.mdpi.com/journal/mathematics

www.ingramcontent.com/pod-product-compliance
Lightning Source LLC
LaVergne TN
LVHW070733100526
838202LV00013B/1224